McGraw-Hill

Welcome to *My Math* — your very own math book! You can write in it – in fact, you are encouraged to write, draw, circle, explain, and color as you explore the exciting world of mathematics. Let's get started. Grab a pencil and finish each sentence.

My name is ____________________.

My favorite color is ____________________.

My favorite hobby or sport is ____________________.

My favorite TV program or video game is

____________________.

My favorite class is ____________________.

Math, of course!

Bothell, WA • Chicago, IL • Columbus, OH • New York, NY

connectED.mcgraw-hill.com

STEM McGraw-Hill is committed to providing instructional materials in Science, Technology, Engineering, and Mathematics (STEM) that give all students a solid foundation, one that prepares them for college and careers in the 21st century.

Send all inquiries to:
McGraw-Hill Education
STEM Learning Solutions Center
8787 Orion Place
Columbus, OH 43240

ISBN: 978-0-02-115024-3 *(Volume 1)*
MHID: 0-02-115024-9

Printed in the United States of America.

22 23 24 25 QSX 22 21 20 19 18

Our mission is to provide educational resources that enable students to become the problem solvers of the 21st century and inspire them to explore careers within Science, Technology, Engineering, and Mathematics (STEM) related fields.

Meet The Artists!

Caelyn Cochran

The Angles of Ballet I was inspired by the gesture mannequin in art class. I thought of how the body moves in different ways and how ballet dancers create different poses. *Volume 1*

Syed Amaan Rahman

Step by Step - Stairway to Learning To me, math means learning in a step-by-step process. I drew consecutive numbers and math signs marching up a staircase and into light to symbolize the process of learning math and beyond with endless possibilities. *Volume 2*

Other Finalists

Jeannine Demarzo
White Cake Recipe

Nate Mitchell
Pixel Man

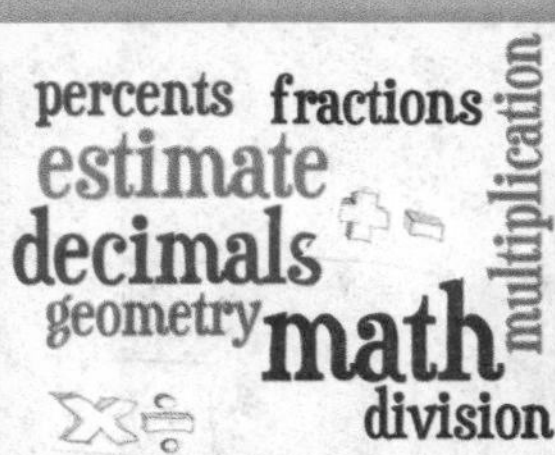

Rachel Harvey
3-D Word Collage

Dante Washington
Insect Mandala 6

Kathleen Saier
Mathematical Language of Computers

Phobe Hanscom, Hope Sternenberg, Haylee Harvey
Sponge Math

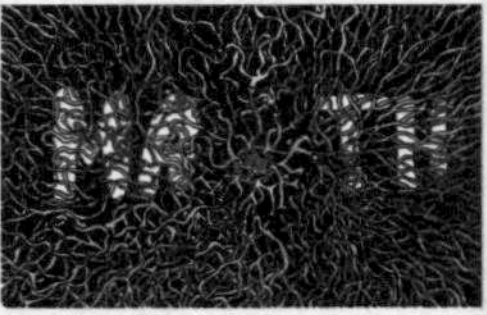

Mazi Blankenship, Abby Gunther, Zach Ginn
Mathematical Quaqmire

Abigail Africa
Math is a Fact of Life

Madeline Murphy
Math Collage

Jessica Li
All the Missing Pieces

Find out more about the winners and other finalists at www.MHEonline.com.

We wish to congratulate all of the entries in the 2011 *McGraw-Hill My Math* "What Math Means To Me" cover art contest. With over 2,400 entries and more than 20,000 community votes cast, the names mentioned above represent the two winners and ten finalists for this grade.

** Please visit mhmymath.com for a complete list of students who contributed to this artwork.*

it's all at
connectED.mcgraw-hill.com

Go to the Student Center for your eBook, Resources, Homework, and Messages.

McGraw-Hill My Math: Student Center

Search

Favorites | McGraw-Hill My Math: Student Center | Tools

Hello, Student | Home | ConnectED | Help | Logout

McGraw-Hill My Math

Search | Standards

Student Center

Home | Homework | Resources

Chapter 5: Multiply with Two-Digit Numbers

Lesson 1: Multiply by Tens

Open eBook

Multiply with Two-Digit Numbers > Multiply by Tens

Number and Operations in Base Ten

Animals in MY world

(l)Meinrad Riedo/imagebroker RF/age fotostock; (r)Sylvia Bors/The Image Bank/Getty Images

Homework

Due: Monday, October 7, 2011
Assignment Title

You have unread teacher comments.

More

Messages

Monday, October 7, 2011
Don't forget to study for the test on Friday.

More

The McGraw-Hill Companies

Terms of Use | Privacy Policy | Technical Support | Minimum System Requirements

Copyright© The McGraw-Hill Companies, Inc.

Write your Username

Password

Get your resources online to help you in class and at home.

Vocab

Find activities for building vocabulary.

Watch

Watch animations of key concepts.

Tools

Explore concepts with virtual manipulatives.

Check

Self-assess your progress.

eHelp

Get targeted homework help.

Games

Reinforce with games and apps.

Tutor

See a teacher illustrate examples and problems.

GO mobile

Scan this QR code with your smart phone* or visit mheonline.com/stem_apps.

*May require quick response code reader app.

Available on the App Store

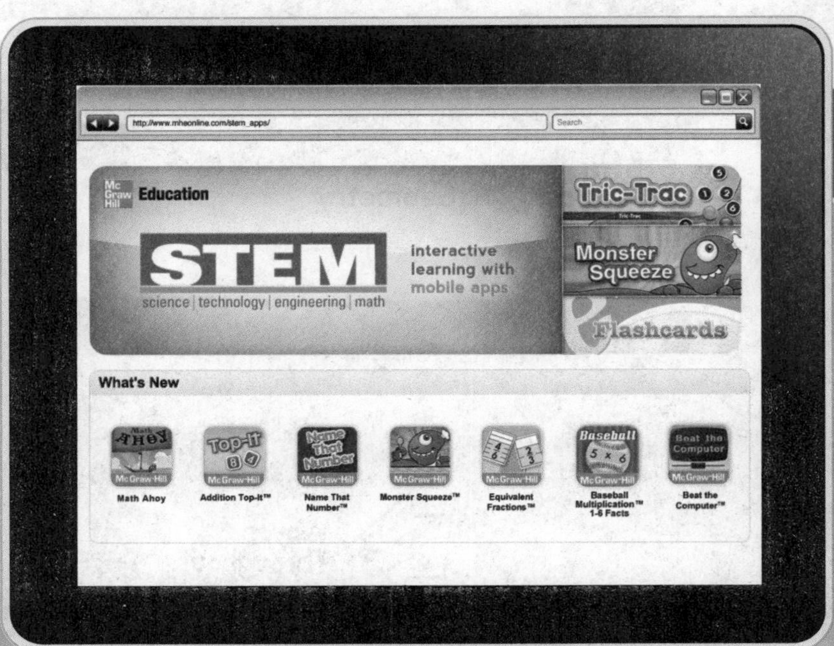

Contents in Brief

Organized by Domain

Standards for Mathematical PRACTICE → *Woven Throughout*

Chapter 1 Place Value

ESSENTIAL QUESTION
How does the position of a digit in a number relate to its value?

Getting Started

Lessons and Homework

Wrap Up

Are you ready for the great outdoors?

Look for this!
Click online and you can watch videos that will help you learn the lessons.

connectED.mcgraw-hill.com

Chapter 2 Multiply Whole Numbers

ESSENTIAL QUESTION
What strategies can be used to multiply whole numbers?

Taking care of my pets!

connectED.mcgraw-hill.com

Chapter 3 Divide by a One-Digit Divisor

ESSENTIAL QUESTION
What strategies can be used to divide whole numbers?

Let's divide and conquer these chores!

Look for this! eHelp
Click online and you can get more help while doing your homework.

connectED.mcgraw-hill.com

Chapter 4 Divide by a Two-Digit Divisor

ESSENTIAL QUESTION
What strategies can I use to divide by a two-digit divisor?

Getting Started

Lessons and Homework

Wrap Up

Look for this!
Click online and you can find tools that will help you explore concepts.
Tools

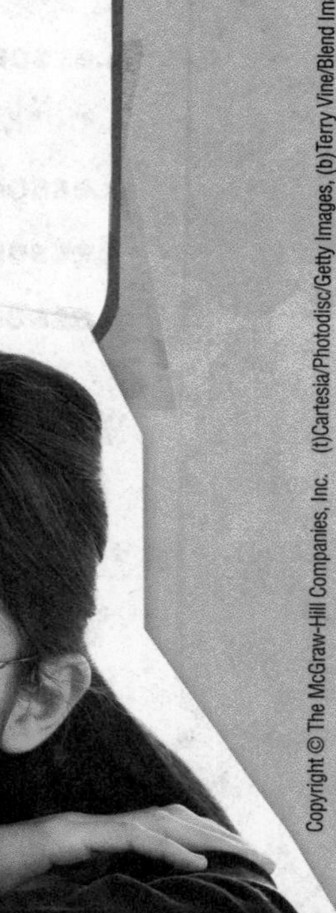

connectED.mcgraw-hill.com

Chapter 5 Add and Subtract Decimals

ESSENTIAL QUESTION
How can I use place value and properties to add and subtract decimals?

Getting Started

Lessons and Homework

Wrap Up

Technology on the go!

connectED.mcgraw-hill.com

Chapter 6

Multiply and Divide Decimals

ESSENTIAL QUESTION
How is multiplying and dividing decimals similar to multiplying and dividing whole numbers?

Getting Started

Lessons and Homework

Wrap Up

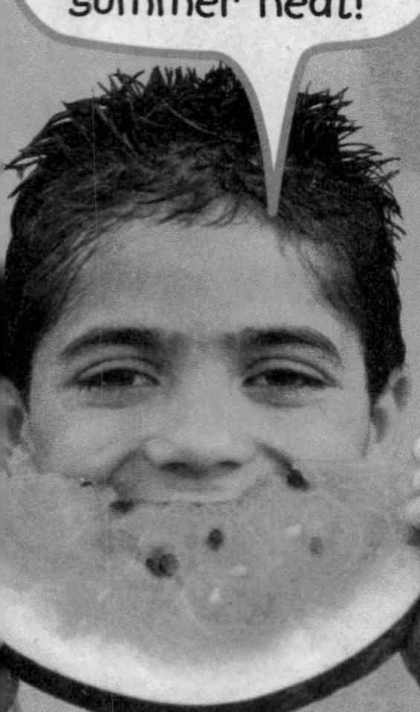

Chapter 7 Expressions and Patterns

ESSENTIAL QUESTION
How are patterns used to solve problems?

Getting Started

Lessons and Homework

Wrap Up

Look for this!
Click online and you can watch a teacher solving problems.

Chapter

8 Fractions and Decimals

ESSENTIAL QUESTION
How are factors and multiples helpful in solving problems?

Getting Started

Lessons and Homework

Wrap Up

Look for this! Vocab
Click online and you can find activities to help build your vocabulary.

connectED.mcgraw-hill.com

Chapter 9 Add and Subtract Fractions

ESSENTIAL QUESTION
How can equivalent fractions help me add and subtract fractions?

Getting Started

Lessons and Homework

Wrap Up

Chapter 10 Multiply and Divide Fractions

ESSENTIAL QUESTION
What strategies can be used to multiply and divide fractions?

What's cooking in the kitchen?

Getting Started

Lessons and Homework

Wrap Up

connectED.mcgraw-hill.com

Chapter 11 Measurement

ESSENTIAL QUESTION
How can I use measurement conversions to solve real-world problems?

Getting Started

Lessons and Homework

Wrap Up

Look for this!
Click online and you can check your progress.

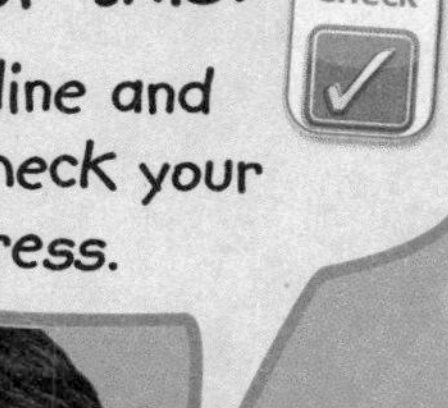

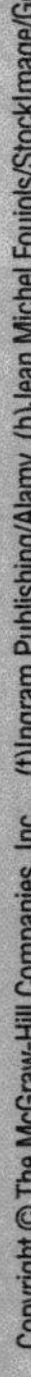

Chapter 12 Geometry

ESSENTIAL QUESTION
How does geometry help me solve problems in everyday life?

Getting Started

Lessons and Homework

Wrap Up

Math is an adventure!

connectED.mcgraw-hill.com

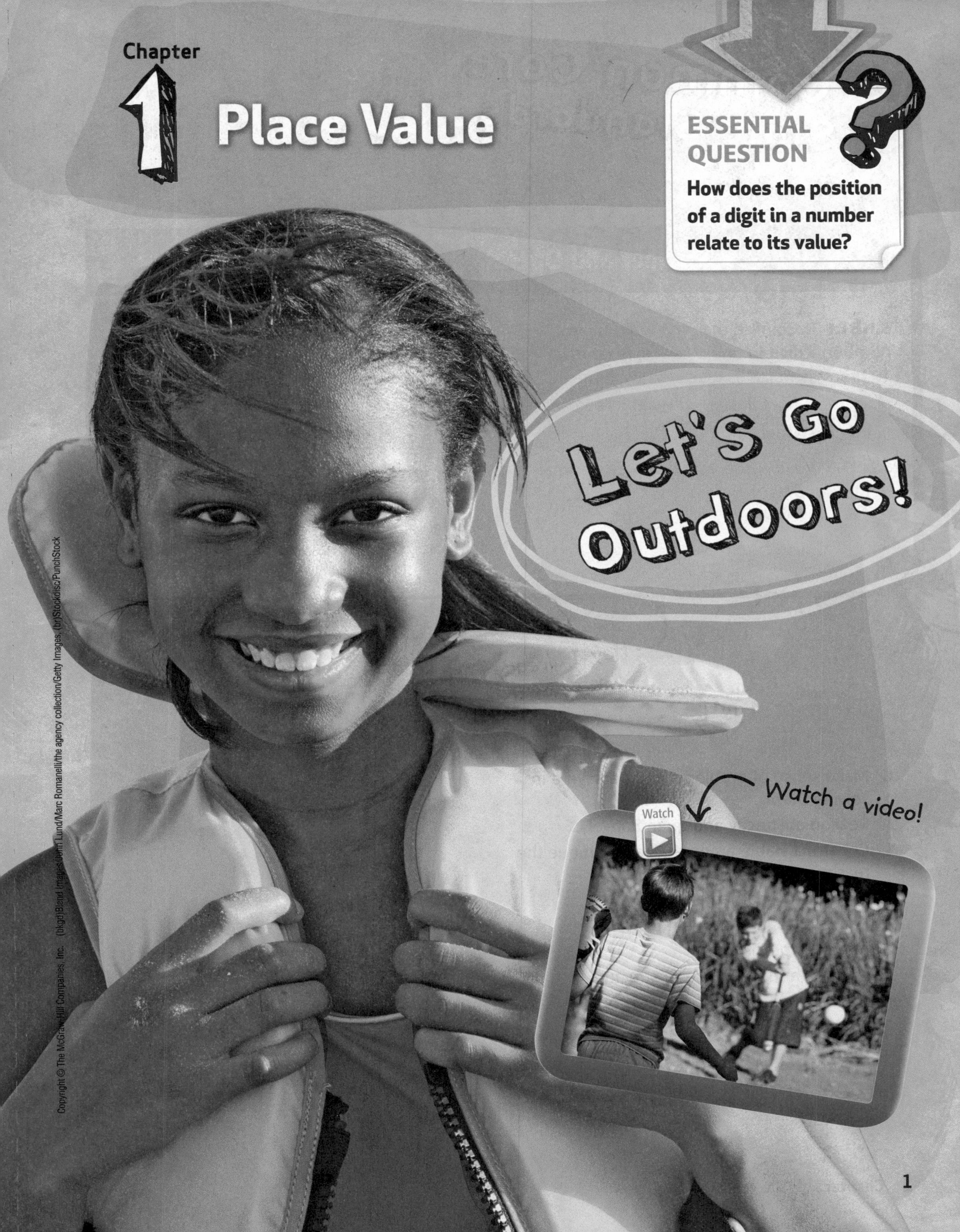
Chapter 1
Place Value
ESSENTIAL QUESTION
How does the position of a digit in a number relate to its value?
Let's Go Outdoors!
Watch a video!
Watch

MY Common Core State Standards

Number and Operations in Base Ten

CCSS

5.NBT.1 Recognize that in a multi-digit number, a digit in one place represents 10 times as much as it represents in the place to its right and $\frac{1}{10}$ of what it represents in the place to its left.

5.NBT.3 Read, write, and compare decimals to thousandths.

5.NBT.3a Read and write decimals to thousandths using base-ten numerals, number names, and expanded form, e.g., $347.392 = 3 \times 100 + 4 \times 10 + 7 \times 1 + 3 \times \left(\frac{1}{10}\right) + 9 \times \left(\frac{1}{100}\right) + 2 \times \left(\frac{1}{1000}\right)$.

5.NBT.3b Compare two decimals to thousandths based on meanings of the digits in each place, using >, =, and < symbols to record the results of comparisons.

Hey, I already know some of these!

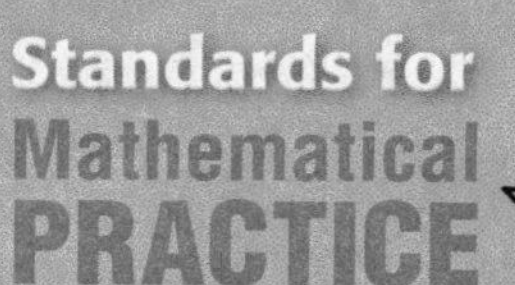

1. Make sense of problems and persevere in solving them.
2. Reason abstractly and quantitatively.
3. Construct viable arguments and critique the reasoning of others.
4. Model with mathematics.
5. Use appropriate tools strategically.
6. Attend to precision.
7. Look for and make use of structure.
8. Look for and express regularity in repeated reasoning.

= focused on in this chapter

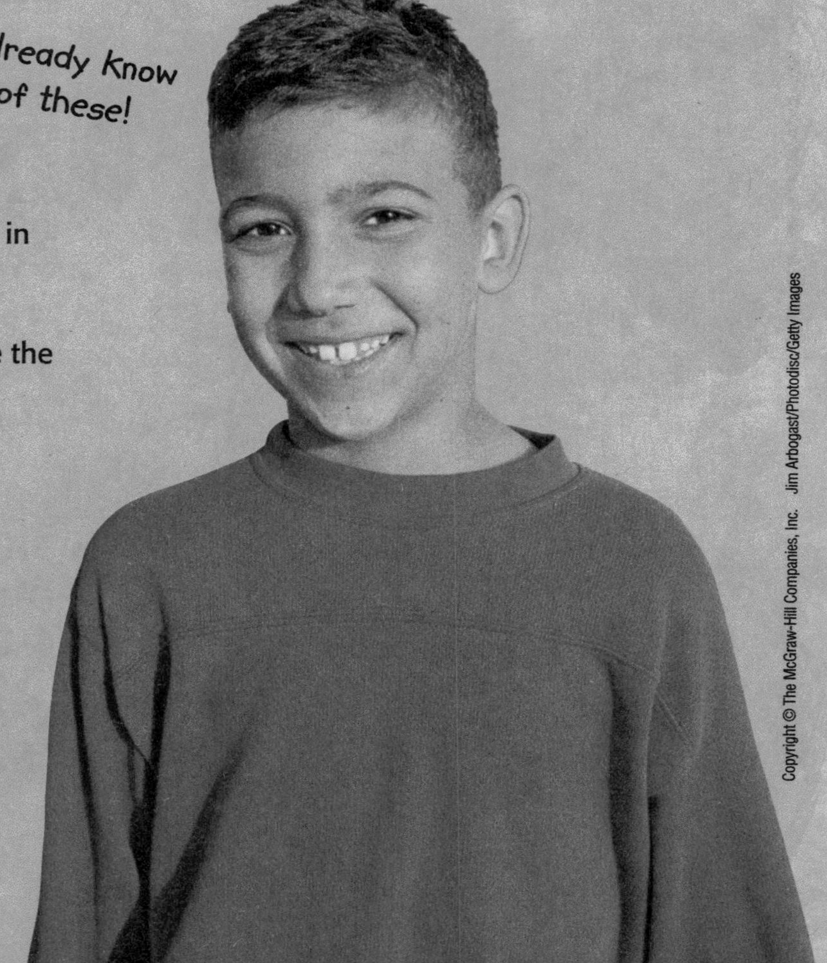

Name ....

Am I Ready?

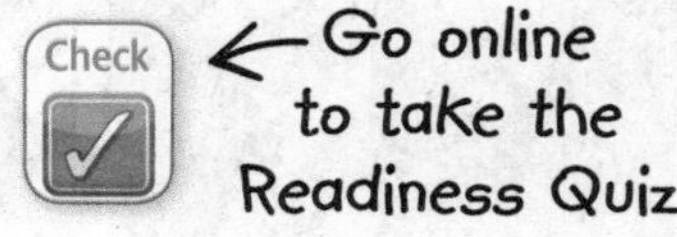

Write each number in word form.

1. 8 eight

2. 23 twenty three

3. 15 fifteen

4. 160 one hundred sixty

Write the number that represents each point on the number line.

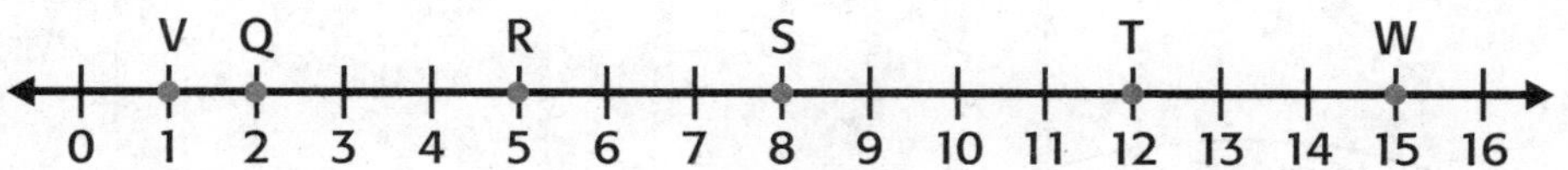

5. *Q* 2

6. *S* 8

7. *R* 5

8. *T* 12

9. *V* 1

10. *W* 15

Write each sentence using the symbols <, >, or =.

11. 8 is less than 12.

8 < 12

12. 24 is greater than 10.

24 > 10.

13. The high temperature yesterday was 64°F. The high temperature today is 70°F. Write 64 is less than 70 using the symbols <, >, or =.

64 < 70

How Did I Do? **Shade the boxes to show the problems you answered correctly.**

1	2	3	4	5	6	7	8	9	10	11	12	13

Name

MY Math Words

Vocab

Review Vocabulary

comma hundreds hundred thousands ones

tens ten thousands thousands

Making Connections

Use the review words to complete each section of the bubble map.

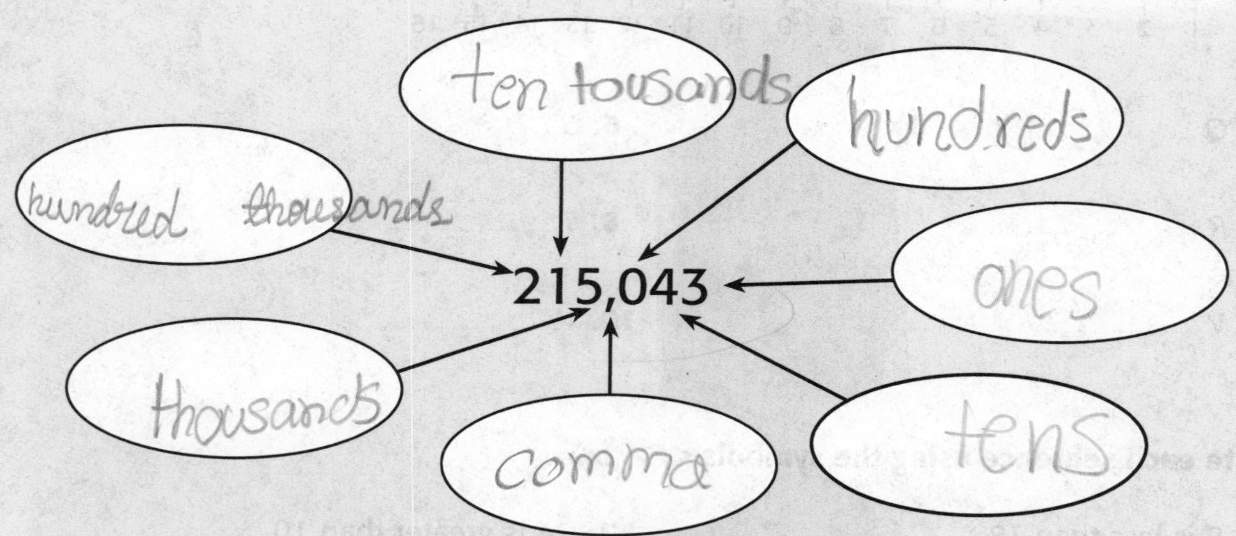

Describe how commas are used in writing greater numbers.

After ever three digit there is a comma.

Example:- 107,698

MY Vocabulary Cards

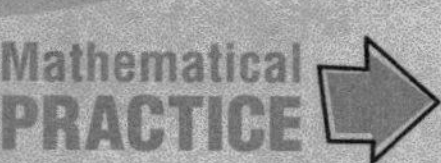

Lesson 1-3

decimal point

1.378 5.0 6.78

Lesson 1-3

decimal

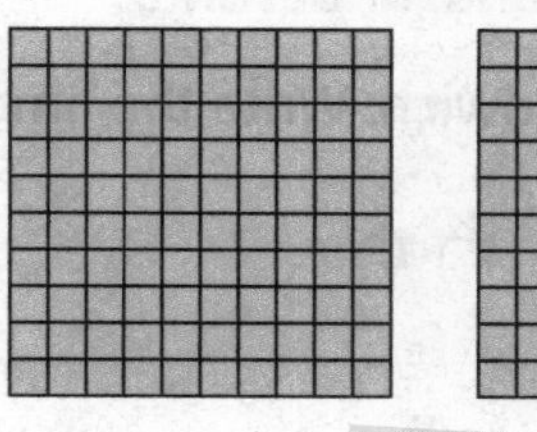

1.75

Lesson 1-7

equivalent decimals

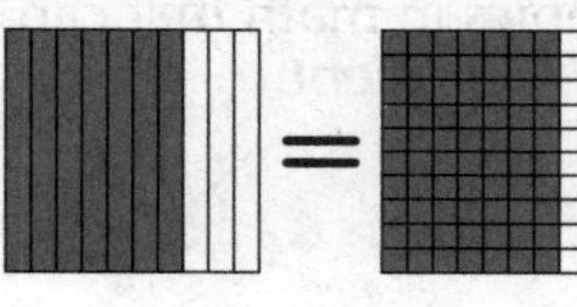

$\frac{7}{10} = \frac{70}{100}$

0.7 = 0.70

Lesson 1-1

expanded form

12,002,060 = 1 × 10,000,000 + 2 × 1,000,000 + 2 × 1,000 + 6 × 10

Lesson 1-1

period

Millions Period			Thousands Period			Ones Period		
Millions			Thousands			Ones		
hundreds	tens	ones	hundreds	tens	ones	hundreds	tens	ones
6	5	0,	0	8	4,	9	7	0

Lesson 1-1

place

Millions Period			Thousands Period			Ones Period		
Millions			Thousands			Ones		
hundreds	tens	ones	hundreds	tens	ones	hundreds	tens	ones
6	5	0,	0	8	4,	9	7	0

Lesson 1-1

standard form

3,000 + 400 + 90 + 1 = 3,491

standard form

Lesson 1-1

place value

Millions Period			Thousands Period			Ones Period		
Millions			Thousands			Ones		
hundreds	tens	ones	hundreds	tens	ones	hundreds	tens	ones
	5	0,	0	8	0,	0	0	0

5 = ten millions
8 = ten thousands

1 → $\frac{1}{4}$, 1/2, $\frac{3}{4}$

Ideas for Use

- Design a crossword puzzle. Use the definition for each word as the clues.
- Work with a partner to name the part of speech of each word. Consult a dictionary to check your answers.

A number that has a digit in the tenths place, hundredths place, and/or beyond.

Erin ate $\frac{3}{4}$ of a sandwich. Write the amount eaten as a decimal.

0.75

A period separating the ones and the tenths in a decimal number.

How does a number with a decimal point differ from a number without one?

A way of writing a number as the sum of the values of its digits.

***Expand* can mean "to stretch or unfold." How can this definition help you remember this definition?**

Decimals that have the same value.

Name two topics in math that can be described as *equivalent.*

The position of a digit in a number.

What is a synonym for *place?*

The name given to each group of three digits in a place-value chart.

What is the meaning of *period* in language arts?

The value given to a digit by its place in a number.

Write a six-digit number. Then write the place value for each digit.

The usual way of writing a number that shows only its *digits,* no words.

Write a different meaning of the word *standard.*

MY Vocabulary Cards

Mathematical PRACTICE

Lesson 1-1

place-value chart

Millions Period			Thousands Period			Ones Period		
Millions			Thousands			Ones		
hundreds	tens	ones	hundreds	tens	ones	hundreds	tens	ones
6	5	0,	0	8	4,	9	7	0

Ideas for Use

- Use a blank card to write this chapter's essential question. Use the back of the card to write or draw examples that help you answer the question.
- Use blank cards to review key concepts from the chapter. Write a few study tips on the back of each card.

A chart showing the value of each number in a multi-digit whole number

Use the space below to write a place-value chart for 7,932,035.

MY Foldable

FOLDABLES® Follow the steps on the back to make your Foldable.

9 ___ ___ , ___ ___ ___ , ___ ___ ___

hundred millions

8 ___ , ___ ___ ___ , ___ ___ ___

ten millions

7 , ___ ___ ___ , ___ ___ ___

millions

6 ___ ___ , ___ ___ ___

hundred thousands

5 ___ , ___ ___ ___

ten thousands

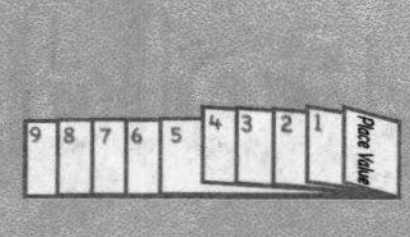

Place Value

1

ones

2 ___

tens

3 ___ ___

hundreds

4 , ___ ___ ___

thousands

Name Krishna

Place Value Through Millions

Lesson 1

ESSENTIAL QUESTION
How does the position of a digit in a number relate to its value?

A **place-value chart** shows the value of the digits in a number. In greater numbers, each group of three digits is separated by commas, and is called a **period**.

Let's GO!

Math in My World

Example 1

The distance from the Earth to the Sun is 92,955,793 miles. Use the place-value chart to list the value of each digit.

1 Complete the place-value chart.

Millions Period			Thousands Period			Ones Period		
hundreds	tens	ones	hundreds	tens	ones	hundreds	tens	ones
0	9	2,	9	5	5,	7	9	3

2 List the values of each digit.

9 × 10,000,000 → 90,000,000

2 × 1,000,000 → 2,000,000

9 × 100,000 → 900,000

5 × 10,000 → 50,000

5 × 1,000 → 5,000

7 × 100 → 700

9 × 10 → 90

3 × 1 → 3

The digit 5 in the ten thousands place is 10 times greater than the digit 5 in the thousands place.

A digit in one **place**, or **place value**, represents 10 times as much as it represents in the place to its right and $\frac{1}{10}$ of what it represents in the place to its left.

The **standard form** of a number is the usual or common way to write a number using digits. The **expanded form** of a number is a way of writing a number as the sum of the values of its digits. The places with zero as a digit are not included in the expanded form.

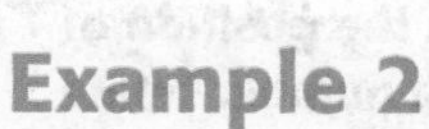

The human eye blinks an average of 5,500,000 times a year. Write 5,500,000 in word form and expanded form.

1 Write the number in the place-value chart.

Millions Period			Thousands Period			Ones Period		
hundreds	tens	ones	hundreds	tens	ones	hundreds	tens	ones
		5,	5	0	0,	0	0	0

2 Write the number in word form.

five million, *five* hundreds *thousand*

3 Write the number in expanded form.

five million: 5 × 1,000,000

five hundred thousand: 5 × 100,000

In expanded form, 5,500,000 =

5 ×  +  × 

Guided Practice

Write the value of the highlighted digit.

1. 469,999 tens
2. 35,098,098 ten thousand
3. Circle the digit in the ten thousands place.

   1, 2 3 5, 9 8 0

Name ....

Independent Practice

Write the value of the highlighted digit.

4. 3,132,685 hundreds **5.** 5,309,573 hundreds thousand **6.** 1,309,841 thousands

Write each number in word form and expanded form.

7. 5,901,452 Five million nine hundred and one thousand four hundred fifty tow

5000000 + 900000 + 1000 + 400 + 50 + 2

8. 309,099,990 three hundred and nine million ninty nine thousand, nine hundred ninety

300000020 + 9000000 + 90000 + 9000 + 900 + 90

Write each number in standard form and expanded form.

9. *eighty-three million, twenty-three thousand, seven*

83,023,007

10. *three hundred four million, eight hundred thousand, four hundred*

304,800,400

Use the place-value chart for Exercises 11 and 12.

Millions Period			Thousands Period			Ones Period		
hundreds	tens	ones	hundreds	tens	ones	hundreds	tens	ones
		5,	9	0	1,	4	5	2

11. The 9 is in the hundreds place.

12. The 1 has a value of 1 × 1000.

Problem Solving

13. In a recent year, the population of the United States was about 304,967,000. Write the population in word form.

three hundred four million nine hundred sixty seven thousand

14. The land area of Florida is $1 \times 100{,}000 + 3 \times 10{,}000 + 9 \times 1{,}000 + 8 \times 100 + 5 \times 10 + 2 \times 1$ square kilometers. Write the area in standard form and word form.

100000 + 30000 + 9000 + 800 + 50 + 2
one hundred thirty nine thousand eight hundred fifty two

15. **Mathematical PRACTICE 6** **Explain to a Friend** The amount of time that American astronauts have spent in space is about 13,507,804 minutes. Is the number read as *thirteen million, fifty-seven thousand, eight hundred four*? Explain to a friend or classmate.

it is not fifty seven thousand but five hundred seven thousand.

Out of this world!

HOT Problems

16. **Mathematical PRACTICE 2** **Use Number Sense** Write the number with the least value using the digits 1 through 9. Use each digit only once.

19

17. **Building on the Essential Question** Explain how you know what number is missing in the equation $3{,}947 = 3{,}000 + \blacksquare + 40 + 7$.

The answer is 3900 you can get the answer by adding 3000 + 40 + 7 then subtracting it from 3947.

Name

MY Homework

Lesson 1
Place Value Through Millions

Homework Helper

Need help? connectED.mcgraw-hill.com

Use the place-value chart to write 12,498,750 in word form and expanded form.

1 Write the number in the place-value chart.

Millions Period			Thousands Period			Ones Period		
hundreds	tens	ones	hundreds	tens	ones	hundreds	tens	ones
	1	2,	4	9	8,	7	5	0

2 Write the number in word form.

twelve million, four hundred ninety-eight thousand, seven hundred fifty

3 Write the number in expanded form.

$1 \times 10{,}000{,}000 + 2 \times 1{,}000{,}000 + 4 \times 100{,}000 +$

$9 \times 10{,}000 + 8 \times 1{,}000 + 7 \times 100 + 5 \times 10$

Practice

Write the value of the highlighted digit.

1. 1,283,479 ten thousands

2. 50,907,652 hundred thousand

3. 318,472,008 hundred million

4. Write 103,727,495 in word form and expanded form.

One hundred three million seven hundred twenty seven thousand four hundred ninty five.

Problem Solving

5. Hanna stated that 11,760,825 people saw the Miami Heat play last season. Chris wants to be sure he heard the number correctly. Write 11,760,825 in word form and expanded form for Chris.

eleven million seven hundred sixty thousand eight hundred twenty five

10000000 + 1000000 + 700000 + 60000 + 800 + 20 + 5

6. Mathematical **PRACTICE** 3 **Find the Error** American car makers produce 5,650,000 cars each year. In a report, Ben wrote that Americans made 6,550,000 cars. What mistake did Ben make? How can he fix it?

The misitake ben made is he switched the frist two numbers

Vocabulary Check 

Match the vocabulary word with its definition.

7. standard form	each group of three digits on a place-value chart
8. period	a way of writing a number as the sum of the values of its digits
9. expanded form	the usual or common way to write a number using digits

Test Practice

10. A popular pirate movie made $135,634,554 in sales during one weekend. What is the value of the highlighted digit?

Ⓐ $30,000 Ⓒ $300,000

Ⓑ $3,000,000 Ⓓ $30,000,000

Name Krishna

Compare and Order Whole Numbers Through Millions

Lesson 2

ESSENTIAL QUESTION
How does the position of a digit in a number relate to its value?

To compare numbers, you can use place value and the symbols <, >, and =.

Words	Symbol
is greater than	>
is less than	<
is equal to	=

Math in My World

Example 1

The table shows the two largest oceans in the world. Which ocean has a greater area?

Ocean	Approximate Area (square miles)
Atlantic Ocean	33,420,160
Pacific Ocean	64,186,600

Write the numbers in the place-value chart.

Millions Period			Thousands Period			Ones Period			
hundreds	tens	ones	hundreds	tens	ones	hundreds	tens	ones	
	3	3,	4	2	0,	1	6	0	← Atlantic Ocean
	6	4,	1	8	6,	6	0	0	← Pacific Ocean

Begin at the greatest place. Compare the digits.

6 (>) 3

Since 6 is greater than 3, then 64,186,600 > 33,420,160.

So, the pacific ocean has a greater area.

Example 2 Tutor

The table shows the area in square miles in different countries. Use place value to order the countries from *greatest* area to *least* area.

Country Areas	
Country	**Area (square miles)**
Argentina	1,068,296
Australia	2,967,893
India	1,269,338
Norway	125,181

1 Line up the ones place. Compare the digits in the greatest place.

2 > 1

So, Australia has the greatest area.

1,068,296
2,967,893
1,269,338
125,181

2 Compare the digits in the next place.

2 > 0.

So, India has the next greatest area.

3 Since 1,068,296 > 125,181, the area of Argentina is greater than the area of Norway.

The order from *greatest* area to *least* area is Australia, India, Argentine, and Norway.

Talk MATH

When ordering whole numbers, explain what to do when the digits in the same place have the same value.

Guided Practice Check

Write <, >, or = in each  **to make a true sentence.**

1. 655,543 (>) 556,543

2. 10,027,301 (<) 10,207,301

3. Order the numbers 145,099; 154,032; 145,004; and 159,023 from *greatest* to *least*.

159025, 154032, 145099, 145004

Name

Independent Practice

Write <, >, or = in each ◯ to make a true sentence.

4. 462,211 ◯ 426,222

5. 42,235,909 ◯ 42,324,909

6. 20,318,523 ◯ 21,318,724

7. 96,042,317 ◯ 96,042,317

8. 132,721,424 ◯ 132,721

9. 152,388,000 ◯ 152,388,010

10. 113,222,523 ◯ 113,333,523

11. 767,676,767 ◯ 676,767,676

Order the numbers from *greatest* to *least*.

12. 138,023; 138,032; 139,006; 183,487

13. 3,452,034; 4,935,002; 34,035,952; 34,530,953

14. 731,364,898; 731,643,898; 73,264,898; 731,643,989

15. 395,024,814; 593,801,021; 395,021,814; 39,021,814

Order the numbers from *least* to *greatest*.

16. 85,289,688; 85,290,700; 85,285,671; 85,301,001

17. 32,356,800; 33,353,800; 32,937,458; 33,489,251

18. 2,009,146; 2,037,579; 2,006,981; 2,011,840

19. 854,236,100; 855,963,250; 855,903,675; 854,114,370

Problem Solving

20. Rank the following states from *least* to *greatest* population.

State Population	
State	**Population**
Alabama	4,627,851
Colorado	4,861,515
Mississippi	2,918,785
Ohio	11,466,917

21. Order the cars from most expensive to least expensive.

Most Expensive Cars	
Car	**Price ($)**
Bugatti Veyron 16.4	1,192,057
Leblanc Mirabeau	645,084
Pagani Zonda Roadster	667,321
Saleen S7	555,000

My Work!

HOT Problems

22. Mathematical PRACTICE 2 **Reason** Write three numbers that are greater than 75,300,000 but less than 75,400,000.

23. **Building on the Essential Question** How do you compare whole numbers through the millions?

Name Krishna

MY Homework

Lesson 2

Compare and Order Whole Numbers Through Millions

Homework Helper

Need help? connectED.mcgraw-hill.com

Order the following numbers from *least* to *greatest*.

84,189,688; 85,290,700; 58,285,671; 80,301,785

1. Line up the ones place. Compare the digits in the greatest place.

   5 < 8
   So, 58,285,671 is the least.

84,189,688
85,290,700
58,285,671
80,301,785

2. Compare the digits in the next place.
   80,301,785 < 84,189,688

3. 85,290,700 > 84,189,688
   So, 85,290,700 is the greatest number.

The order from least to greatest is 58,285,671; 80,301,785; 84,189,688; and 85,290,700.

Practice

Write <, >, or = in each ◯ to make a true sentence.

1. 67,982,001 (>) 67,892,001

2. 100,542,089 (<) 105,042,098

3. 1,986,034 (>) 1,896,075

4. 12,165,982 (<) 12,178,983

5. 239,742,005 (<) 289,650,010

6. 1,652,985 (>) 1,563,218

Order the numbers from *greatest* to *least*.

7. 3,356,000; 2,359,412; 2,937,158; 3,368,742

3368742, 3356000, 2937158, 2359412

8. 2,009,832; 2,103,425; 2,009,604; 2,112,300

2112300, 2103425, 2009832, 2009604

Order the numbers from *least* to *greatest*.

9. 14,258,123; 14,259,688; 14,256,001; 14,258,252

10. 574,210,033; 574,211,874; 574,198,852; 874,210,089

Problem Solving

11. Madison wants to know which sports are most popular. The list below shows how many kids play each sport. Order the sports from most players to least players to help show Madison which sports are most popular.

Soccer: 3,875,026 Surfing: 250,982
Baseball: 900,765 Basketball: 2,025,351

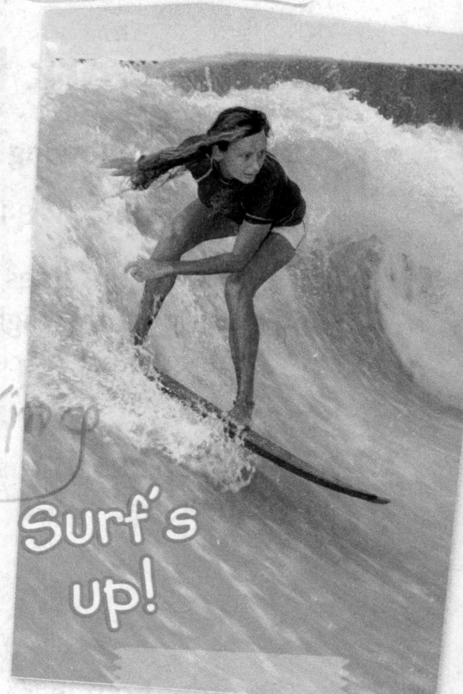

12. Andrea wants to live in the city with the most people. She read that New York City has 8,008,278 people and that Seoul, South Korea has 10,231,217 people. In which city does Andrea want to live?

13. The Denver Mint made 2,638,800 nickels. The Philadelphia Mint made 2,806,000 nickels. Which mint made more nickels?

14. Mathematical PRACTICE 3 **Draw a Conclusion** In 1950, bike stores sold about 205,850 bikes per year per store. In 2000, bike stores sold about 185,000 bikes per year per store. Is the number of bikes being sold getting larger or smaller?

Test Practice

15. Which set of numbers are in order from *greatest* to *least*?

Ⓐ 74,859,623; 74,759,458; 74,905,140; 73,569,991

Ⓑ 74,905,140; 74,859,623; 74,759,458; 73,569,991

Ⓒ 73,569,991; 74,759,458; 74,859,623; 74,905,140

Ⓓ 74,905,140; 74,759,458; 74,859,623; 73,569,991

Name

Hands On
Model Fractions and Decimals

Lesson 3

ESSENTIAL QUESTION
How does the position of a digit in a number relate to its value?

Numbers that have digits in the tenths place, hundredths place, and/or beyond are called **decimals**. A **decimal point** is used to separate the ones place from the tenths place.

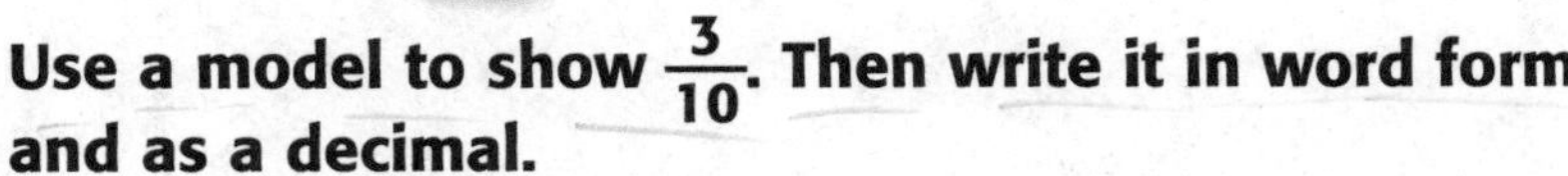

Draw It

Tools

Use a model to show $\frac{3}{10}$. Then write it in word form and as a decimal.

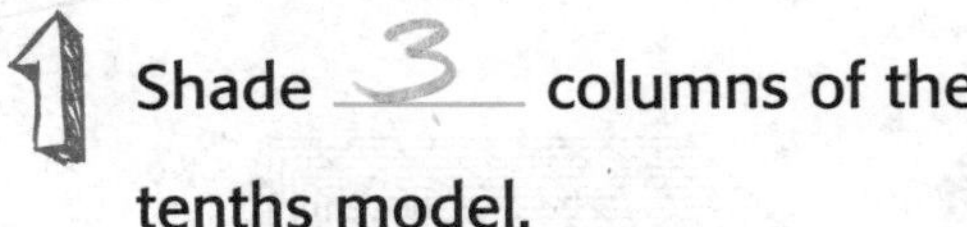

1 Shade 3 columns of the tenths model.

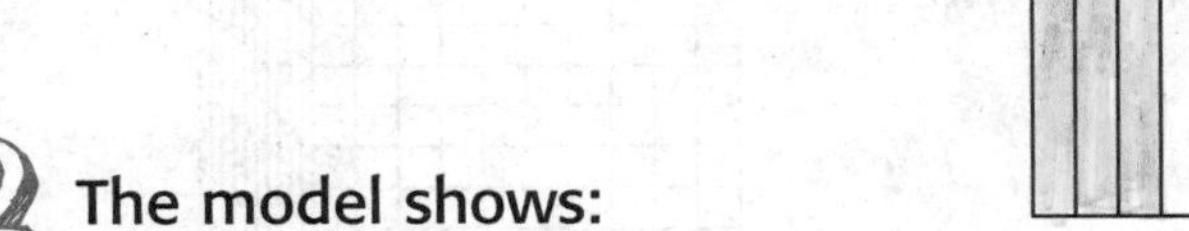

2 The model shows:

Word form: 3 *tenths*

Decimal: 0. 3

Ones	Tenths
0	3

Try It

Use a model to show $\frac{9}{100}$. Then write it in word form and as a decimal.

1 Shade 9 of the small squares on the hundredths model.

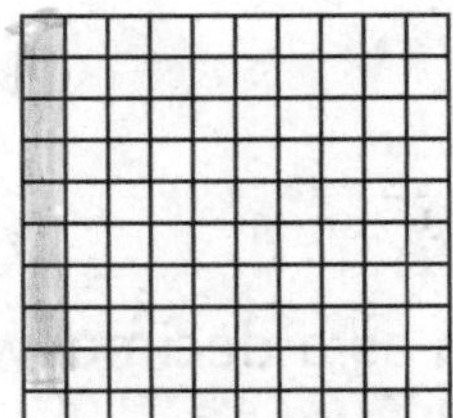

2 The model shows:

Word form: 9 *hundredths*

Decimal: 0. 0 9

Ones	Tenths	Hundredths
0	0	9

Try It

Use a model to show $\frac{34}{100}$. Then write it in word form and as a decimal.

1 Shade 34 of the small squares.

2 The model shows:

Word form: 4 *hundredths*

Notice that there are 3 tenths

and 4 hundredths shaded.

Decimal: 0. 3 4

Ones	Tenths	Hundredths
0	. 3	4

Talk About It

1. The model at the right shows a thousandths cube. What fraction of the model is shaded? Then write it as a decimal.

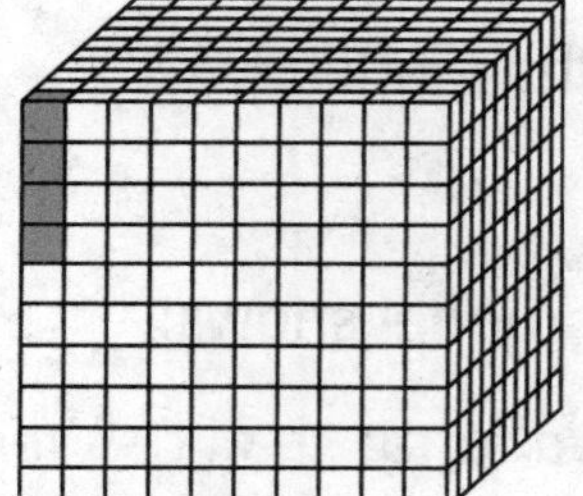

four of the fraction is shaded

0.004

2. Mathematical PRACTICE 5 **Use Math Tools** Shade the model to show $\frac{80}{100}$. Then write the fraction in word form and as a decimal.

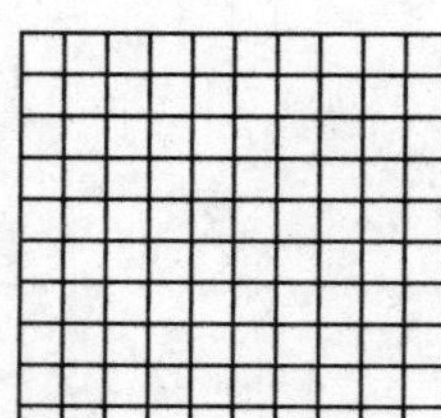

3. Explain why $\frac{45}{100}$ is written as a decimal with a 4 in the tenths place and a 5 in the hundredths place.

Name

Practice It

Shade the model to show each fraction. Write each fraction in word form.

4. $\frac{7}{10}$

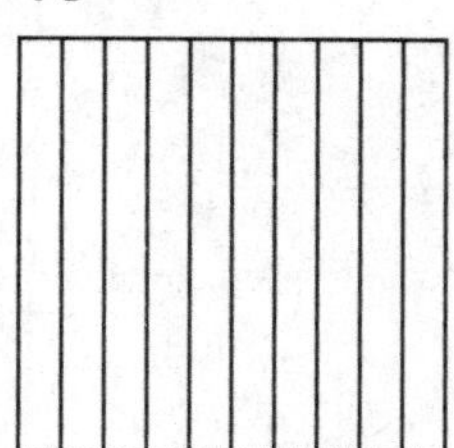

5. $\frac{9}{10}$

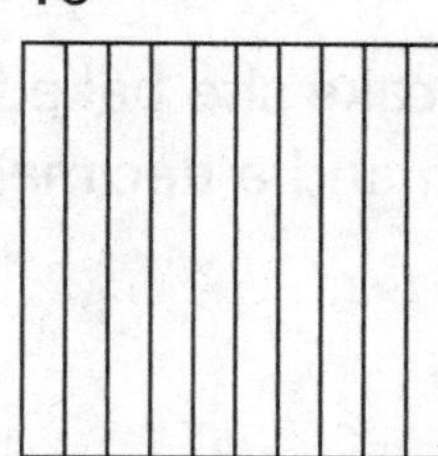

6. $\frac{4}{10}$

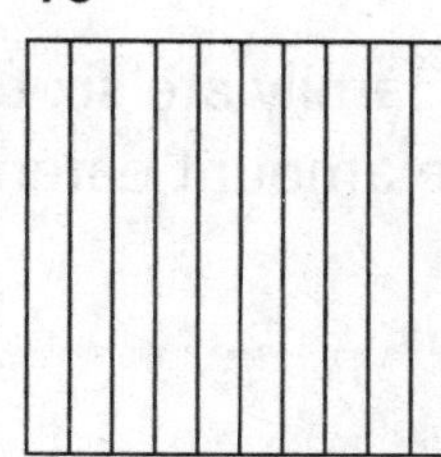

7. $\frac{5}{100}$

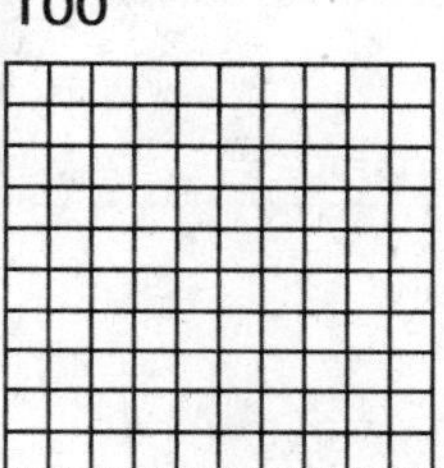

8. $\frac{63}{100}$

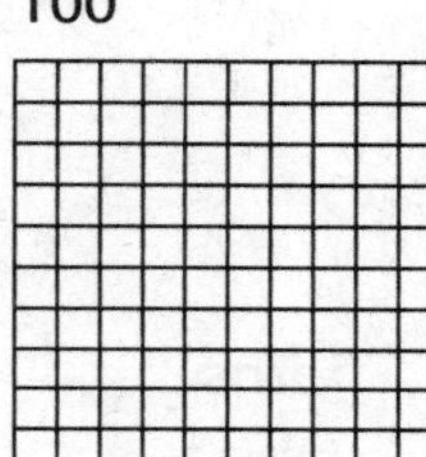

9. $\frac{21}{100}$

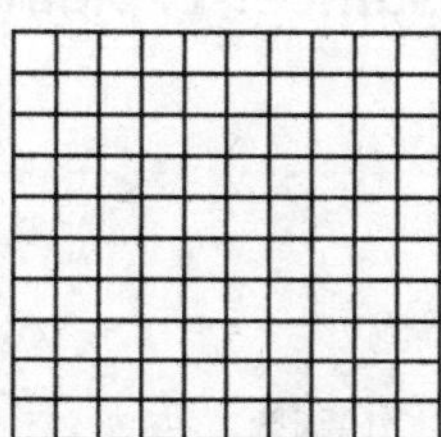

Write the decimal for each model.

10.

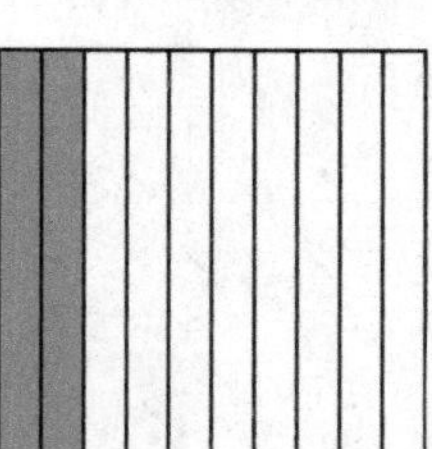

11.

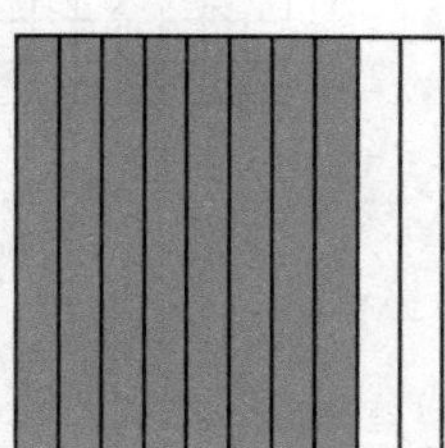

12.

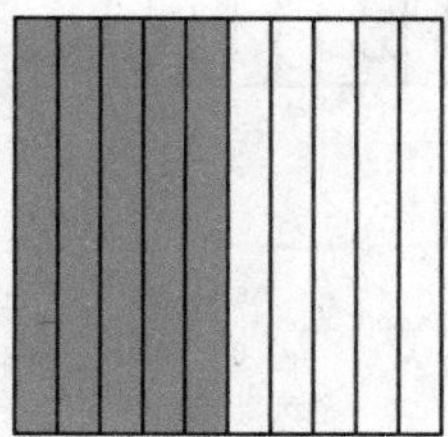

13.

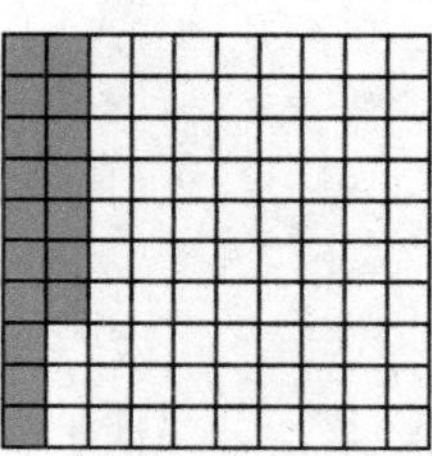

14.

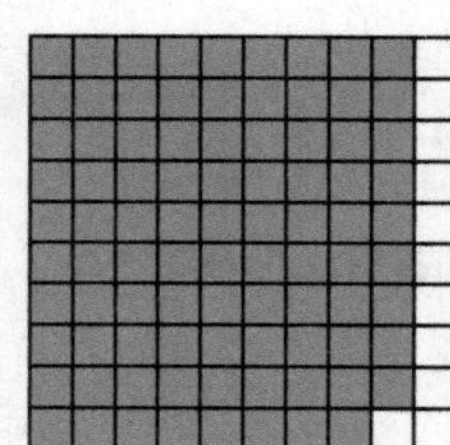

15.

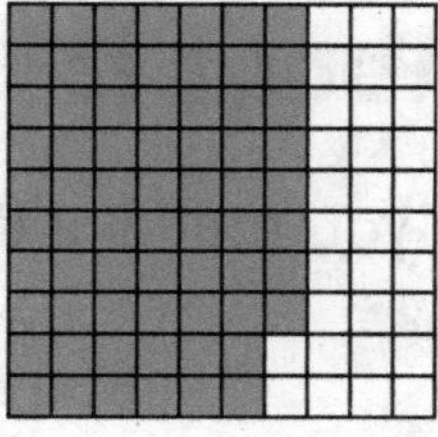

Apply It

16. Dewayne waters his plant with 0.76 quart of water every other day. Write the amount of water he uses in word form and as a fraction.

17. Jessica's family ate six-tenths of the cake she baked. Write the amount eaten as a fraction and a decimal.

18. Mathematical PRACTICE 3 **Draw a Conclusion** Mei rode her bike $\frac{7}{10}$ mile to school. Her friend Julie walked 0.07 mile to school. Did the two girls travel the same distance? Explain.

19. Mathematical PRACTICE 2 **Use Number Sense** Write the decimal and fraction for each model. Are the two models equal? Explain.

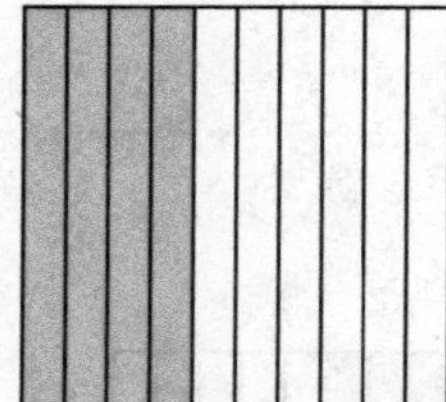

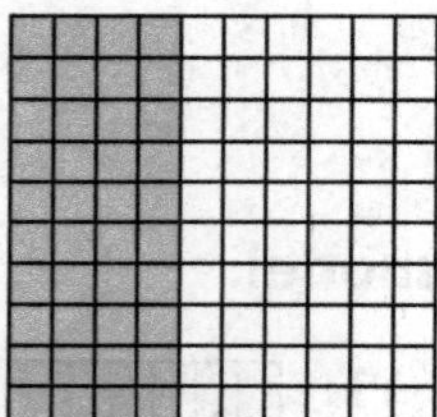

Write About It

20. How do you model a decimal using a fraction?

Name

MY Homework

Lesson 3

Hands On: Model Fractions and Decimals

Homework Helper

Need help? connectED.mcgraw-hill.com

Use a model to show $\frac{57}{100}$. Then write it in word form and as a decimal.

1 Shade 57 of the small squares.

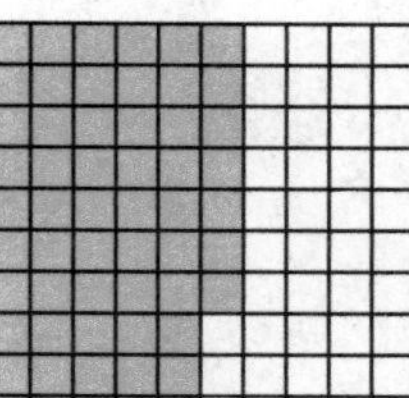

2 The model shows:

Word form: *fifty-seven hundredths*
Decimal: 0.57

Practice

Shade the model to show each fraction. Write each fraction in word form.

1. $\frac{4}{10}$

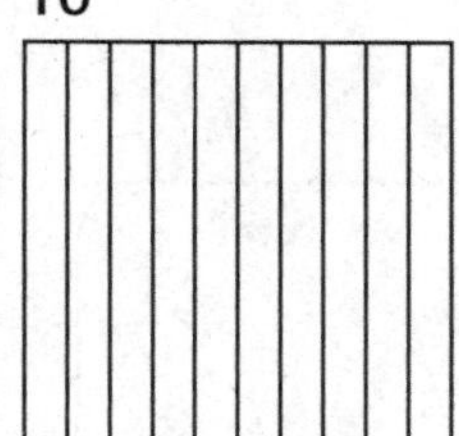

____

2. $\frac{8}{10}$

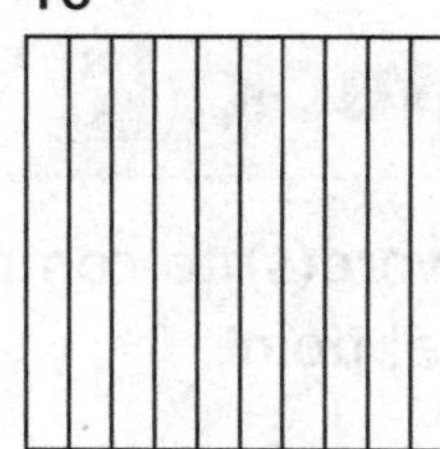

____

3. $\frac{45}{100}$

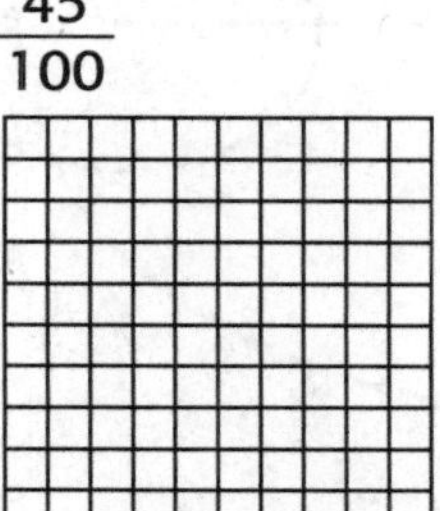

____

4. $\frac{76}{100}$

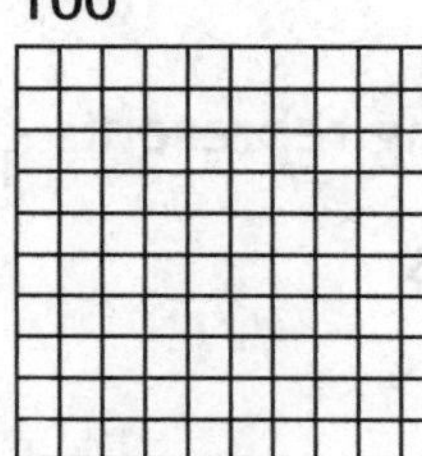

____

Write the decimal for each model.

5. 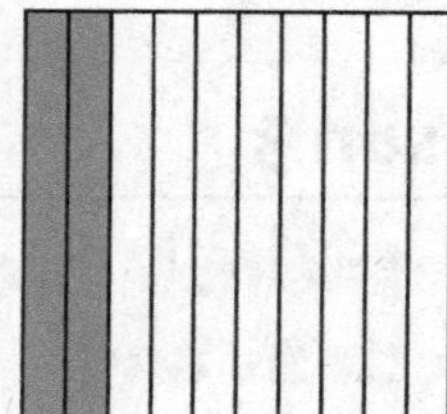

6. 

7. 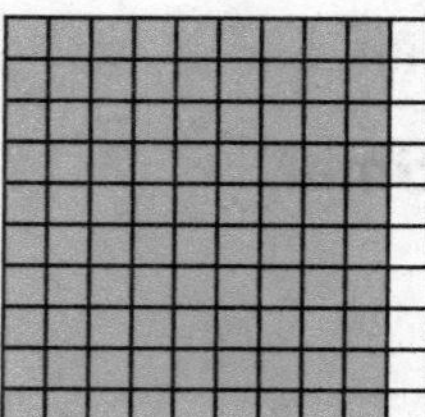

8. 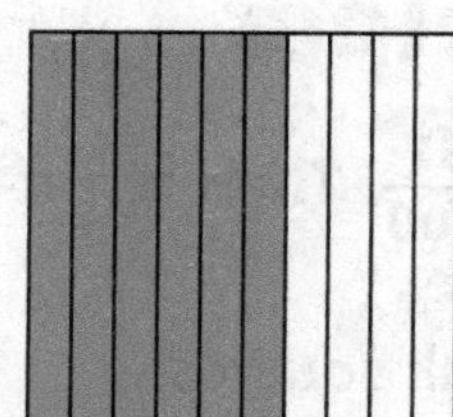

Problem Solving

9. **Mathematical PRACTICE 2** **Use Number Sense** The largest butterfly in the world is found in Papua, New Guinea. The female of the species weighs about $\frac{9}{10}$ ounce. Use a decimal to write the female's weight.

Vocabulary Check

10. Choose the correct word(s) to complete the sentence below.
decimal   decimal point

A ______________ is used to separate the ones place from the tenths place.

11. Which of the following represents $\frac{32}{100}$? Circle it.

thirty-two hundredths

0.032

0.23

thirty-two tenths

Name

Represent Decimals

Lesson 4

ESSENTIAL QUESTION
How does the position of a digit in a number relate to its value?

Math in My World

Example 1

A bee hummingbird weighs only about $\frac{56}{1,000}$ of an ounce. Represent this fraction as a decimal. Then write it in word form.

1 The model represents thousandths by showing one thousand small cubes.

Shade ______ of the small cubes.

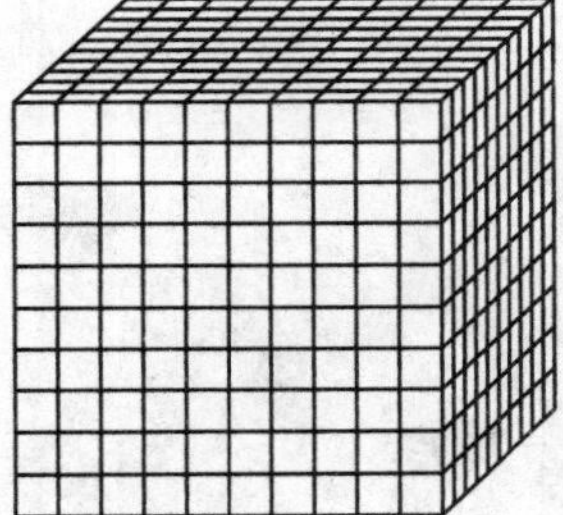

2 The fraction names thousandths, so there should be three digits to the right of the decimal point.

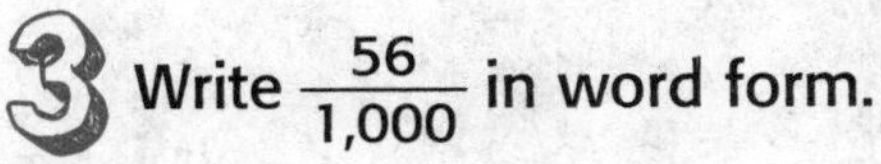

$\frac{56}{1,000}$ = 0. ☐ ☐ ☐

3 Write $\frac{56}{1,000}$ in word form.

$\frac{56}{1,000}$ ← fifty-six ← thousandths

So, $\frac{56}{1,000}$ is ______________________.

Example 2 

Model $\frac{35}{100}$. Then write it in word form and as a decimal.

 Shade ______ of the small squares.

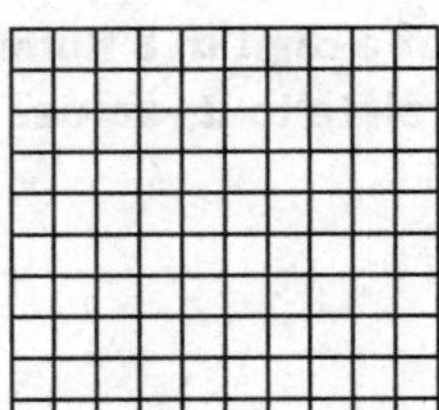

2 The fraction names *thirty-five* ____________, so there should be two digits to the right of the decimal point.

So, $\frac{35}{100}$ is *thirty-five hundredths* and 0.☐☐.

Guided Practice 

Shade the model. Then write each fraction as a decimal.

1. $\frac{2}{10}$

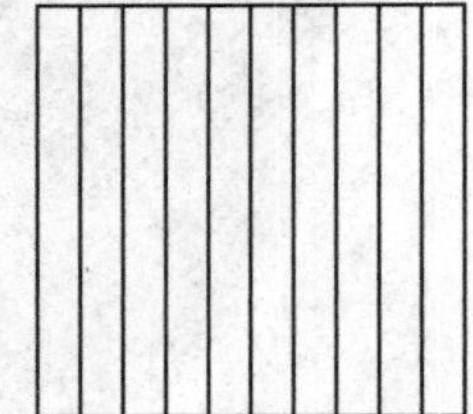

Decimal: ____________

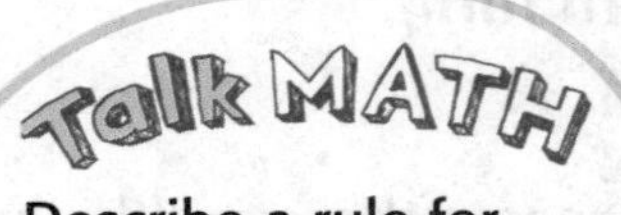

Describe a rule for writing fractions like $\frac{8}{100}$ and $\frac{32}{1,000}$ as decimals.

2. $\frac{58}{100}$ 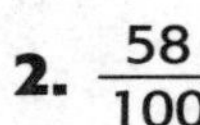

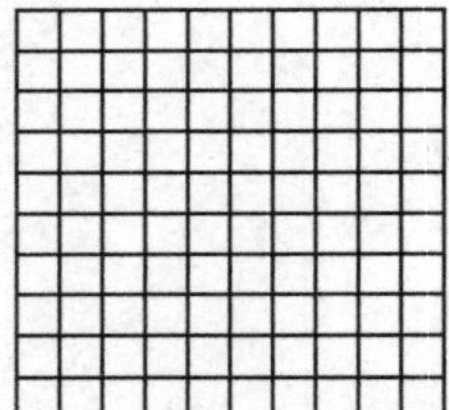

Decimal: ____________

3. $\frac{95}{1,000}$

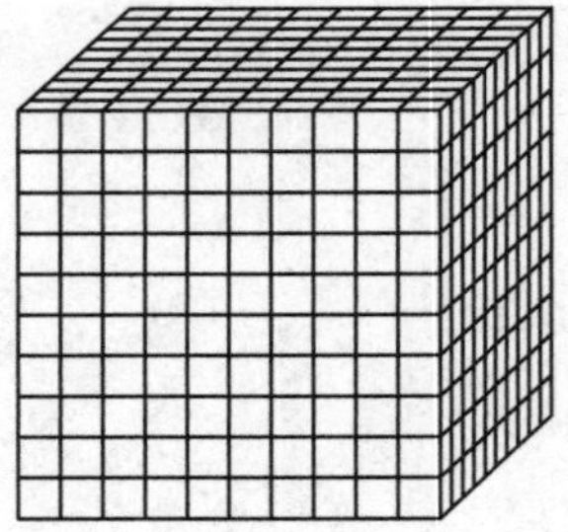

Decimal: ____________

Name ....

Independent Practice

Shade the model. Then write each fraction in word form and as a decimal.

4. $\frac{3}{10}$

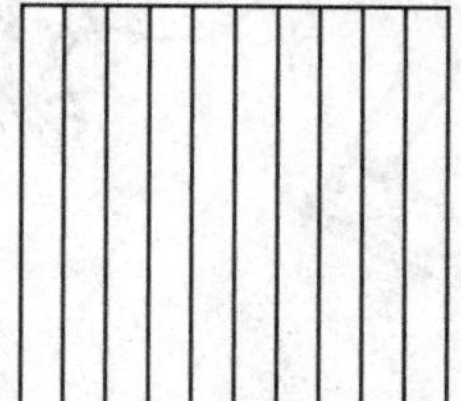

Word form: ______________________

Decimal: ______________________

5. $\frac{86}{100}$

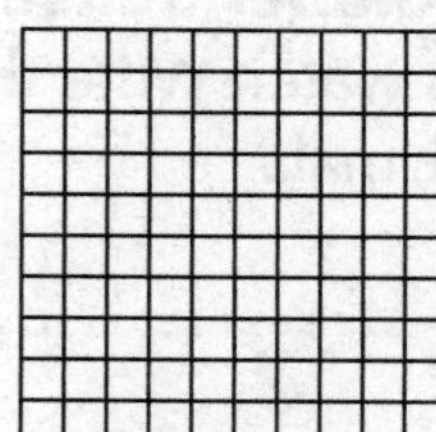

Word form: ______________________

Decimal: ______________________

6. $\frac{99}{100}$

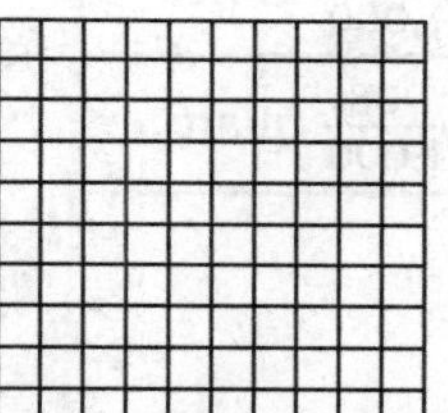

Word form: ______________________

Decimal: ______________________

7. $\frac{51}{1,000}$

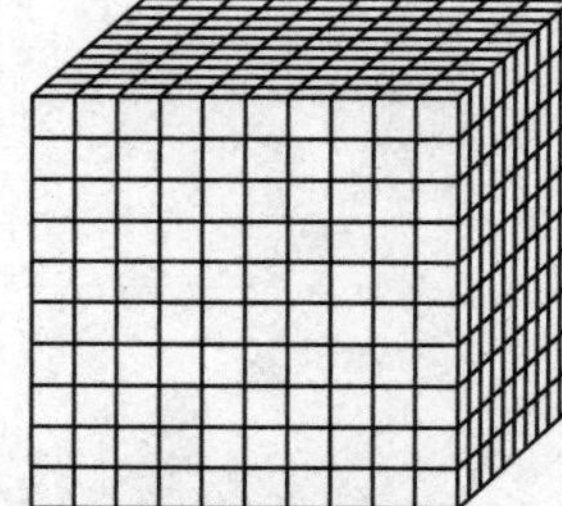

Word form: ______________________

Decimal: ______________________

8. $\frac{22}{1,000}$

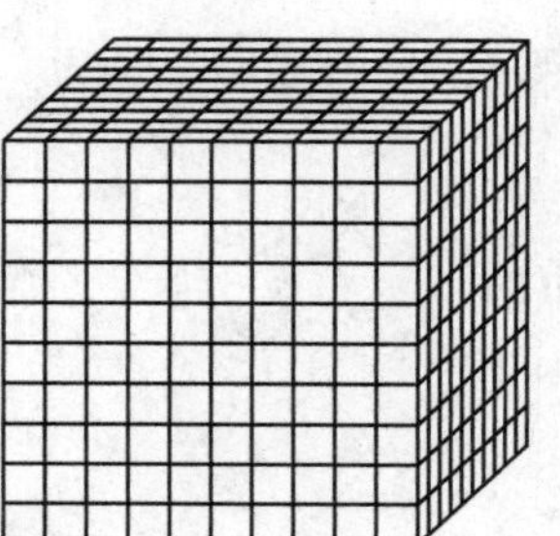

Word form: ______________________

Decimal: ______________________

9. $\frac{1}{1,000}$

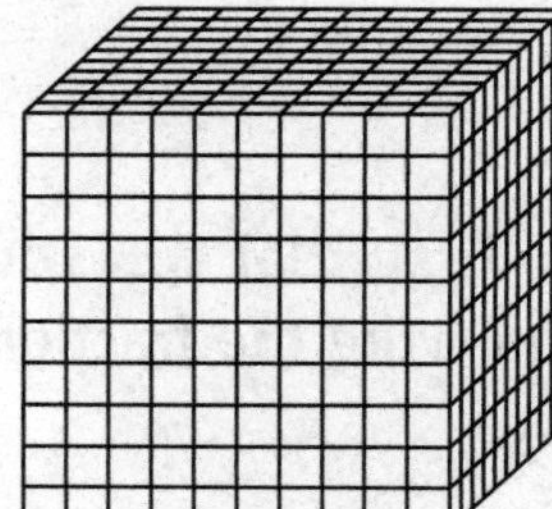

Word form: ______________________

Decimal: ______________________

Problem Solving

10. A runner decreased his time by $\frac{5}{100}$ second. Express the decrease as a decimal.

____

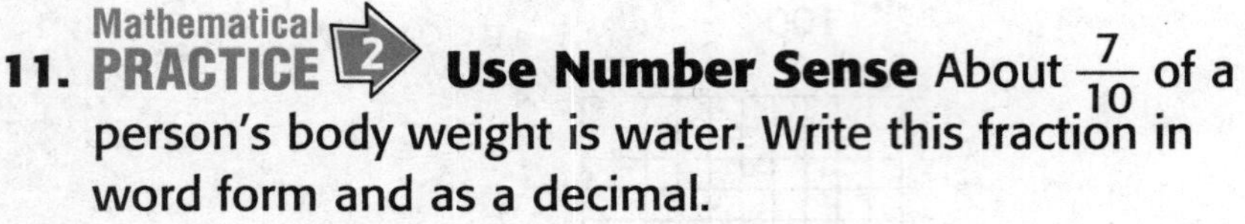

11. Mathematical PRACTICE 2 **Use Number Sense** About $\frac{7}{10}$ of a person's body weight is water. Write this fraction in word form and as a decimal.

____

Write the customary measure for each metric measure as a decimal.

Metric Measure	Customary Measure
1 kilometer	$\frac{62}{100}$ mile
1 millimeter	$\frac{4}{100}$ inch
1 gram	$\frac{35}{1,000}$ ounce
1 liter	$\frac{908}{1,000}$ quart

12. 1 kilometer = ______________

13. 1 millimeter = ______________

14. 1 gram = ______________

15. 1 liter = ______________

HOT Problems

16. Mathematical PRACTICE 3 **Find the Error** Dylan is writing $\frac{95}{1,000}$ as a decimal. Find his mistake and correct it.

$\frac{95}{1,000} = 0.950$

____

____

17. **Building on the Essential Question** How can the word form of a fraction help you write the fraction as a decimal?

____

____

____

Name

MY Homework

Lesson 4
Represent Decimals

Homework Helper

Need help? connectED.mcgraw-hill.com

Model $\frac{98}{1,000}$. Then write it in word form and as a decimal.

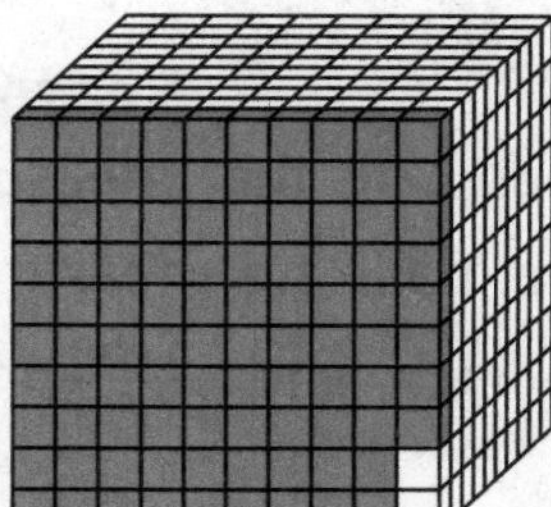

1. The model shows 98 of the small cubes are shaded.

2. The fraction names *ninety-eight thousandths*, so there should be three digits to the right of the decimal point.

So, $\frac{98}{1,000} = 0.098$.

Practice

Shade the model. Then write each fraction in word form and as a decimal.

1. $\frac{7}{10}$

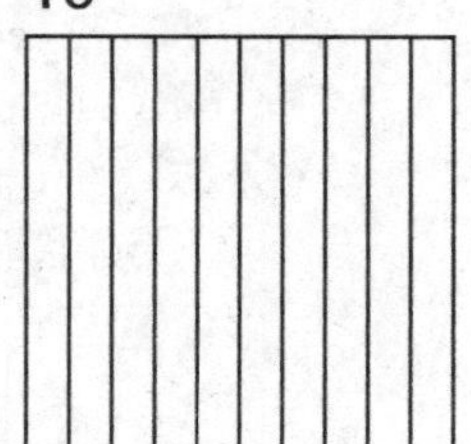

2. $\frac{62}{100}$

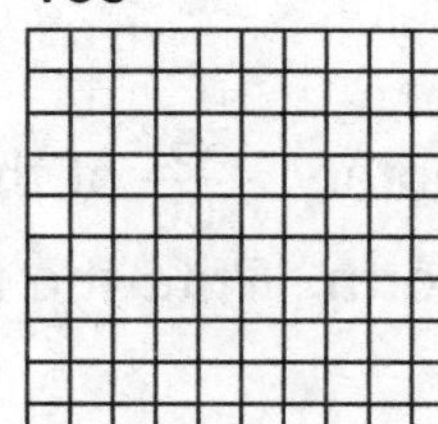

3. $\frac{91}{1,000}$

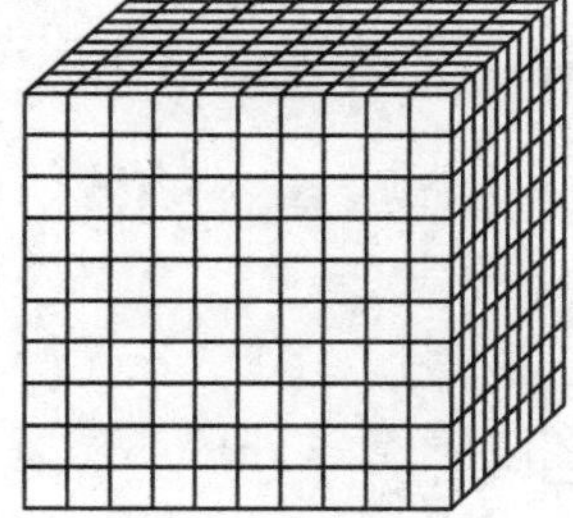

4. $\frac{75}{1,000}$

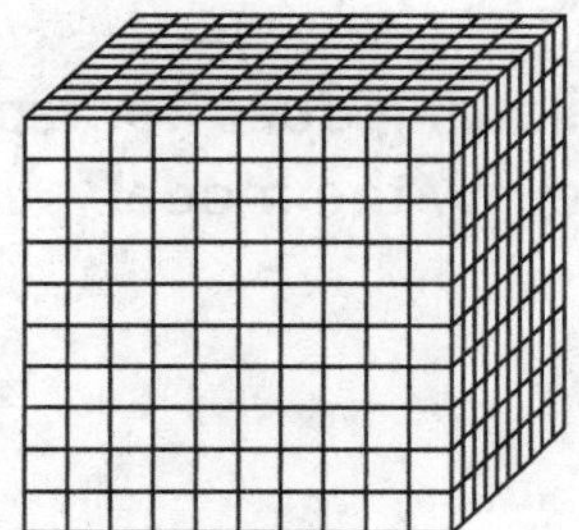

Write each fraction as a decimal.

5. $\frac{15}{100} =$ ______ **6.** $\frac{129}{1,000} =$ ______ **7.** $\frac{17}{100} =$ ______

8. $\frac{8}{10} =$ ______ **9.** $\frac{815}{1,000} =$ ______ **10.** $\frac{2}{10} =$ ______

Problem Solving

11. Mathematical PRACTICE 2 **Reason** Trudy is making a picture frame and needs nails that measure 0.375 inch. At the hardware store, nails are measured in fractions of an inch: $\frac{125}{1,000}$ inch, $\frac{25}{100}$ inch, and $\frac{375}{1,000}$ inch. Which of these nails should she buy?

______

12. It rained 16 hundredths of an inch on Tuesday. Write this amount as a decimal and a fraction.

______

13. At Richardson Elementary, $\frac{35}{100}$ of the buses were late because of a snowstorm. Write the fraction as a decimal.

______

Test Practice

14. Which of the following does *not* represent the number given by the model?

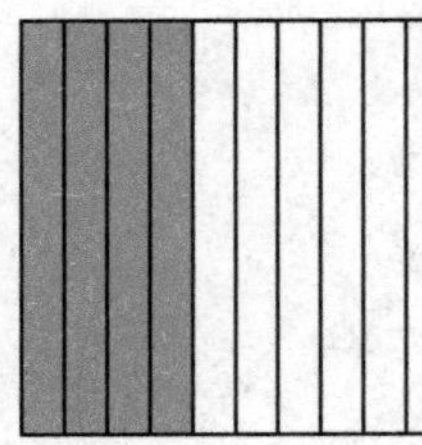

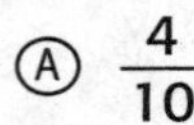

Ⓐ $\frac{4}{10}$

Ⓑ *forty tenths*

Ⓒ 0.4

Ⓓ *four tenths*

Check My Progress

Vocabulary Check

Choose the correct word(s) to complete each sentence.

decimal **expanded form** **period** **standard form**

1. Each group of three digits on a place-value chart is called a(n) ________________.

2. ________________ is the usual or common way to write a number using digits.

3. A(n) ________________ is a number that has a digit in the tenths place, hundredths place, and/or beyond.

4. A way of writing a number as the sum of the values of its digits is called ________________.

Concept Check

Name the place of the highligted digit. Then write the value of the digit.

5. 42,924,603 ________________

6. 953,187 ________________

7. Write 13,180,000 in expanded form.

____

8. Write 4,730,000 in word form.

____

Write <, >, or = in each ◯ to make a true sentence.

9. 84 ◯ 90 **10.** 542 ◯ 524 **11.** 925 ◯ 1,024 **12.** 6,123 ◯ 6,231

Shade the model. Then write each fraction in word form and as a decimal.

13. $\frac{1}{10}$ **14.** $\frac{85}{100}$ **15.** $\frac{39}{1,000}$

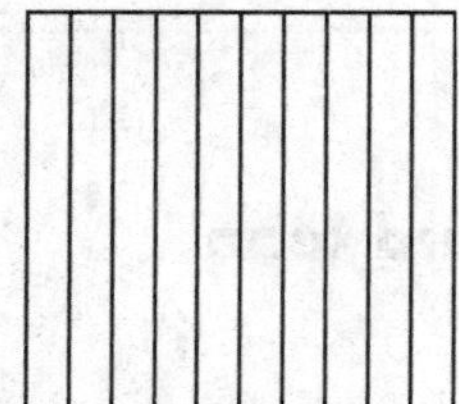

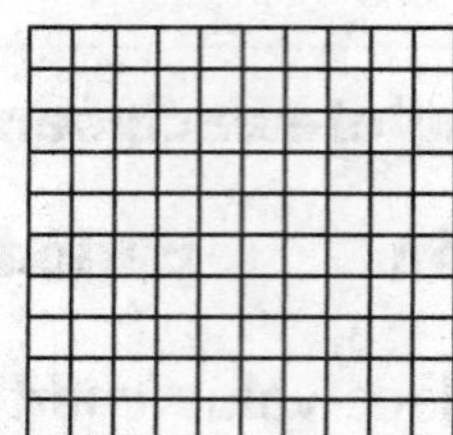

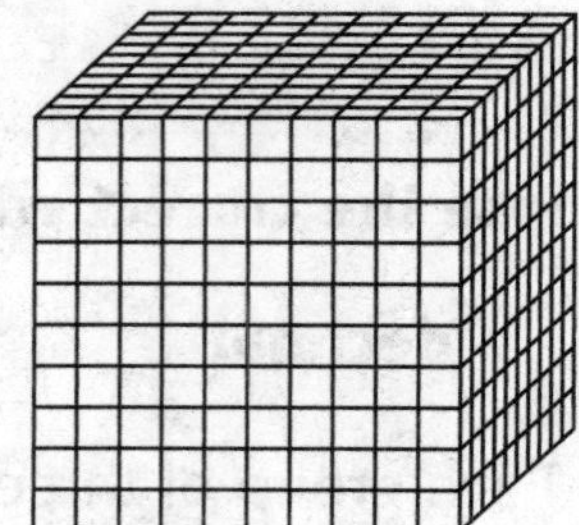

______________ ______________ ______________

______________ ______________ ______________

Problem Solving

16. The attendance at Friday's baseball game was 45,673. Sunday's game attendance was 45,761. Which game had a greater attendance?

____

17. The shortest fish ever recorded is the dwarf goby, found in the Indo-Pacific. The female of this species is about $\frac{35}{100}$ inch long. Use a decimal to write the female's length.

____

Test Practice

18. Which decimal represents the shaded part of the figure?

Ⓐ 0.0052 Ⓒ 0.52

Ⓑ 0.052 Ⓓ 5.2

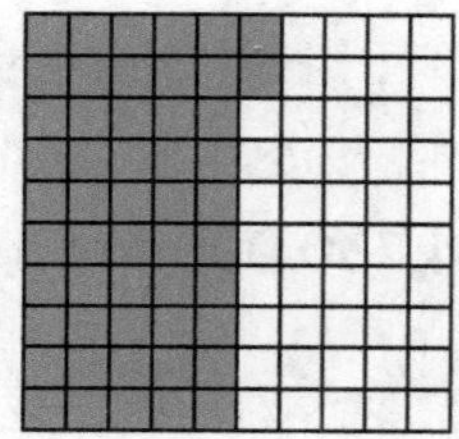

Name

Hands On
Understand Place Value

Lesson 5

ESSENTIAL QUESTION
How does the position of a digit in a number relate to its value?

Draw It

Use models to describe the relationship between the value of the digits in the decimal 0.77 and their place-value position.

1 Use models.

Shade the model to show 0.77. Write the decimal in word form and standard form.

Word form: *seventy-*______ *hundredths*

Standard form: 0.77

The place-value chart that you used for whole numbers can be extended to include decimals.

Write the decimal 0.77 in the place-value chart.

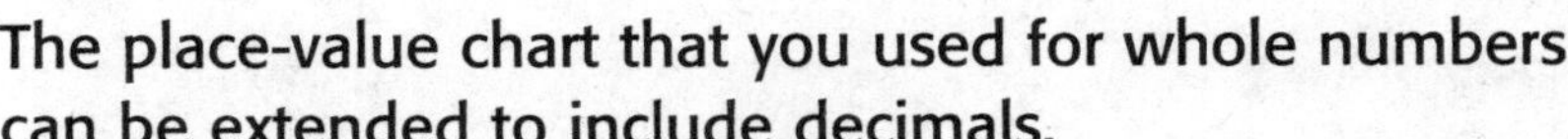

Tens	Ones	Tenths	Hundredths	Thousandths
	.			

2 Describe the relationship.

The value of the digit 7 in the tenths place is 0.7 or ______ *tenths.*

The value of the digit 7 in the hundredths place is 0.07 or

______ *hundredths.*

Use a calculator to find 0.7 ÷ 0.07. ______

The value of the digit 7 in the tenths place is ______ times as much as the value of the digit 7 in the hundredths place.

The value of the digit 7 in the hundredths place is $\frac{\square}{\square}$ times as much as the value of the digit 7 in the tenths place.

Try It

Describe the relationship between the value of the digits in the decimal 0.027 and their place-value position.

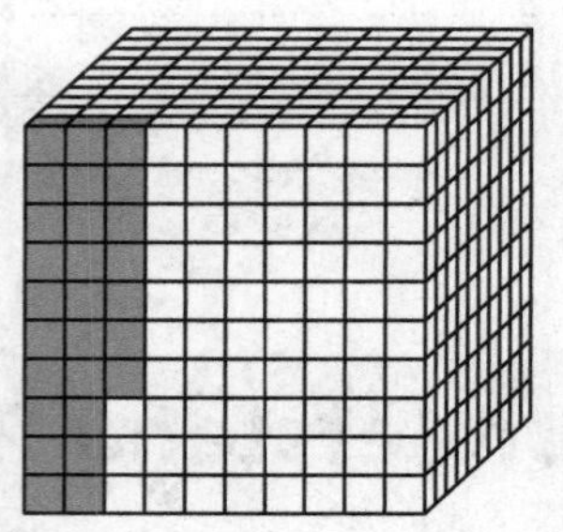

Use a place-value chart.

Write the decimal in the place-value chart.

Tens	Ones	Tenths	Hundredths	Thousandths
	•			

2 Describe the relationship.

The digit in the thousandths place is ______.

It has a value of ______.

If this digit were to move to the hundredths place,

it would have a value of ______.

If this digit were to move to the tenths place, it would have a value of ______.

The digit in each place has a value that is $\frac{1}{10}$ times as much as it has in the place to its left.

Talk About It

The decimal model shows 0.033. Use this decimal to answer Exercises 1–4.

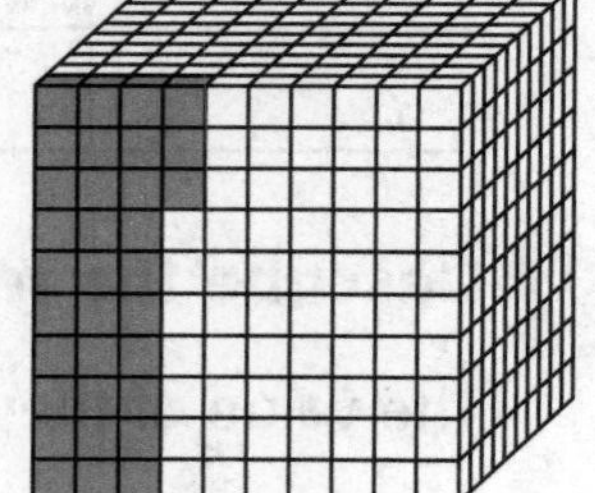

1. What is the value of the 3 in the hundredths place?

____

2. What is the value of the 3 in the thousandths place?

____

3. The digit in the hundredths place is how many times as much as the digit in the thousandths place?

____

4. Mathematical PRACTICE 2 **Use Number Sense** How many times as great is the value of the digit in the thousandths place as in the hundredths place?

____

Name ....

CCSS

Practice It

Use each model to write a decimal in standard form and word form. Then complete each sentence.

5. 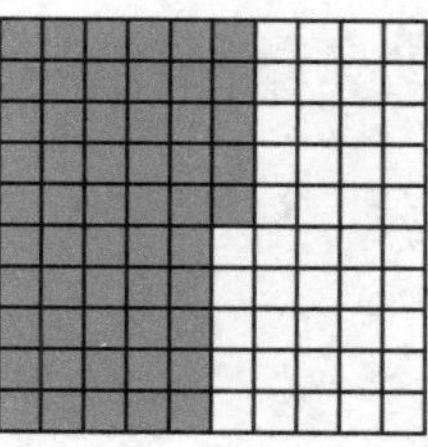

The value of the digit in the tenths place is ________ times as much as the digit in the hundredths place.

____

6. 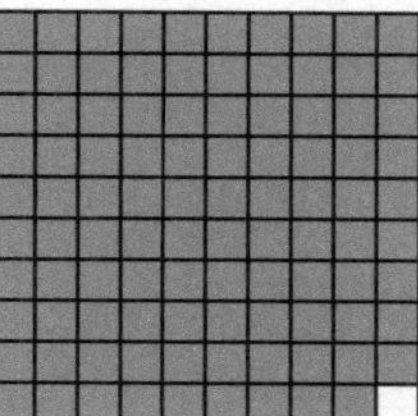

The value of the digit in the hundredths place is ________ times as much as the digit in the tenths place.

____

7. 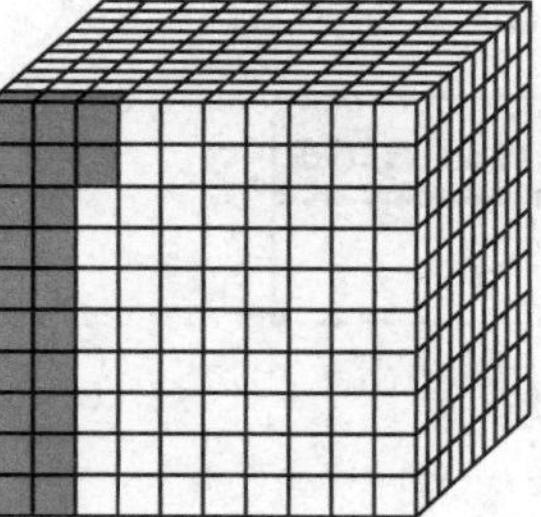

The value of the digit in the thousandths place is ________ times as much as the digit in the hundredths place.

____

8. 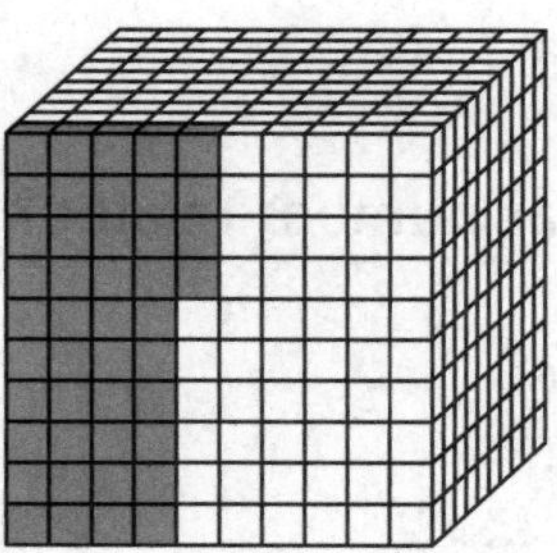

The value of the digit in the hundredths place is ________ times as much as the digit in the thousandths place.

____

Apply It

9. Micah asked for directions to a local park. He was told to go 0.33 mile south to get to the park. The value of the digit in the tenths place is how many times as much as the value of the digit in the hundredths place?

10. Mathematical PRACTICE 2 **Use Number Sense** In science class, Abigail's frog weighed 0.88 kilogram. What is the value of the digit in the tenths place?

11. Mathematical PRACTICE 8 **Look for a Pattern** For the decimal 0.555, the value of the digit in the tenths place is how many times as much as the value of the digit in the thousandths place?

12. Look at the place-value chart. What happens to the value of the digit 2 as it moves places to the left from the thousandths place?

Tens	Ones	Tenths	Hundredths	Thousandths
	0 .	2	2	2

Write About It

13. How are the place-value positions to either side of a particular number related?

Name

MY Homework

Lesson 5
Hands On: Understand Place Value

Homework Helper

Need help? connectED.mcgraw-hill.com

1 The model shows a decimal. Write the decimal.

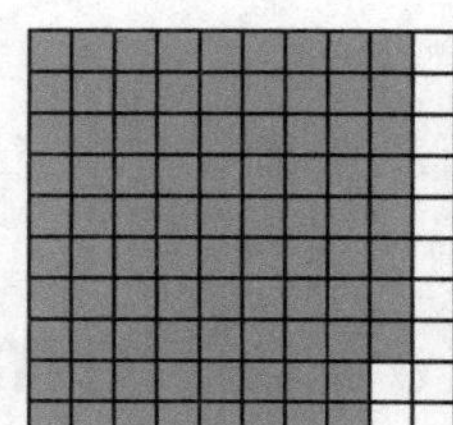

Word form: *eighty-eight hundredths*

Standard form: 0.88

2 Write the decimal in the place-value chart.

Tens	Ones	Tenths	Hundredths	Thousandths
	0 .	8	8	

3 The value of the digit 8 in the tenths place is 0.8.

The value of the digit 8 in the hundredths place is 0.08.

The value of the digit 8 in the tenths place is 10 times as much as the value of the digit 8 in the hundredths place.

The value of the digit 8 in the hundredths place is $\frac{1}{10}$ times as much as the value of the digit 8 in the tenths place.

Practice

1. Use the model to write a decimal in standard form and word form. Then complete the sentence.

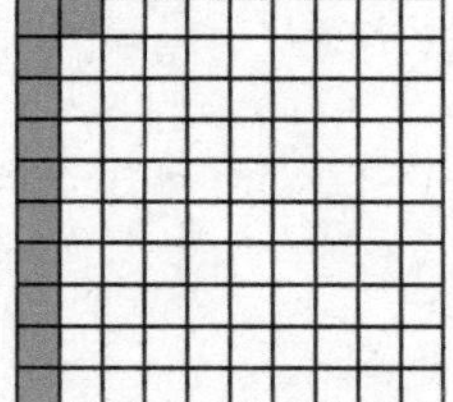

The value of the digit in the hundredths place is

_______ times as much as the digit in the tenths place.

2. Use the model to write a decimal in standard form and word form. Then complete the sentence.

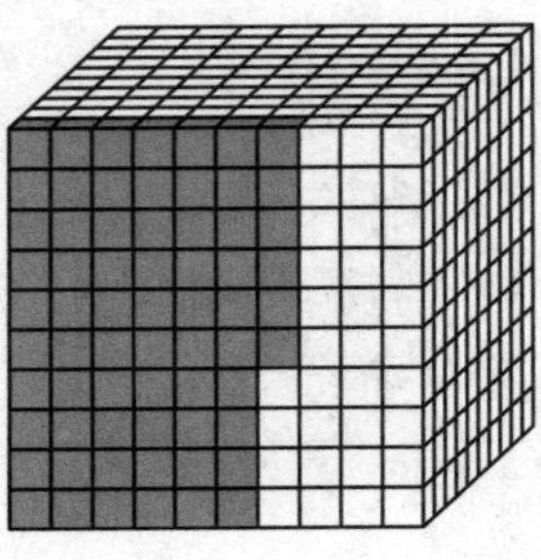

The value of the digit in the thousandths place is ________ times as much as the digit in the hundredths place.

____

Problem Solving

3. **Michael bought 0.44 pound of sliced turkey. What is the value of the digit in the hundredths place?**

____

The value of the digit in the hundredths place is how many times as much as the value of the digit in the tenths place?

____

4. A small piece of metal weighs 0.77 gram. What is the value of the digit in the tenths place?

____

The digit in the tenths place is how many times as much as the value of the digit in the hundredths place?

____

5. Mathematical PRACTICE 2 **Use Number Sense** Write a decimal where the value of a digit is $\frac{1}{10}$ as much as a digit in another place.

____

Name ....

Place Value Through Thousandths

Lesson 6

ESSENTIAL QUESTION
How does the position of a digit in a number relate to its value?

A decimal can be greater than one. For example, 1.5 is greater than one because there is a non-zero digit in the ones place.

Math in My World

Example 1

Five tree taps produce enough maple sap to make 1 gallon, or about 3.79 liters, of syrup. Read and write the number of liters in word form.

 Write the number in the place-value chart.

Tens	Ones	Tenths	Hundredths	Thousandths
	.			

 The place value of the last digit, 9, is __________.

Use the word *and* for the decimal point.

Word form: __________ *and seventy-nine* __________

Example 2

Circle the digit in the thousandths place. Then write the value of the digit.

0.247

1 The thousandths place is __________ places to the right of the decimal place. Circle the digit.

2 The digit has a value of __________ thousandths.

Example 3 

Write *five and six hundred fourteen thousandths* in standard form and expanded form.

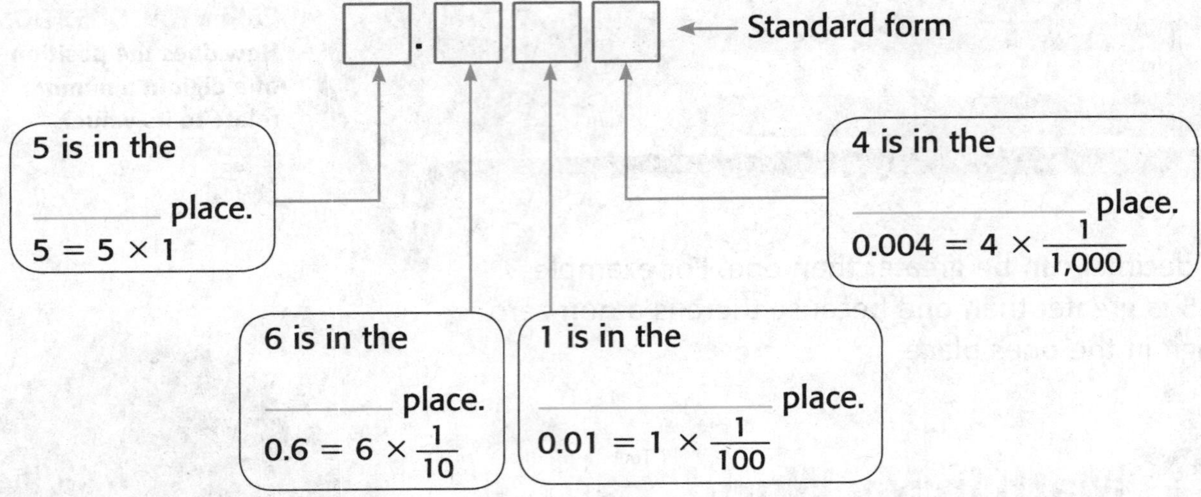

So, in expanded form, $5.614 = 5 \times 1 + \left(6 \times \frac{1}{10}\right) + \left(1 \times \frac{1}{100}\right) + \left(4 \times \frac{1}{1{,}000}\right)$.

Guided Practice 

1. Circle the digit in the tenths place. 6.14

2. Circle the digit in the hundredths place. 4.036

Write each number in standard form.

3. 5 and 87 hundredths

____

4. $2 \times 10 + 6 \times 1 + \left(9 \times \frac{1}{10}\right) + \left(1 \times \frac{1}{100}\right) + \left(4 \times \frac{1}{1{,}000}\right)$

____

5. Write 19.4 in expanded form. Then write in word form.

____

____

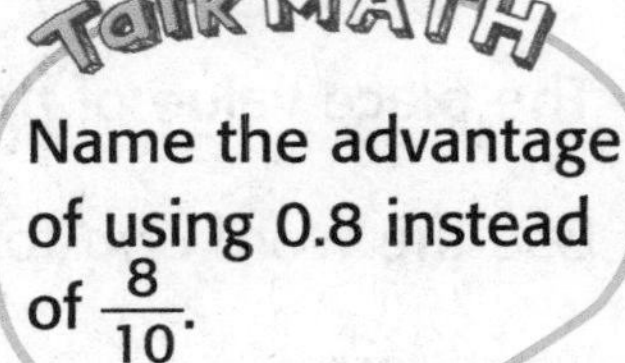

Name ....

Independent Practice

Name the place of the highlighted digit. Then write the value of the digit.

6. 63.47 ____________

7. 9.56 ____________

8. 4.072 ____________

9. 81.453 ____________

10. 1.608 ____________

11. 7.017 ____________

Write each number in standard form.

12. *thirteen and nine tenths* ____________

13. *fifty and six hundredths* ____________

14. $1 \times 10 + 1 \times 1 + \left(9 \times \frac{1}{10}\right) + \left(2 \times \frac{1}{100}\right) + \left(3 \times \frac{1}{1,000}\right)$ ____________

15. $7 \times 10 + \left(1 \times \frac{1}{10}\right) + \left(5 \times \frac{1}{1,000}\right)$ ____________

16. *five and three thousandths* ____________

17. $6 \times 10 + 4 \times 1 + \left(4 \times \frac{1}{10}\right) + \left(1 \times \frac{1}{100}\right) + \left(8 \times \frac{1}{1,000}\right)$ ____________

Write each number in expanded form. Then write in word form.

18. 0.917

____

____

19. 69.409

____

____

CCSS

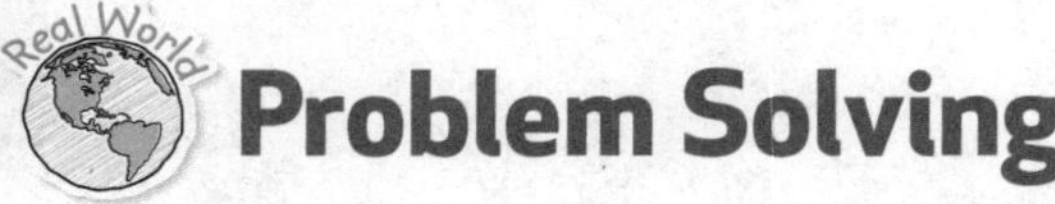

Problem Solving

20. A baseball player had a batting average of 0.334 for the season. Write this number in expanded form.

21. There were three and five hundredths inches of rain yesterday. Write this number in standard form.

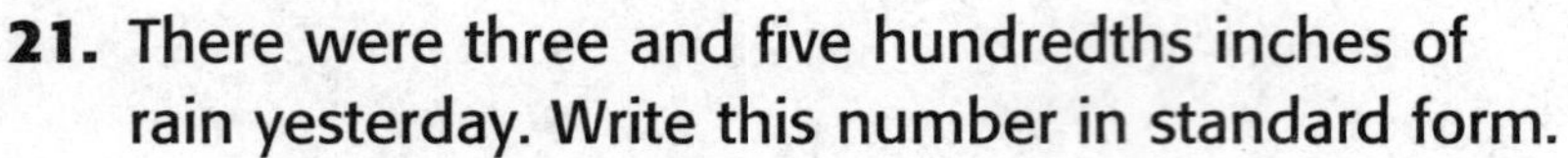

22. An athlete completes a race in 55.72 seconds. How many times greater is the digit in the tens place than the digit in the ones place?

23. The table shows the amount of salt that remains when a cubic foot of water evaporates. Read each number that describes the amount of salt. Then write each number in words.

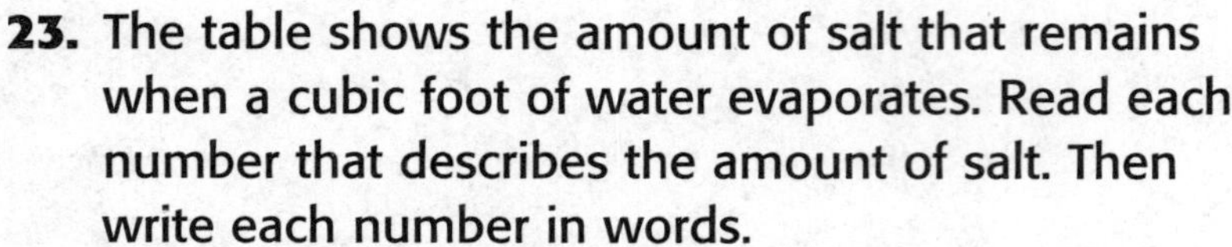

Salt Comparison	
Source of Water	**Amount of Salt**
Atlantic Ocean	2.2 pounds
Lake Michigan	0.01 pound

HOT Problems

24. Mathematical PRACTICE 3 **Which One Doesn't Belong?** Circle the decimal that does not belong with the other three.

five and thirty-nine hundredths	5.39	$5 \times 1 + \left(3 \times \frac{1}{10}\right) + \left(9 \times \frac{1}{100}\right)$	5 and 39 tenths

25. **Building on the Essential Question** How is place value used to read decimals?

Name

MY Homework

Lesson 6
Place Value Through Thousandths

Homework Helper

Need help? connectED.mcgraw-hill.com

Write *six and seven hundred eighty-two thousandths* in standard form and expanded form.

Standard form
6.782

6 is in the ones place.
$6 = 6 \times 1$

7 is in the tenths place.
$0.7 = 7 \times \frac{1}{10}$

8 is in the hundredths place.
$0.08 = 8 \times \frac{1}{100}$

2 is in the thousandths place.
$0.002 = 2 \times \frac{1}{1{,}000}$

So, in expanded form, $6.782 = 6 \times 1 + \left(7 \times \frac{1}{10}\right) + \left(8 \times \frac{1}{100}\right) + \left(2 \times \frac{1}{1{,}000}\right)$.

Practice

Name the place of the highlighted digit. Then write the value of the digit.

1. 35.052

2. 5.654

3. 4.95

Write each number in standard form.

4. *thirty-four and twelve hundredths*

5. $2 \times 10 + 4 \times 1 + \left(7 \times \frac{1}{10}\right) + \left(4 \times \frac{1}{100}\right) + \left(5 \times \frac{1}{1{,}000}\right)$

Write each number in expanded form. Then write in word form.

6. 23.5

7. 164.38

8. 209.106

Problem Solving

9. Mathematical PRACTICE 4 **Model Math** When measuring board footage for some exotic woods, a carpenter must use 1.25 inches for thickness rather than 1 inch in her calculations. Write 1.25 in expanded form.

10. The summer camp Jessica attends is exactly four hundred twenty-three and four tenths miles from her home. Write *four hundred twenty-three and four tenths* in standard form.

Test Practice

11. Which statement is true regarding the value of the digit in the tenths place of the decimal 19.993?

Ⓐ It is 10 times as great as the value of the digit in the ones place.

Ⓑ It is 10 times as great as the value of the digit in the thousandths place.

Ⓒ It is $\frac{1}{10}$ as great as the value of the digit in the ones place.

Ⓓ It is $\frac{1}{10}$ as great as the value of the digit in the tens place.

Name

Compare Decimals

Lesson 7

ESSENTIAL QUESTION
How does the position of a digit in a number relate to its value?

Math in My World

Example 1

Luis downloaded two songs onto his MP3 player. Which song is longer?

Song	Length (min)
1	3.6
2	3.8

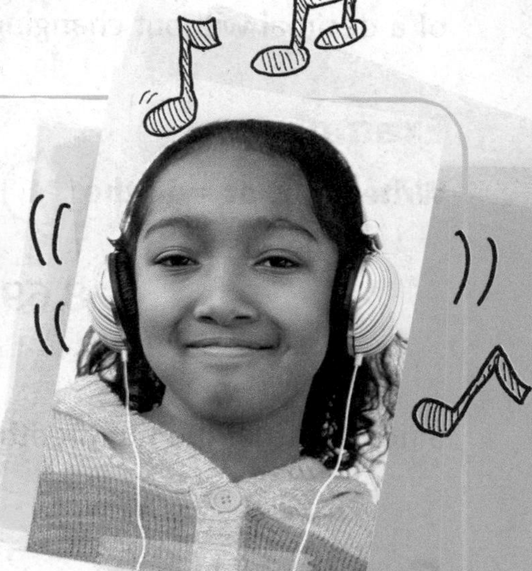

One Way **Use a number line.**

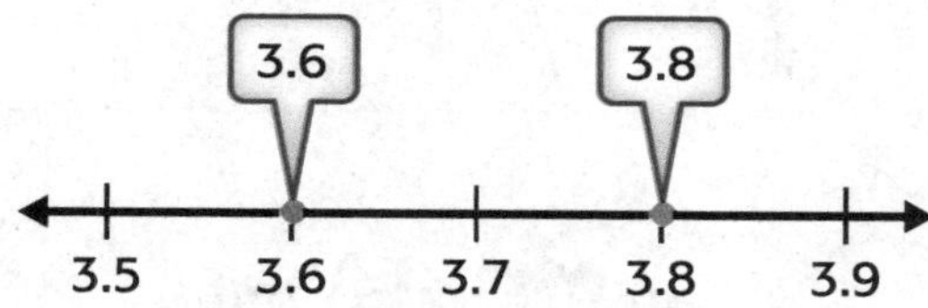

Numbers to the right are greater than numbers to the left.

Since 3.8 is to the right of 3.6, 3.8 ◯ 3.6.

Another Way **Line up the decimal points.**

1. Compare the digits in the greatest place.

   The ones digits are the ______.

3.6
3.8

2. Continue comparing until the digits are different. In the tenths place, 8 ◯ 6.

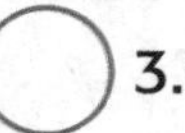

So, 3.8 ◯ 3.6. Song ______ is longer.

Decimals that have the same value are **equivalent decimals**.

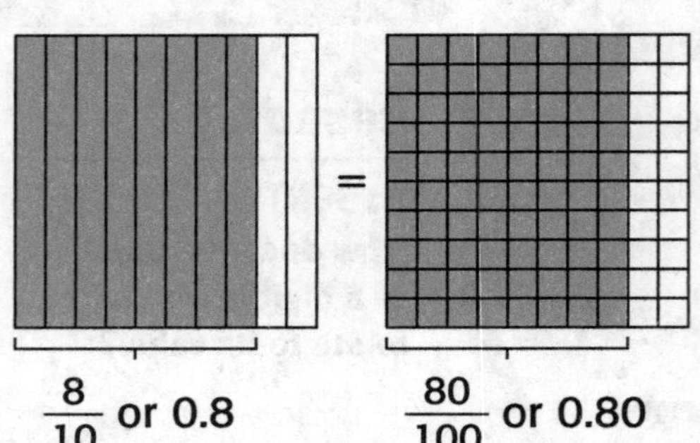

The shaded part of each model is the same. So, 0.8 = 0.80.

$\frac{8}{10}$ or 0.8 $\qquad$ $\frac{80}{100}$ or 0.80

The models show you can annex, or place zeros, to the right of a decimal without changing its value.

Example 2

Write <, >, or = in the ◯ below to make a true sentence.

8.69 ◯ 8.6___

Annex a zero to the right of 8.6 so that it has the same number of decimal places as 8.69.

Since 9 > 0 in the hundredths place, 8.69 ◯ 8.6.

Guided Practice

Plot each decimal on the number line.

Write <, >, or = in each ◯ to make a true sentence.

1. 0.5 ◯ 0.7

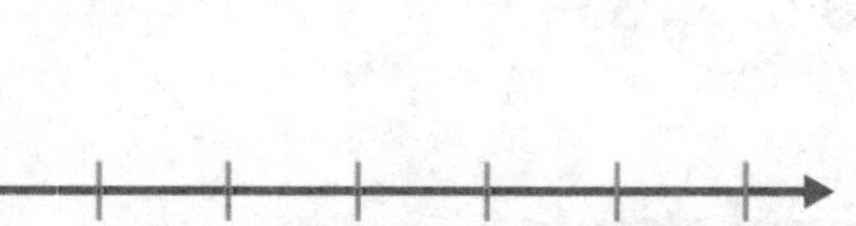

2. 4.40 ◯ 4.44

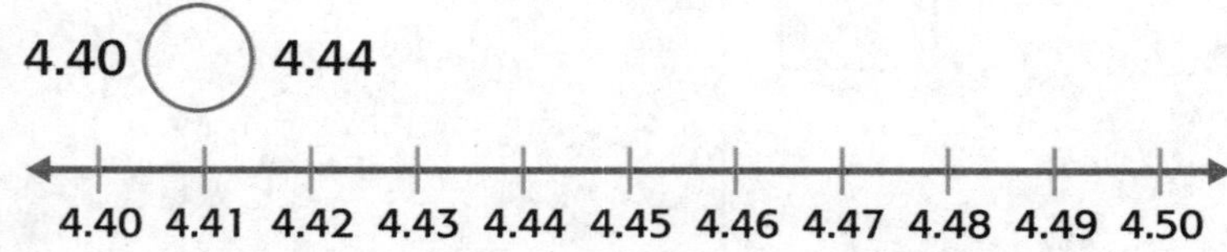

Talk MATH

Describe how you know if two decimals are equivalent.

Name ....

Independent Practice

Plot each decimal on the number line. Write <, >, or = in each  **to make a true sentence.**

3. 4.4 ◯ 4.1

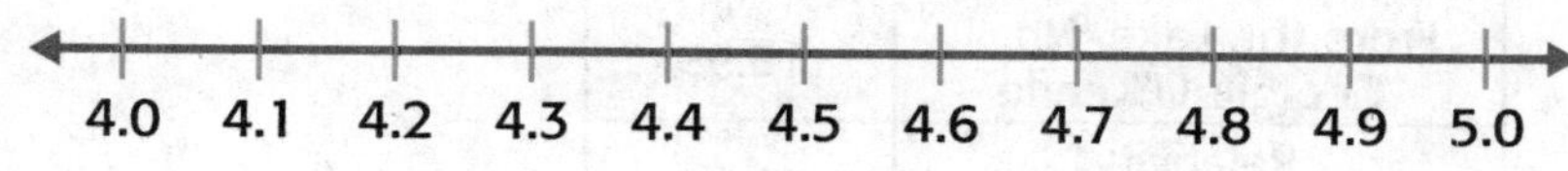

4. 0.37 ◯ 0.39

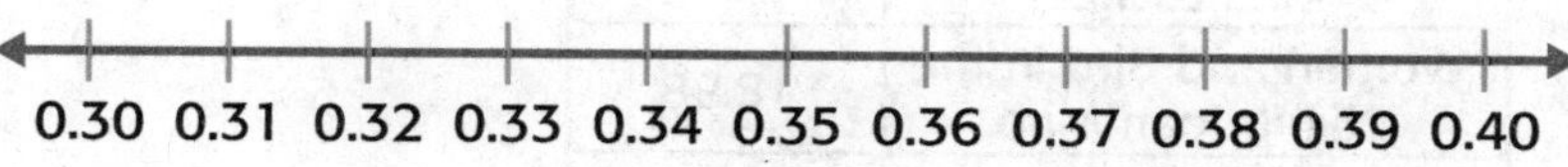

5. 0.57 ◯ 0.65

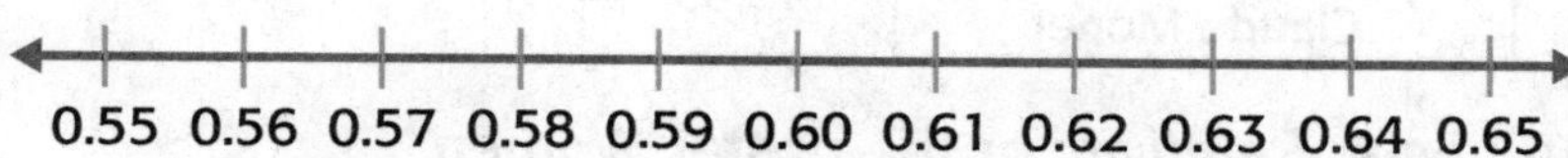

Write <, >, or = in each ◯ to make a true sentence.

6. 2.15 ◯ 2.150	**7.** 0.006 ◯ 0.1	**8.** 0.652 ◯ 0.647
9. 0.09 ◯ 0.001	**10.** 7.31 ◯ 7.304	**11.** 2.800 ◯ 2.8
12. 0.5 ◯ 0.7	**13.** 0.62 ◯ 0.26	**14.** 3.7 ◯ 3.70

Problem Solving

For Exercises 15–17, use the table that shows the cost of posters of famous works of art.

Poster Prices	
Poster	**Cost ($)**
From the Lake, No. 1, Georgia O'Keeffe	16.99
Relativity, M.C. Escher	11.49
Women and Bird in the Night, Joan Miro	18.98
Waterlillies, Claude Monet	15.99

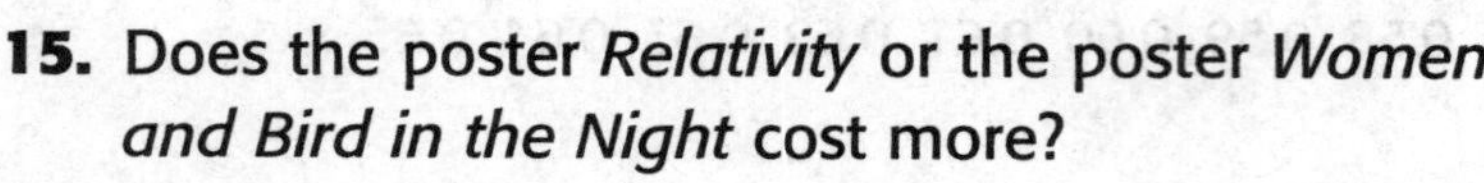

15. Does the poster *Relativity* or the poster *Women and Bird in the Night* cost more?

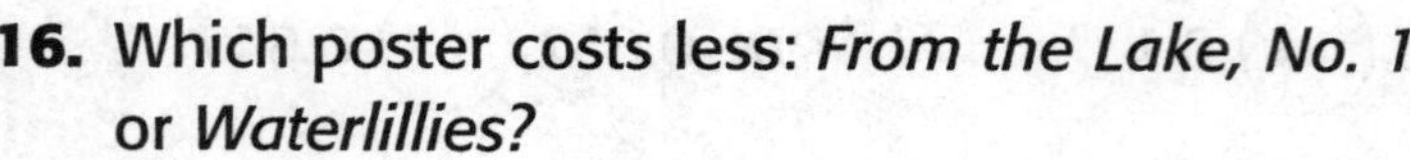

16. Which poster costs less: *From the Lake, No. 1* or *Waterlillies?*

17. Which poster costs less than *Waterlillies?*

HOT Problems

18. Mathematical PRACTICE 6 **Explain to a Friend** How many times greater is 46 than 0.46? Explain to a classmate.

19. **Building on the Essential Question** What are the similarities and differences between comparing whole numbers and comparing decimals?

Name

MY Homework

Lesson 7
Compare Decimals

Homework Helper

eHelp

Need help? connectED.mcgraw-hill.com

Compare 59.296 and 59.6.

1. Line up the decimal points. Annex zeros where necessary.

59.296
59.600

2. Compare the digits in the greatest place. The tens and ones digits are each the same.

Annex 2 zeros so that the numbers have the same number of decimal places.

3. Continue comparing until the digits are different. In the tenths place, 2 < 6.

So, 59.296 < 59.6.

Practice

Write <, >, or = in each ◯ to make a true sentence.

1. 3.976 ◯ 4.007

2. 89.001 ◯ 89.100

3. 126.698 ◯ 126.689

4. 5.05 ◯ 5.050

5. 9.087 ◯ 9.807

6. 3.674 ◯ 6.764

7. 0.256 ◯ 0.256

8. 2.7 ◯ 2.82

9. 6.030 ◯ 6.03

10. 7.89 ◯ 7.189

11. 12.54 ◯ 1.254

12. 0.981 ◯ 2.3

Problem Solving

13. In January, the average low temperature in Montreal, Quebec, Canada, is 5.2°F. The average low temperature in Cape Town, South Africa, is 60.3°F. Which city is warmer in January?

____

14. In one year Detroit, Michigan, recorded 30.9 inches of snow and Chicago, Illinois, recorded 39.9 inches of snow. Which city had more snow?

____

15. Mathematical **PRACTICE** 6 **Explain to a Friend** George was weighed at the doctor's office. The scale read 67.20 pounds. The doctor wrote 67.2 pounds on George's chart. Did the doctor make a mistake? Explain to a friend.

____

16. The two fastest times in the past 20 years for the girls' 200-meter run at Clarksville Elementary School are 27.97 seconds and 27.93 seconds. At yesterday's track meet, Claire ran 27.99 seconds. Was her time faster than either of the two fastest? Explain.

____

Vocabulary Check

17. Write if the following statement is true or false.

Equivalent decimals are decimals that have the same value. ________

Test Practice

18. Which of the following symbols make the statement below true?

98.546 ◯ 98.654

Ⓐ <   Ⓒ =

Ⓑ >   Ⓓ ≥

Name ____________________

Order Whole Numbers and Decimals

Lesson 8

ESSENTIAL QUESTION
How does the position of a digit in a number relate to its value?

Math in My World

Example 1

The table shows the cost to build three National Football League stadiums. Order the costs of the stadiums from *greatest* to *least*.

Cost to Build (millions $)	
INVESCO Field Englewood, Co	364.2
Ford Field Detroit, MI	430.0
Qwest Field Seattle, WA	350.0

One Way **Use place value.**

1 Line up the numbers by their decimal points. →

364.2
430.0
350.0

2 Compare the digits in the greatest place. ______ > 3

3 Compare the digits in the next greatest place.

______ > 5 and 5 > ______

Another Way **Use a number line.**

Place dots on the number line to represent the approximate locations of the decimals.

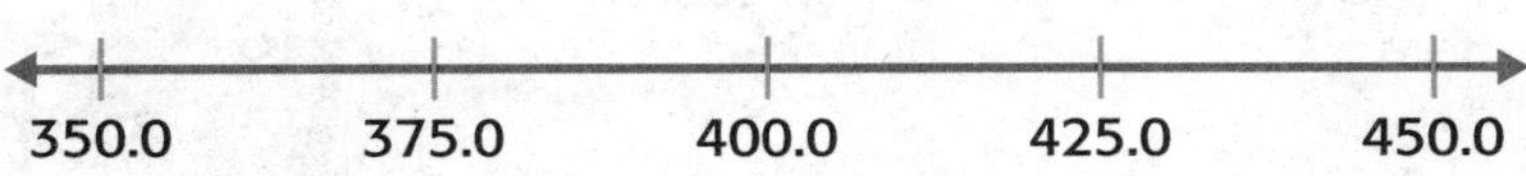

So, the costs, in millions of dollars, from greatest to least are ______, ______, and ______.

Example 2 Tutor

Four trees in a state forest had heights of 22.65, 23.8, 22, and 23.25 feet. Order the heights from *least* to *greatest*.

To annex a zero means to add a zero at the end of a number.

1. Line up the numbers by their decimal points.

2. Annex zeros so that all numbers have the same final place value.

22.65
23.80
22.00
23.25

3. Compare the digits using place value.

The least number is ________.

The greatest number is ________.

The heights, in feet, from least to greatest are ________, ________, ________, and ________.

Guided Practice

Order each set of numbers from *least* to *greatest*.

1. weight in kilograms of a dog: 56.7, 64.3, 59.0, 64.5

____

2. rainfall in inches: 0.76, 0.09, 0.63, 0.24

____

3. height of flowers in inches: 8.9, 8.59, 8.705, 8.05

____

4. The lengths of insects in centimeters are 1.35, 0.9, 1.48, and 1.8. Order the sizes of the insects from greatest to least.

____

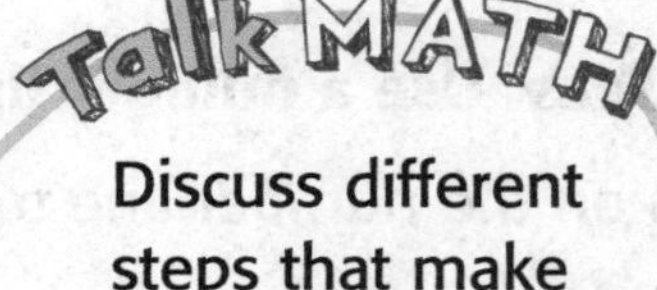

Discuss different steps that make ordering numbers easier.

Name

Independent Practice

Order each set of numbers from *least* to *greatest*.

5. cost of cellphones: $98.75, $114.99, $105.99

6. temperatures in °F: 106.3, 99.8, 101.1, 110.5

7. distance in light years: 4.2, 6.0, 4.3, 7.7

8. heights of buildings in meters: 419.7, 346.5, 178.3, 527.3

9. kilometers ran: 4.9, 3.7, 3.4, 4.2

Order each set of numbers from *greatest* to *least*.

10. cost of snacks: $2.43, $2.34, $2.05, $2.18

11. masses of bottles in grams: 9.14, 7.99, 9.02, 8.95, 8.91

12. race times in seconds: 43.789, 67.543, 86.347, 78.432, 34.678

13. heights of trees in meters: 9.8, 10, 10.2, 9.6, 11

14. weights of dogs in pounds: 25.4, 26.2, 26, 25.8, 27

Problem Solving

Mathematical PRACTICE 3 Draw a Conclusion **The table shows facts about snakes common to the United States.**

Snake	Average Adult Body Length (cm)	Average Baby Body Length (cm)
Copperhead	63.5	27.9
Western Cottonmouth	91.25	21.5
Timber Rattlesnake	121.6	29.5
Queen Snake	61	15.2

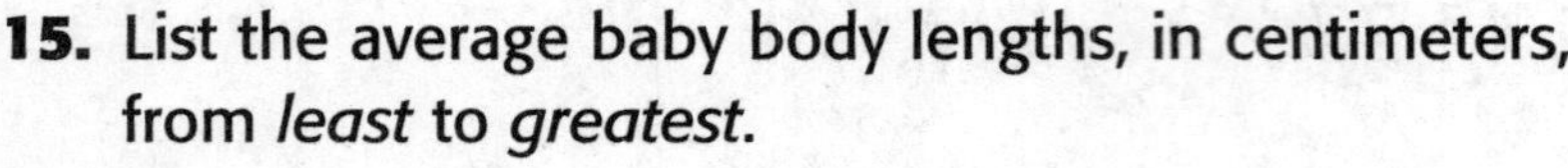

15. List the average baby body lengths, in centimeters, from *least* to *greatest*.

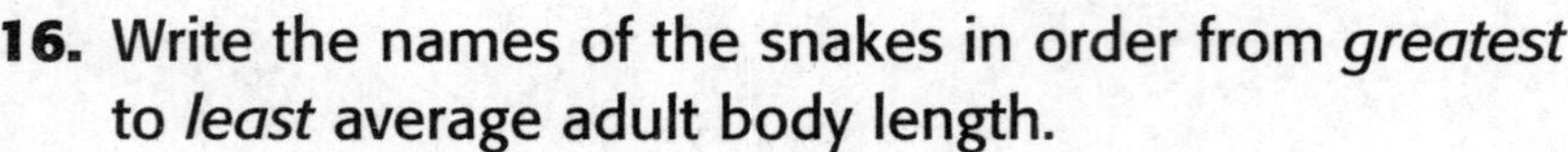

16. Write the names of the snakes in order from *greatest* to *least* average adult body length.

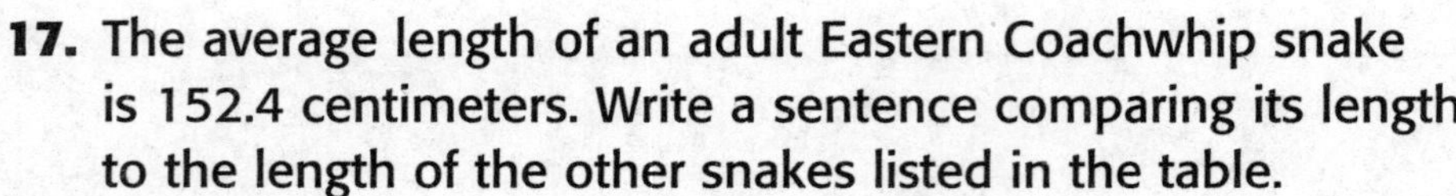

17. The average length of an adult Eastern Coachwhip snake is 152.4 centimeters. Write a sentence comparing its length to the length of the other snakes listed in the table.

HOT Problems

18. **Mathematical PRACTICE 1 Make Sense of Problems** Write an ordered list of five numbers whose values are between 50.98 and 51.6. Tell whether your list is from least to greatest or greatest to least.

19. **Building on the Essential Question** How does comparing numbers help to order numbers?

Name ....

MY Homework

Lesson 8
Order Whole Numbers and Decimals

Homework Helper

Need help? connectED.mcgraw-hill.com

Order the set of numbers 9.275, 8.950, and 9.375 from *least* to *greatest*.

Use place value.

1 Line up the numbers by their decimal points.

9.275
8.950
9.375

2 Compare the digits in the greatest place. 9 > 8

3 Compare the digits in the next place. 3 > 2

So, the order from least to greatest is 8.950, 9.275, and 9.375.

Practice

Order each set of numbers from *least* to *greatest*.

1. 17.639, 3.828, 45.947

____

2. 890.409, 890.904, 809.904

____

Order each set of numbers from *greatest* to *least*.

3. 2.654, 2.564, 2.056, 2.465

____

4. 1.11, 0.111, 1.01, 1.001

____

Problem Solving

5. Mathematical PRACTICE 2 **Use Number Sense** The table shows the heights of four students. Arrange the students in order from shortest to tallest.

Student Heights	
Name	**Height (in.)**
Kim	56.03
Alexa	56.3
Roy	56.14
Tom	57.1

____

6. Lauren spent $3.26 for lunch on Tuesday. She spent $1.98 on Wednesday and $2.74 on Thursday. Order the prices from greatest to least.

____

7. The four fastest times in a race were 27.08 seconds, 27.88 seconds, 27.8 seconds, and 26.78 seconds. Order these times from least to greatest.

____

Test Practice

8. Four boxes to be mailed are weighed at the post office. Box A weighs 8.25 pounds, Box B weighs 8.2 pounds, Box C weighs 8.225 pounds, and Box D weighs 8.05 pounds. Which box is the heaviest?

   Ⓐ Box A

   Ⓑ Box B

   Ⓒ Box C

   Ⓓ Box D

Name ....

Real World

Problem-Solving Investigation

STRATEGY: Use the Four-Step Plan

Lesson 9

ESSENTIAL QUESTION
How does the position of a digit in a number relate to its value?

Learn the Strategy

Victor spent $61 on some sandpaper for his model cars. He bought 2 packages of the smallest-grain sandpaper and spent the rest on the largest-grain sandpaper. How many packages of the largest-grain sandpaper did he buy?

Size of Grain (cm)	Cost per Package ($)
0.003	13
0.011	7
0.001	20

1 Understand

What facts do you know?

A total of ________ was spent. ________ packages of the smallest-grain sandpaper were purchased.

What do you need to find?

the number of packages of the largest grain sandpaper he bought

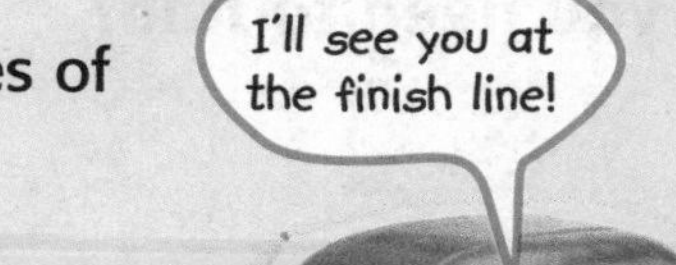

2 Plan

To solve this problem, I can work backward.

3 Solve

The smallest size is ________ centimeter. Victor spent 2 × ________ or ________.

Subtract to find the remaining amount spent: $61 − ________ = ________.

The largest size is 0.011 centimeter. Each package costs $7. Divide.

Victor bought $21 ÷ $7, or ________ packages of the largest grain sandpaper.

4 Check

Does your answer make sense? Explain.

____

Practice the Strategy

Luisa bought a roll of ribbon. She used 34 inches on each of two gifts. Then she used 13 inches on a scrapbook page. There are 39 inches left. How many inches did she start with?

1 Understand

What facts do you know?

____

____

What do you need to find?

____

2 Plan

____

____

3 Solve

4 Check

Is your answer reasonable? Explain.

____

Name

Apply the Strategy

Solve each problem using the four-step plan.

1. The table shows the number of ounces of butter Marti used in different recipes. She has 6 ounces of butter left. How many ounces of butter did she have at the beginning?

Recipe	Ounces of Butter
Pie	4
Cookie	8
Pasta	6

2. At the end of their 3-day vacation, the Palmers traveled a total of 530 miles. On the third day, they drove 75 miles. On the second day, they drove 320 miles. How many miles did they drive the first day?

3. Mathematical PRACTICE 1 **Keep Trying** You divide a number by 3, add 6, then subtract 7. The result is 4. What is the number?

4. Mr. Toshio lent out 11 rulers at the beginning of class, collected 4 rulers in the middle of class, and gave out 7 at the end of class. He had 18 at the end of the day. How many rulers did he start with?

5. The Math Club is selling gift wrap for a fundraiser. They sold all 45 rolls of solid wrapping paper at $4 each and rolls of patterned wrapping paper at $5 each. If they made $265, how many rolls of patterned wrapping paper did they sell?

Review the Strategies

Use any strategy to solve each problem.

- Use the four-step plan.
- Make a table.
- Act it out.

6. Mathematical PRACTICE 1 **Make a Plan** Ming-Li spent $15 at the movies. She then earned $30 babysitting. She spent $12 at the bookstore. She now has $18 left. How much money did Ming-Li have to begin with?

____

7. Mr. Jenkins bought pavers for some landscaping projects. He used 120 for a small patio, 86 for the border of a flower bed, and 70 for a wall. He has 24 pavers left. How many pavers did Mr. Jenkins buy?

____

8. The table shows the number of candy bars the cheerleaders sold each week. They have 9 candy bars left. How many candy bars did they have to sell to begin with?

Week	Candy Bars Sold
1	117
2	130
3	83

____

9. You multiply a number by 3, subtract 6 and then add 2. The result is 20. What is the number?

____

10. Dimitrius returned 5 library books last week, returned 3 books this week, and then checked out 8 more books. He now has 12 library books. How many books did he have before last week?

____

MY Homework

Lesson 9
Problem Solving: Use the Four-Step Plan

Homework Helper

Need help? connectED.mcgraw-hill.com

Dana walks every day. She walked 3 miles on Tuesday, 5 miles on Wednesday, and 8 miles on Thursday. After Thursday's walk, she had walked a total of 21 miles for the week. How many miles did she walk on Monday?

1 Understand

What facts do you know?

Dana walked 3 miles on Tuesday, 5 miles on Wednesday, and 8 miles on Thursday. Dana walked a total of 21 miles from Monday to Thursday.

What do you need to find?

the number of miles Dana walked on Monday

2 Plan

I can solve the problem by adding 3, 5, and 8 and then subtracting the sum from 21.

3 Solve

$$21 - (3 + 5 + 8) = 21 - 16$$
$$= 5 \text{ miles}$$

So, Dana walked 5 miles on Monday.

4 Check

Does your answer make sense? Explain.

Yes. $5 + 3 + 5 + 8 = 21$

Problem Solving

Solve each problem using the four-step plan.

1. The table shows the number of vehicles washed at a car wash fundraiser over the weekend. If there were a total of 94 vehicles, how many were washed on Saturday?

Day	Vehicles
Fri.	27
Sat.	?
Sun.	34

2. The volleyball team sold 16 items on the first day of a bake sale, 28 items the second day, and 12 items the last day. There were 4 items left that had not been purchased. How many total items were for sale at the bake sale?

3. Ricardo lost 6 golf balls while playing yesterday. He bought a box of 12 golf balls, then lost 4 on the course today. He now has 18 golf balls. How many golf balls did Ricardo have to begin with?

4. The distance between Cincinnati, Ohio, and Charlotte, North Carolina, is about 336 miles. The distance between Cincinnati and Chicago, Illinois, is about 247 miles. If Perry drove from Charlotte to Chicago by way of Cincinnati, find the distance he drove.

5. Mathematical PRACTICE 1 **Plan Your Solution** Adison earned $25 mowing her neighbor's lawn. Then she loaned her friend $18, and got $50 from her grandmother for her birthday. She now has $86. How much money did Adison have to begin with?

Review

Vocabulary Check

Choose the correct word(s) to complete each sentence.

decimal	**decimal point**	**equivalent decimals**	**expanded form**
period	**place value**	**place-value chart**	**standard form**

1. Decimals that have the same value are ______________________.

2. ______________________ is a system for writing numbers. In this system, the position of a digit determines its value.

3. The usual or common way to write a number is called ______________________.

4. The way of writing a number as the sum of the values of its digits is called ______________________.

5. A ______________________ is a number that has a digit in the tenths place, hundredths place, and/or beyond.

6. The ______________________ is a period separating the ones and the tenths in a decimal number.

7. A ______________________ is a chart that shows the value of the digits in a number.

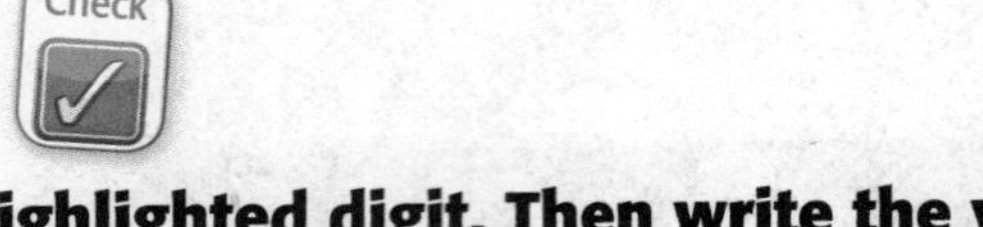

Concept Check

Name the place of the highlighted digit. Then write the value of the digit.

8. 195,489

____

9. 6,720,341

____

Write each number in standard form.

10. 94 million, 237 thousand, 108 ______________

11. $8 \times 1{,}000{,}000 + 5 \times 10{,}000 + 2 \times 1{,}000 + 6 \times 100$ ______________

Shade the model to show each fraction. Write each fraction as a decimal.

12. $\frac{19}{100}$ 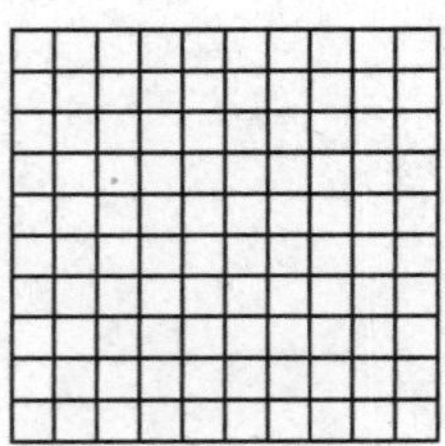 ________

13. $\frac{8}{10}$ 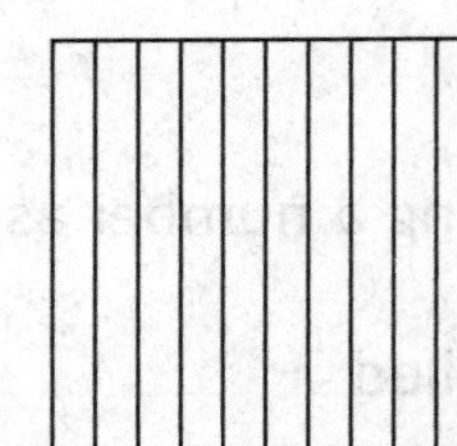 ________

Write each number in standard form and expanded form.

14. *five and nine tenths*

____

15. *seven hundred twelve thousandths*

____

Write $<$, $>$, or $=$ in each ◯ to make a true sentence.

16. 14,589 ◯ 14,985 **17.** 506,789 ◯ 505,789 **18.** 8,913 ◯ 8,931

19. 0.49 ◯ 0.71 **20.** 9.02 ◯ 9.020 **21.** 0.843 ◯ 0.846

22. Order the set of numbers from *least* to *greatest*.

13.84, 13.097, 12.655, 13.6

____

Problem Solving

23. One cup is equal to $\frac{5}{10}$ pint. Write this fraction as a decimal.

____

24. Charles is moving from Springfield, which has 482,653 people, to Greenville, which has 362,987 people. Is he moving to a city with a greater or smaller population? Explain.

____

25. The Pacific Ocean has *sixty-four million, one hundred eighty-six thousand, three hundred square miles.* How many square miles is the Pacific Ocean in standard form?

____

Test Practice

26. Which number results in a true sentence in 475 $<$ _____?

Ⓐ 473

Ⓑ 474

Ⓒ 475

Ⓓ 476

Reflect

Chapter 1

Answering the ESSENTIAL QUESTION

Use what you learned about place value to complete the graphic organizer.

Example with Whole Numbers

ESSENTIAL QUESTION

How does the position of a digit in a number relate to its value?

Example with Decimals

Now reflect on the ESSENTIAL QUESTION  **Write your answer below.**

Chapter

Multiply Whole Numbers

ESSENTIAL QUESTION

What strategies can be used to multiply whole numbers?

Taking Care of My Pets

Watch

Watch a video!

MY Common Core State Standards

Number and Operations in Base Ten

5.NBT.2 Explain patterns in the number of zeros of the product when multiplying a number by powers of 10, and explain patterns in the placement of the decimal point when a decimal is multiplied or divided by a power of 10. Use whole-number exponents to denote powers of 10.

5.NBT.5 Fluently multiply multi-digit whole numbers using the standard algorithm.

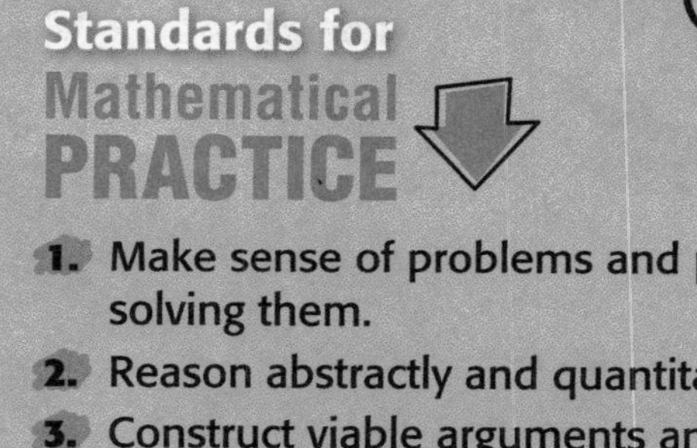

1. Make sense of problems and persevere in solving them.
2. Reason abstractly and quantitatively.
3. Construct viable arguments and critique the reasoning of others.
4. Model with mathematics.
5. Use appropriate tools strategically.
6. Attend to precision.
7. Look for and make use of structure.
8. Look for and express regularity in repeated reasoning.

= focused on in this chapter

I'll be able to get this – no problem!

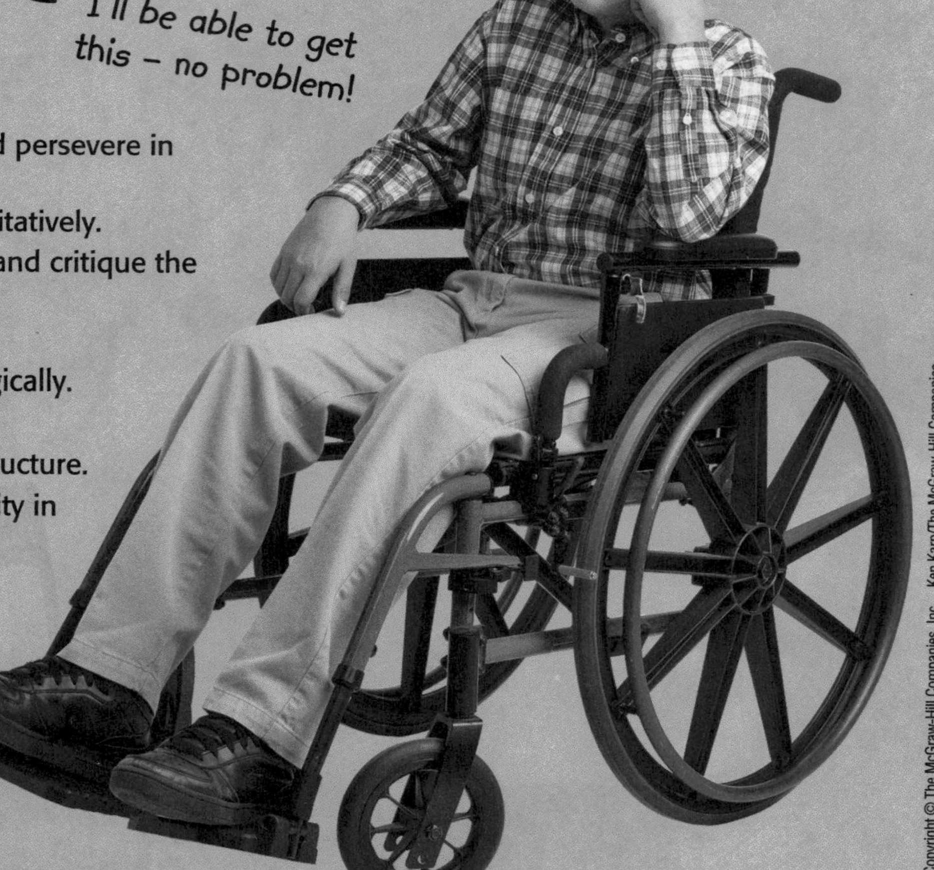

Name ....

Am I Ready?

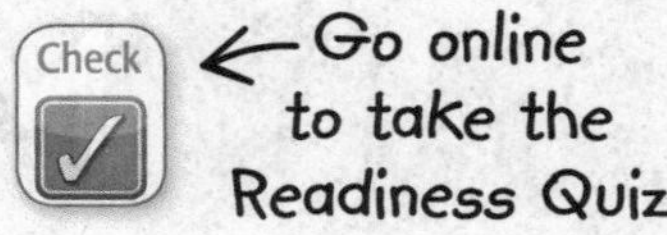

Write all of the factors of each number.

1. 8 ____, ____, ____, ____

2. 11 ____, ____

3. 6 ____, ____, ____, ____

4. 15 ____, ____, ____, ____

5. 32 ____, ____, ____, ____, ____, ____

6. 10 ____, ____, ____, ____

Write each repeated addition sentence as a multiplication sentence.

7. $5 + 5 + 5 + 5 = 20$

____

8. $8 + 8 + 8 = 24$

____

9. $21 + 21 = 42$

____

10. $6 + 6 + 6 + 6 + 6 = 30$

____

11. $13 + 13 + 13 = 39$

____

12. $7 + 7 + 7 + 7 + 7 + 7 = 42$

____

Multiply.

13. $6 \times 3 =$ ______

14. $1 \times 8 =$ ______

15. $7 \times 8 =$ ______

16. $4 \times 10 =$ ______

17. Nikki purchased three used books at a garage sale for $5 each. Find the total cost for all three books.

____

How Did I Do?

Shade the boxes to show the problems you answered correctly.

1	2	3	4	5	6	7	8	9	10	11	12	13	14	15	16	17

Name

MY Math Words

Vocab

Review Vocabulary

composite numbers   prime numbers

Making Connections

Complete the chart below using the review words.

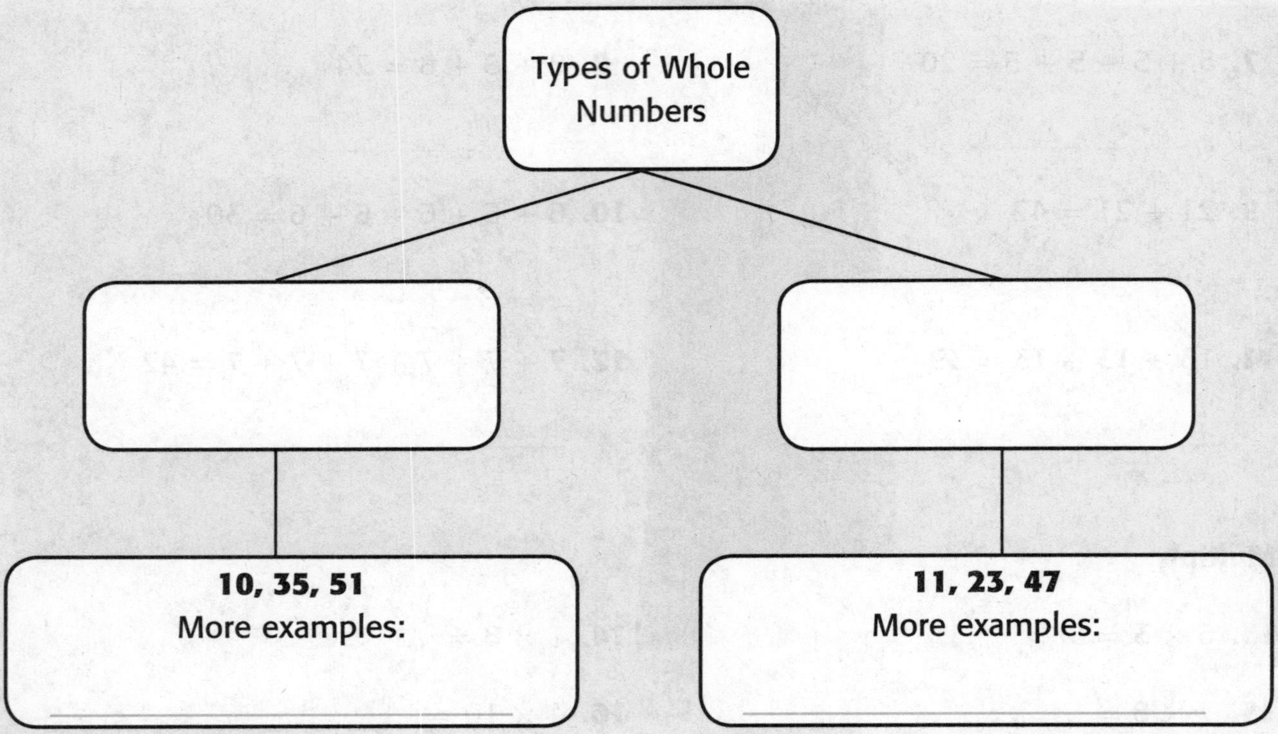

Types of Whole Numbers

10, 35, 51
More examples:

11, 23, 47
More examples:

Explain how you decided to categorize each set of numbers.

MY Vocabulary Cards

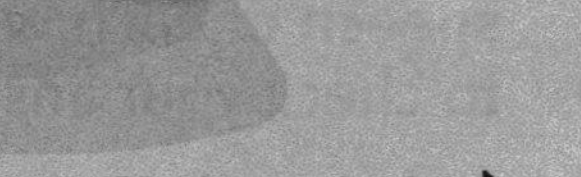

Mathematical PRACTICE

Lesson 2–3

base

$3^3 = 3 \times 3 \times 3 = 27$

$5^4 = 5 \times 5 \times 5 \times 5 = 625$

Lesson 2–8

compatible numbers

$42 \times 7 = 294$

$40 \times 7 = 280$

compatible numbers

Lesson 2–3

cubed

$3^3 = 3 \times 3 \times 3 = 27$

Lesson 2–6

Distributive Property

$4 \times (2 + 7) = (4 \times 2) + (4 \times 7)$

Lesson 2–3

exponent

$5^4 = 5 \times 5 \times 5 \times 5 = 625$

Lesson 2–3

power

$3^4 = 3 \times 3 \times 3 \times 3 = 81$

81 is a power of 3.

Lesson 2–4

power of 10

$10^4 = 10 \times 10 \times 10 \times 10$

$= 10{,}000$

Lesson 2–1

prime factorization

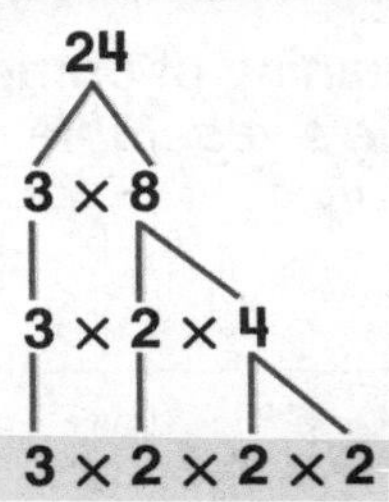

Ideas for Use

- Work with a partner to name the parts of speech of each word. Consult a dictionary to check your answers.
- Write a tally mark on each card every time you read the word in this chapter or use it in your writing. Try to use at least 2 to 3 tally marks for each card.

Numbers in a problem that are easy to work with mentally.

How can the meaning of *compatible* help you remember this definition?

In a power, the number used as a factor.

Consult a dictionary to find another math-related definition of *base.* Describe that meaning in your own words.

To multiply a sum by a number, multiply each addend by the number, and add the products.

How can the meaning of the verb *distribute* help you remember this property?

A number raised to the third power.

Consult a dictionary to describe how *cube* is used in geometry. Write that meaning.

A number obtained by raising a base number to an exponent.

Add a suffix to *power* to form a new word.

In a power, the number of times the base is used as a factor.

What suffix can you add to *exponent* to make a word meaning "growing or increasing very rapidly"? Name the word.

A way of expressing a composite number as a product of its prime factors.

Identify the meaning of prime in this sentence: *Maria's vegetable garden is in its prime in mid-July.*

A number like 10, 100, 1,000 and so on. It is the result of using only 10 as a factor.

Is the number 30 a power of 10? Explain.

MY Vocabulary Cards

Mathematical PRACTICE

Lesson 2–6

property

Commutative Property $25 \times 47 = 47 \times 25$

Identity Property $25 \times 0 = 0$

Distributive Property $5 \times (2 + 7) = (5 \times 2) + (5 \times 7)$

Lesson 2–3

squared

$25^2 = 25 \times 25 = 625$

Ideas for Use

- Use the back of the card to write or draw examples to help you answer the question.
- Use the blank cards to write notes about important concepts in this chapter, such as how powers of 10 apply to multiplication patterns.

A number raised to the second power.

Write a sentence using *square* as a noun.

A rule in mathematics that can be applied to all numbers.

Write a sentence using *property* in a non-mathematical way.

MY Foldable

FOLDABLES® Follow the steps on the back to make your Foldable.

2 × 24

3 × 16

4 × 12

FOLDABLES
Study Organizer

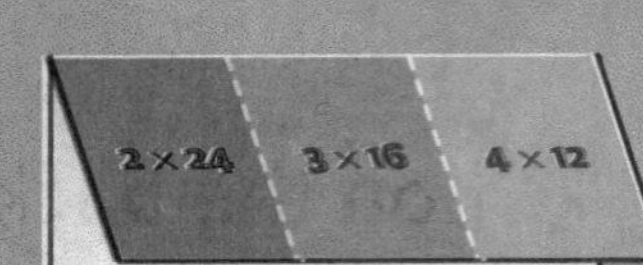
1
2 × 24
3 × 16
4 × 12

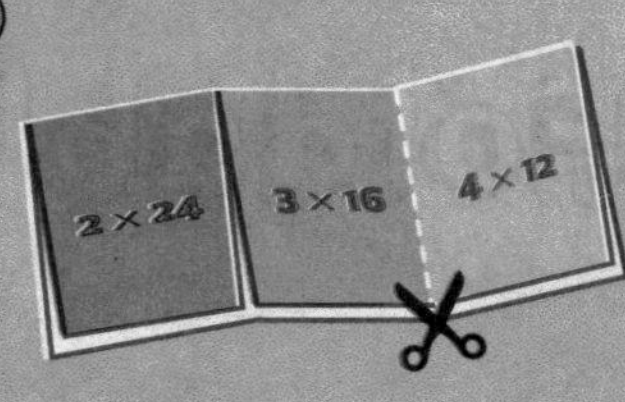
2
2 × 24
3 × 16
4 × 12

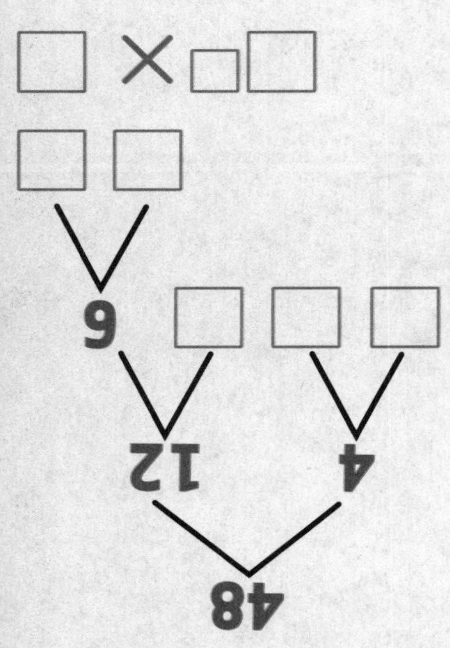
6
12
4
48

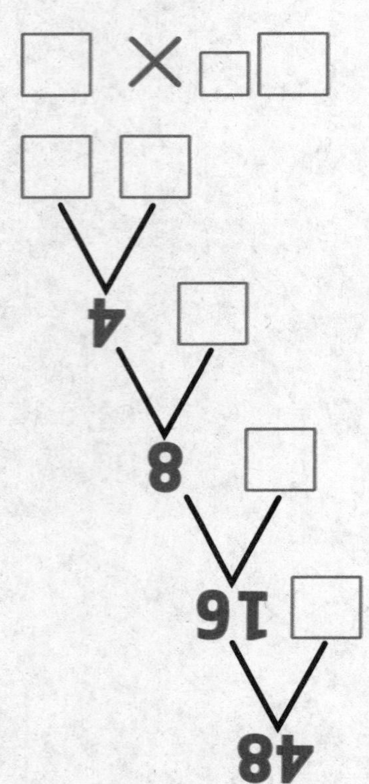
4
8
16
48

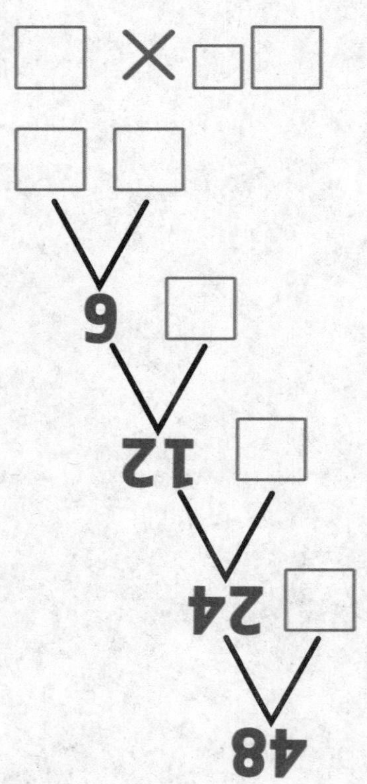
6
12
24
48

Name

Prime Factorization

Lesson 1

ESSENTIAL QUESTION
What strategies can be used to multiply whole numbers?

You can write every composite number as a product of prime factors. This is called the **prime factorization** of a number. A factor tree is a diagram that shows the prime factorization of a composite number.

Math in My World

Example 1

Mr. Dempsey surveyed his class and found that his students have a total of 36 pets. Find the prime factorization of 36.

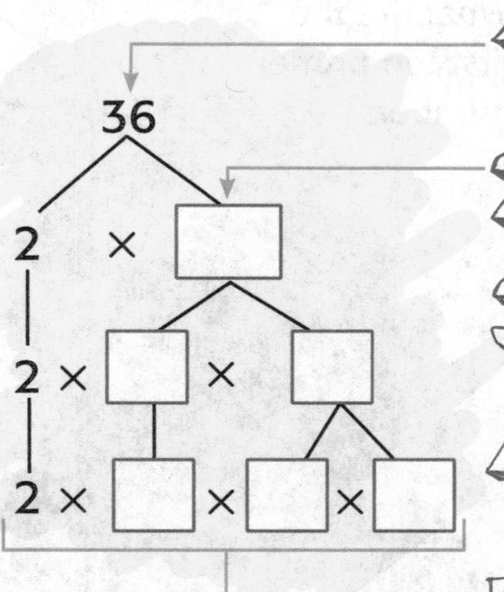

1. Write the number to be factored at the top.
2. Choose any pair of whole number factors of 36.
3. Continue to factor any number that is not prime.
4. Except for the order, the prime factors are the same.
5. Write the prime factors in order from least to greatest.

36
3 ×
3 × ×
3 × × ×

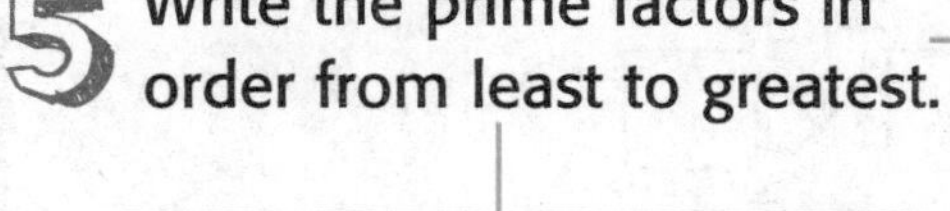

The prime factorization of 36 is ____ × ____ × ____ × ____.

Check Work backward. Multiply all the prime factors in order from left to right. Then compare your product with the composite number.

$2 \times 2 \times 3 \times 3 =$ ______

Example 2 

Find the prime factorization of 24.

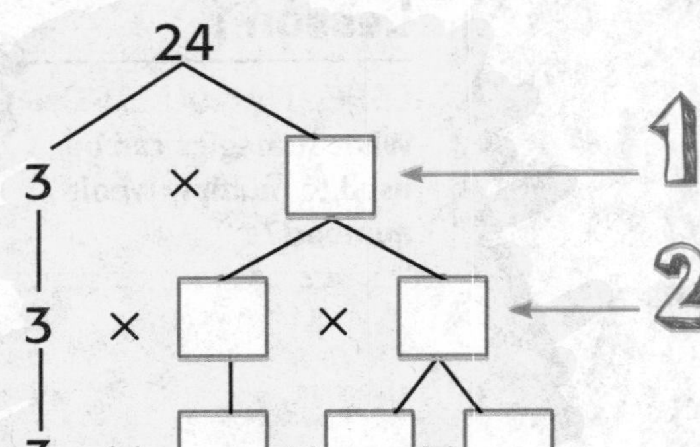

1. You can choose any pair of whole number factors, such as 3 × 8, 4 × 6, or 2 × 12.
2. Continue to factor any number that is not prime.
3. Write the prime factors in order from least to greatest.

The prime factorization of 24 is ___ × ___ × ___ × ___.

Check

Work backward. Multiply all the prime factors in order from left to right. Then compare your product with the composite number.

2 × 2 × 2 × 3 = ______

What are the first ten prime numbers?

Guided Practice 

1. Find the prime factorization of 16.

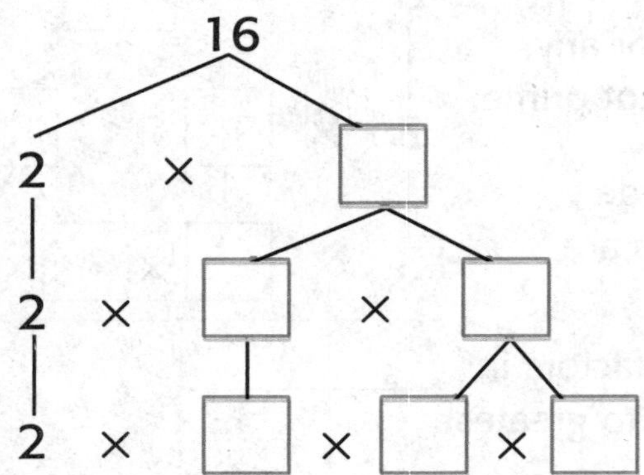

The prime factorization of 16 is

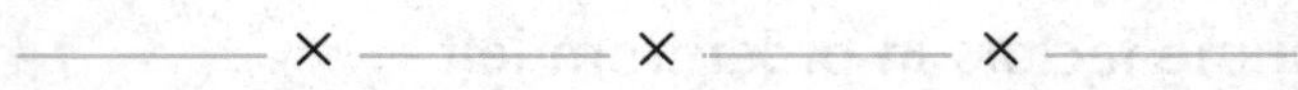

___ × ___ × ___ × ___.

Name

Independent Practice

Find the prime factorization of each number.

2. 63 = ______

3. 18 = ______

4. 40 = ______

5. 75 = ______

6. 27 = ______

7. 32 = ______

8. 49 = ______

9. 44 = ______

Mathematical PRACTICE 2 **Understand Symbols** **Find the missing number.**

10. 104 = 2 × 2 × ■ × 13

■ = ______

11. 55 = ■ × 11

■ = ______

12. 77 = 7 × ■

■ = ______

CCSS

Problem Solving

Use the table for Exercises 13–16 that shows the average weights of popular dog breeds.

Breed	Weight (lb)	Prime Factorization
Cocker Spaniel	20	
German Shepherd	81	
Labrador Retriever	67	
Beagle	25	
Golden Retriever	70	
Siberian Husky	50	

13. Complete the table.

14. Which weight(s) have a prime factorization of exactly three factors?

15. Which weight(s) have a prime factorization with factors that are all the same number?

16. Which breed(s) have weights that are prime numbers?

HOT Problems

17. Mathematical PRACTICE 7 **Identify Structure** Find the prime factorization of 2,800.

18. **Building on the Essential Question** How do factor trees help you find the prime factorization of a number?

Name ______________________

MY Homework

Lesson 1
Prime Factorization

Homework Helper

Need help? connectED.mcgraw-hill.com

Find the prime factorization of 42.

1. You can choose any pair of whole number factors, such as 2 × 21, 3 × 14, or 6 × 7.

2. Continue to factor any number that is not prime.

3. Write the prime factors in order from least to greatest.

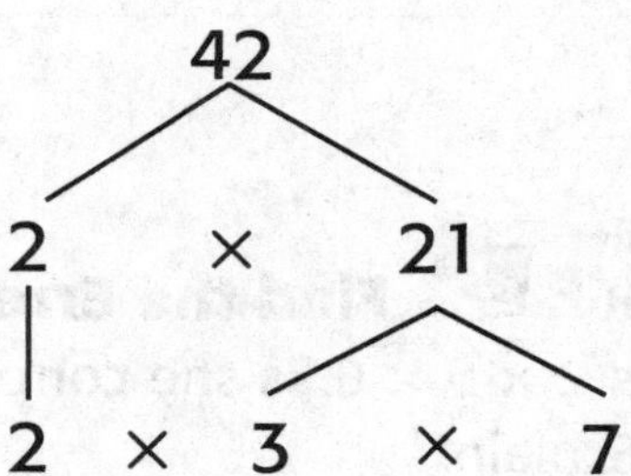

The prime factorization of 42 is 2 × 3 × 7.

Practice

Find the prime factorization of each number.

1. 50 = ______________ **2.** 81 = ______________

3. 65 = ______________ **4.** 28 = ______________

Real World Problem Solving

5. Britney scored an 85 on her last math test. Write the prime factorization of 85.

____

6. Priscilla has 56 stickers in her collection. Write the prime factorization of 56.

____

7. Mathematical PRACTICE 3  **Find the Error** Lainey wrote the prime factorization of 60 as 2 × 5 × 6. Is she correct? If not, what is the prime factorization of 60? Explain.

____

____

Vocabulary Check

Fill in each blank with the correct term or number to complete the sentence.

8. _______________ numbers can be written as a product of _______________ factors. This is called the prime factorization of a number.

Test Practice

9. Josiah had a pot belly pig as a pet that weighed 46 pounds. What is the prime factorization of 46?

Ⓐ 2 × 23 Ⓒ 2 × 2 × 13

Ⓑ 2 × 2 × 11 Ⓓ 3 × 23

Name ....

Hands On
Prime Factorization Patterns

Lesson 2

ESSENTIAL QUESTION
What strategies can be used to multiply whole numbers?

Build It

You can make a pattern using paper and a hole punch. By folding the paper, hole punching it, and counting the holes, you can discover a pattern.

1. Fold a piece of paper in half and make one hole punch. Open the paper.

   How many holes are in the paper? ______

   Find the prime factorization for the number of holes. ______

2. Fold another piece of paper in half twice. Make one hole punch.

   Unfold the paper. How many holes are in the paper? ______

   What is the prime factorization for the number of holes? ______ × ______

3. Complete the table for one, two, and three folds.

Number of Folds	Number of Holes	Prime Factorization
1		
2		
3		

4 What pattern do you notice between the number of factors in each prime factorization and the number of folds?

5 Using the pattern you found in Step 4, complete the table for four and five folds.

Number of Folds	Number of Holes	Prime Factorization
1	2	2
2	4	2×2
3	8	$2 \times 2 \times 2$
4		
5		

My Work!

Talk About It

1. Which prime number did you record in each prime factorization?

2. How many holes will you make if you fold the paper eight times? Write the prime factorization of that number.

3. Mathematical PRACTICE 1 **Make Sense of Problems** How can you check that you have the correct prime factorization?

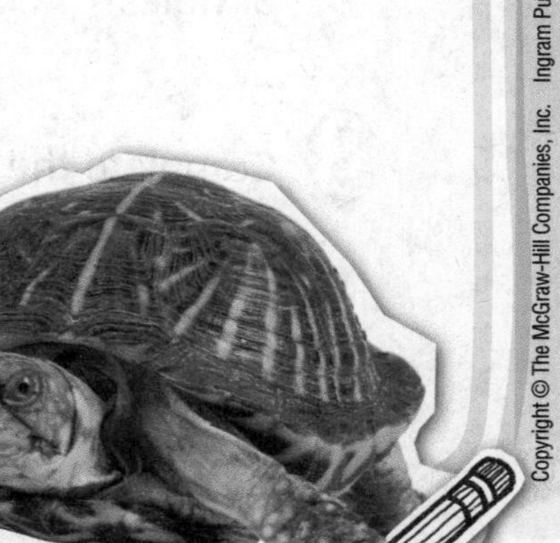

Name ....

CCSS

Practice It

4. Use paper and a hole punch to complete the table below. Start by folding a piece of paper in half and making 3 holes. Use a new piece of paper each time you increase the number of folds.

Number of Folds	Number of Holes	Prime Factorization
1	6	2×3
2	12	$2 \times 2 \times 3$
3		
4		
5		

Find a pattern to complete the tables in Exercises 5–7.

5.

Number of Folds	Number of Holes	Prime Factorization
1	10	2×5
2	20	$2 \times 2 \times 5$
3	40	
4		
5		

6.

Number of Folds	Number of Holes	Prime Factorization
1	14	2×7
2	28	$2 \times 2 \times 7$
3	56	
4		
5		

7.

Number of Folds	Number of Holes	Prime Factorization
1	18	$2 \times 3 \times 3$
2	36	$2 \times 2 \times 3 \times 3$
3	72	
4		
5		

Apply It

Use the information below to solve Exercises 8–11. There is one skin cell used in a science lab. Each day, the skin cell will split into two cells. The next day the cell then splits into two cells again.

Number of Days Passed	Number of Cells
1	2
2	4
3	8
4	16
5	32

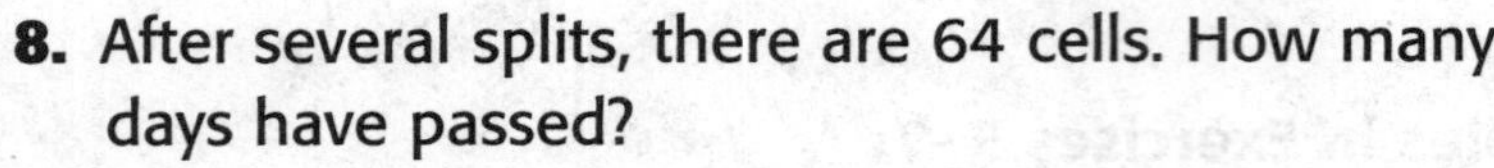

8. After several splits, there are 64 cells. How many days have passed?

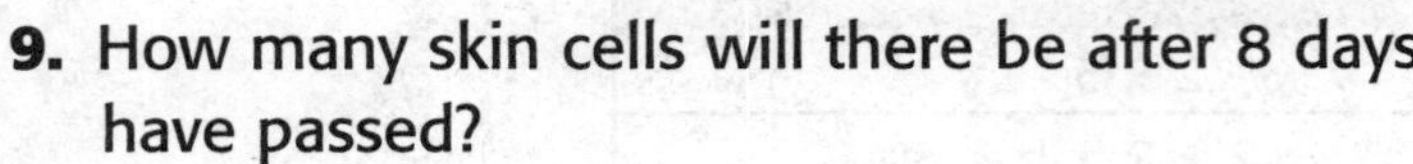

9. How many skin cells will there be after 8 days have passed?

10. How many days would need to pass before there were over 2,000 cells?

11. Mathematical PRACTICE 1 **Make a Plan** After 15 days, there were 32,768 cells. After how many days were there 16,384 cells?

Write About It

12. How can I use patterns to describe relationships?

Name

MY Homework

Lesson 2
Hands On: Prime Factorization Patterns

Homework Helper

Need help? connectED.mcgraw-hill.com

A design with equilateral triangles is shown. The triangle is divided into four equal-sized, smaller triangles as shown. Then each of the four triangles is divided into four equal-sized, smaller triangles. If the pattern continues, how many triangles will there be in Figure 3?

Figure 1

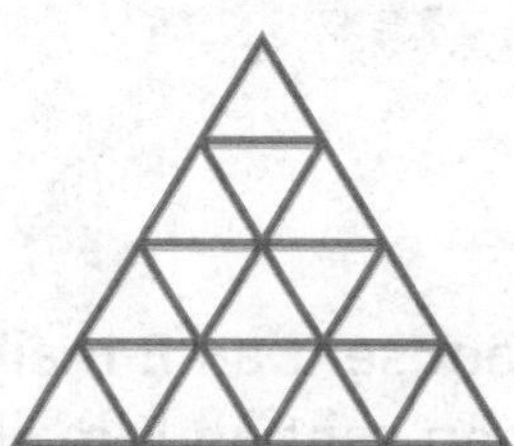
Figure 2

In Figure 1, there are 4 resulting triangles. In Figure 2, there are 16 triangles.

How many triangles will be in Figure 3?

The table shows the Figure numbers, the number of triangles formed, and the prime factorization of the number of triangles.

Figure Number	Number of Triangles Formed	Prime Factorization
1	4	2×2
2	16	$2 \times 2 \times 2 \times 2$
3	64	$2 \times 2 \times 2 \times 2 \times 2 \times 2$

By following the pattern, there are 64 triangles formed in Figure 3.

Practice

1. Complete the table for Figures 4 and 5 if the above pattern continues.

Figure Number	Number of Triangles Formed	Prime Factorization
4		
5		

Problem Solving

2. A population of rabbits triples every month. The population starts with two rabbits. How many rabbits are there after three months?

____

3. Three friends each create 4 bags of starter bread dough. After ten days, each of those four bags is then divided into four more bags of dough. How many days pass before 192 bags of dough are created?

____

4. Sam sent an E-mail to 3 friends on Monday. Each of the friends then sent an E-mail to 3 friends on Tuesday. On Wednesday, each of those friends then sent an E-mail to 3 friends. Write the prime factorization of the number of E-mails that were sent on Wednesday.

____

5. Mathematical PRACTICE 8 **Look for a Pattern** Annie opened a savings account and deposited $10. If the balance in her account doubles each month, what is the account balance after 4 months?

____

6. Elyse folded a piece of paper in half 3 times. She then punched 3 holes in the paper. How many holes are in the paper when she unfolds it?

____

Name

Powers and Exponents

Lesson 3

ESSENTIAL QUESTION
What strategies can be used to multiply whole numbers?

A product of identical factors can be written using an exponent and a base. The **base** is the number used as a factor. The **exponent** indicates how many times the base is used as a factor.

Math in My World

Tutor

Example 1

The number of Calories in six pancakes can be written as 10^3. Write 10^3 as a product of the same factor. Then find the value.

$10 \times 10 \times 10 = 10^3$ ← exponent

3 factors (under $10 \times 10 \times 10$); base (arrow to 10)

$10 \times 10 \times 10 =$ ______

Six pancakes have ______ Calories.

Numbers expressed using exponents are called **powers.** Numbers raised to the second or third power have special names.

Powers	Words
2^5	2 to the fifth power
3^2	3 to the second power or 3 **squared**
10^3	10 to the third power or 10 **cubed**

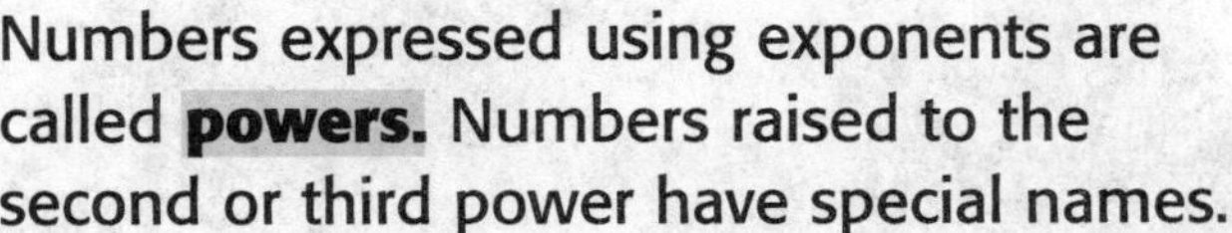

Example 2 

Write $3 \times 3 \times 3 \times 3$ using an exponent.

The base is ______. Since 3 is used as a factor ______ times,

the exponent is ______.

Write as a power. $3 \times 3 \times 3 \times 3 =$ ______

Example 3 

Write the prime factorization of 72 using exponents.

1 Complete the factor tree.

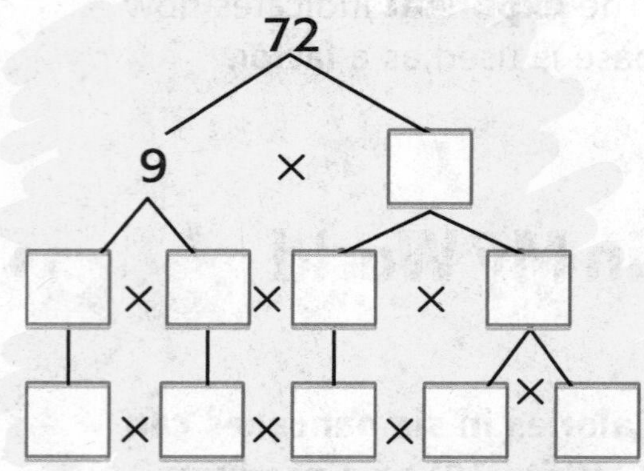

2 Order the factors from least to greatest.

______ × ______ × ______ × ______ × ______

3 Write products of identical factors using exponents.

$2^{\square} \times 3^{\square}$

So, 72 = ______ × ______.

Explain how a factor tree helps you to write the prime factorization of a number using exponents.

Guided Practice 

1. Write $4 \times 4 \times 4 \times 4$ using an exponent.

   The base is ______. Since 4 is used as a factor ______ times, the exponent is ______.

   $4 \times 4 \times 4 \times 4 =$ ______

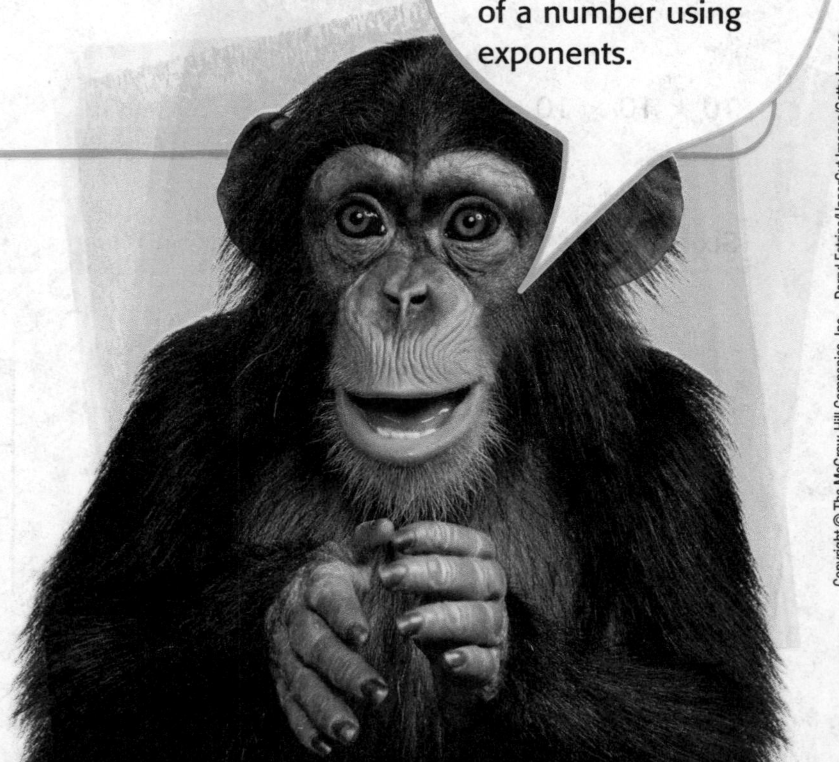

Name ....

Independent Practice

Write each product using an exponent.

2. $10 \times 10 =$ ________

3. $8 \times 8 \times 8 \times 8 =$ ________

4. $3 \times 3 \times 3 \times 3 \times 3 \times 3 =$ ________

5. $5 \times 5 \times 5 \times 5 \times 5 =$ ________

6. $9 \times 9 \times 9 \times 9 =$ ________

7. $1 \times 1 \times 1 \times 1 \times 1 =$ ________

Write each power as a product of the same factor. Then find the value.

8. $10^4 =$ ________________

9. $3^2 =$ ________________

10. $9^3 =$ ________________

11. $6^5 =$ ________________

Write the prime factorization of each number using exponents.

12. $25 =$ ________

13. $56 =$ ________

14. $68 =$ ________

15. $88 =$ ________

CCSS

Problem Solving

16. To find the amount of space a cube-shaped bird cage occupies, find the *cube* of the measure of one edge of the bird cage. Express the amount of space occupied by the bird cage shown as a power. Then find the amount in cubic units.

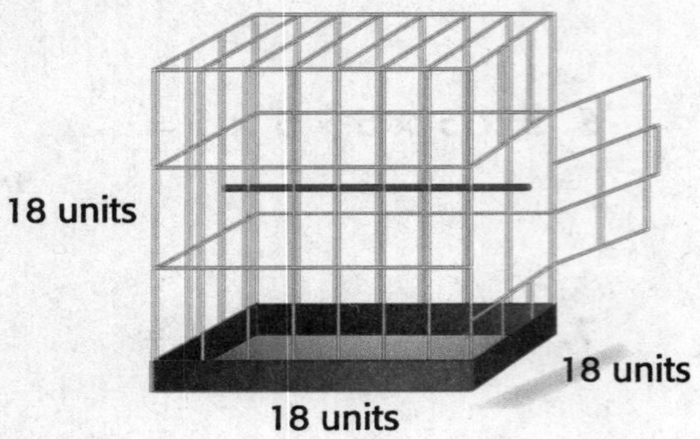

17. A single tusk that weighed just over 2^8 pounds from an African elephant is the largest tooth ever recorded from any modern animal. About how many pounds did the tusk weigh?

HOT Problems

18. Mathematical PRACTICE 2 **Reason** Which is greater: 3^5 or 5^3? Explain your reasoning.

19. **Building on the Essential Question** What does it mean for a product of factors to be expressed using exponents?

Name ....

MY Homework

Lesson 3
Powers and Exponents

Homework Helper

Need help? connectED.mcgraw-hill.com

Write $6 \times 6 \times 6$ using an exponent.

The base is 6. Since 6 is used as a factor three times, the exponent is 3.

So, $6 \times 6 \times 6 = 6^3$.

Practice

Write each product using an exponent.

1. $10 \times 10 \times 10 =$ ______

2. $12 \times 12 =$ ______

Write each power as a product of the same factor. Then find the value.

3. $3^7 =$ ______

4. $10^6 =$ ______

Write the prime factorization of each number using exponents.

5. $20 =$ ______

6. $50 =$ ______

Problem Solving

7. **Mathematical PRACTICE 8** **Look for a Pattern** The Newfoundland is a large breed of dog. It weighs about 10×10 pounds. Write 10×10 using an exponent. Then find the value of the power. How many pounds does the Newfoundland weigh?

8. The area of San Bernardino County, California, the largest county in the United States, is about 3^9 square miles. Write this as an expression. What is the area of San Bernardino County?

Vocabulary Check

Fill in each blank with the correct term or number to complete each sentence.

9. Numbers expressed using exponents are called ________.

10. The exponent indicates how many times the ________ is used as a factor.

Test Practice

11. A 100-pound person on Earth would weigh about $4 \times 4 \times 4 \times 4$ pounds on Jupiter. Evaluate the expression to determine how much a 100-pound person would weigh on Jupiter.

Ⓐ 16 pounds　Ⓒ 256 pounds

Ⓑ 64 pounds　Ⓓ 1,024 pounds

Name ....

Multiplication Patterns

Lesson 4

ESSENTIAL QUESTION
What strategies can be used to multiply whole numbers?

Powers of 10 include numbers like 10, 100, and 1,000, because they can be written as 10^1, 10^2, and 10^3.

Math in My World

Example 1

Pet Paws is ordering more goldfish to sell at their store. Each goldfish costs $2. Use the table to determine the cost of 10, 100, and 1,000 goldfish. Describe the pattern in the number of zeros when multiplying the cost, $2, by powers of ten.

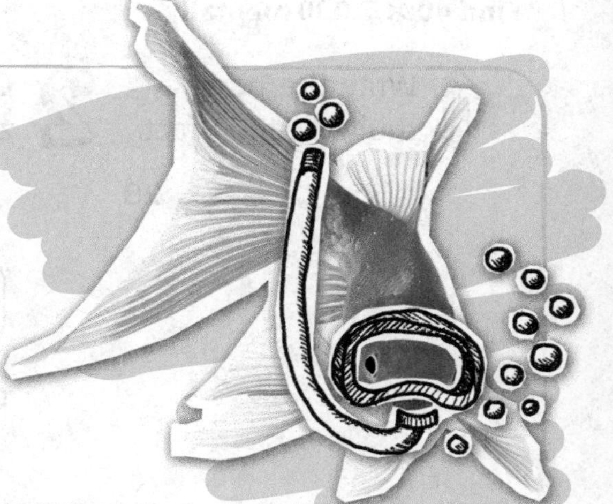

Cost of One Goldfish ($)	Power of Ten	Product	Number of Zeros in Product
2	× 1	2	0
2	× 10		
2	× 100		
2	× 1,000		

The number of zeros in the product increases when the power of ten increases. Each successive power of ten adds

________ zero to the product.

How many zeros are in the product of 7 and 100? ________

How many zeros are in the product of 21 and 10? ________

How many zeros are in the product of 12 and 1,000? ________

Example 2

Find 13×10^2 mentally.

Write 10^2 without exponents.

$13 \times 10^2 =$

$13 \times$ ________

Count the number of zeros in the power of 10.

100 ← ____ zeros

Compare the number of zeros to the exponent of 10^2.

They are the ________.

Write the zeros to the right of 13.

1, 3 ___

So, the product is ________.

Example 3

Find $40 \times 7{,}000$ mentally.

Write the basic multiplication fact.

$4 \times$ ________ $= 28$

Count the number of zeros in each factor.

$40 \times 7{,}000$

1 zero and ____ zeros

There are a total of ________ zeros.

Write the zeros to the right of the product from step 1.

28___, ___

So, the product is ________.

Guided Practice

Find each product mentally.

1. 8×10^2

$10^2 =$ ________

total number of zeros = ________

So, $8 \times 10^2 =$ ________.

2. $14 \times 2{,}000$

basic multiplication fact:

________ × ________ = 28

total number of zeros = ________

So, $14 \times 2{,}000 =$ ________.

Talk MATH

Explain how you could find the product of 29 and 10^3 mentally.

Name

Independent Practice

Let's PRACTICE!

Find each product mentally.

3. $13 \times 1{,}000 =$ ______

4. $37 \times 10^2 =$ ______

5. $9{,}000 \times 3 =$ ______

6. $8 \times 10^3 =$ ______

7. $21 \times 10^1 =$ ______

8. $9 \times 50 =$ ______

Algebra Find the missing number.

9. ■ $\times 10^4 = 70{,}000$

■ = ______

10. $300 \times$ ■ $= 120{,}000$

■ = ______

11. $100 \times$ ■ $= 900$

■ = ______

12. ■ $\times 10^2 = 4{,}400$

■ = ______

Problem Solving

13. Each box contains 10^2 pencils. The school store has 15 boxes of pencils. How many pencils does the school store have?

___

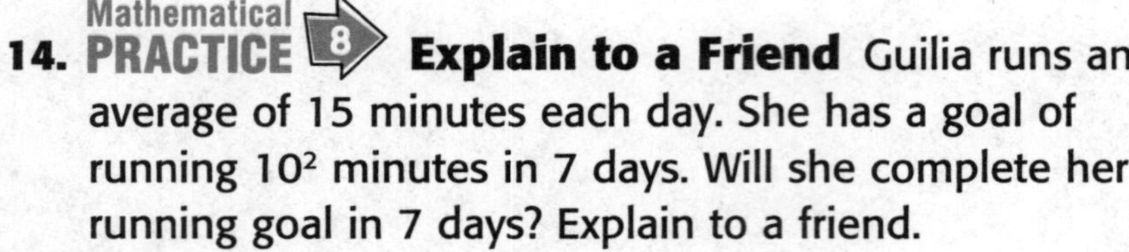

14. Mathematical PRACTICE 8 **Explain to a Friend** Guilia runs an average of 15 minutes each day. She has a goal of running 10^2 minutes in 7 days. Will she complete her running goal in 7 days? Explain to a friend.

___

___

15. A group of friends bought 7 tickets to a dog show for \$30 each. How much did they spend on the tickets?

___

HOT Problems

Mathematical PRACTICE 2 **Use Number Sense Find each missing exponent.**

16. $40 \times 10^{\blacksquare} = 4{,}000$

■ = ___

17. $32 \times 10^{\blacksquare} = 32{,}000$

■ = ___

18. $80{,}000 = 10^{\blacksquare} \times 8$

■ = ___

19. $10{,}000 = 10 \times 10^{\blacksquare}$

■ = ___

20. **Building on the Essential Question** How can patterns be used to determine products of a number and a power of 10?

___

___

___

Name ....

MY Homework

Lesson 4
Multiplication Patterns

Homework Helper

Need help? connectED.mcgraw-hill.com

A truck is loaded with 10^2 boxes of skateboards. Each box weighs 36 pounds. What is the total weight of the boxes?

Find 36×10^2 mentally.

1. 10^2 without exponents is equal to 100.

2. There are 2 zeros in 100.

After placing the zeros to the right of 36, the product is 3,600.

So, the total weight of the boxes is 3,600 pounds.

Practice

Find each product mentally.

1. $70 \times 500 =$ ____________

2. $320 \times 10^2 =$ ____________

3. $56 \times 10^3 =$ ____________

4. $10^2 \times 72 =$ ____________

5. $80 \times 3{,}000 =$ ____________

6. $10^3 \times 31 =$ ____________

Problem Solving

7. To protect themselves from extreme hot or cold temperatures, American Alligators dig burrows in the mud. Suppose there are 20 alligators, each with 50 feet of burrows. What is the total length of all the burrows?

Later GATOR!

____

8. Paulita reads an average of 20 pages each day. She has 6 days to read 10^2 pages. Will she finish her reading in 6 days? Explain.

____

____

9. Mathematical PRACTICE 1 **Make Sense of Problems** Explain how using basic facts can help you find $10 \times 20 \times 30 \times 40$ mentally.

____

____

____

Vocabulary Check

Fill in the blank with the correct term or number to complete the sentence.

10. Numbers like 10, 100, 1,000, and so on are called _______________.

Test Practice

11. A music store sold 10^3 CDs and 10^2 CD players. If each CD costs $12 and each CD player costs $35, what was the store's total earnings?

Ⓐ $15,500 Ⓒ $36,200

Ⓑ $24,500 Ⓓ $47,000

Name

Real World

Problem-Solving Investigation

STRATEGY: Make a Table

Lesson 5

ESSENTIAL QUESTION
What strategies can be used to multiply whole numbers?

Learn the Strategy

In a recent year, about 63 out of every 100 households in the United States owned at least one pet. About how many households out of ten thousand owned at least one pet?

My house is always ready to go!

HOME SWEET HOME

Understand

What facts do you know?

Sixty-three out of every ______ households owned at least one pet.

What do you need to find?

how many households out of ten thousand owned at least one pet

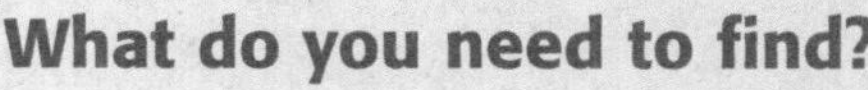

Plan

I will make a ______ to solve the problem.

Solve

Total Number of Households	Number of Households Owning at least one Pet
100	63

×10 ×10 (left); ×10 ×10 (right)

So, about ______ out of ten thousand households owned at least one pet.

Check

Is my answer reasonable? Explain.

Use mental math to multiply. $63 \times 100 =$ ______

Practice the Strategy

Nestor is saving money to buy a new camping tent. Each week he doubles the amount he saved the previous week. If he saves $1 the first week, how much money will Nestor save in 7 weeks?

Understand

What facts do you know?

____

____

What do you need to find?

____

2 Plan

____

3 Solve

4 Check

Is my answer reasonable? Explain.

____

____

Name

Apply the Strategy

Solve each problem by making a table.

1. Betsy is saving to buy a bird cage. She saves $1 the first week, $3 the second week, $9 the third week, and so on. How much money will she save in 5 weeks?

____

2. Kendall is planning to buy a laptop for $1,200. Each month she doubles the amount she saved the previous month. If she saves $20 the first month, in how many months will Kendall have enough money to buy the laptop?

____

3. Mrs. Piant's yearly salary is $42,000 and increases $2,000 per year. Mr. Piant's yearly salary is $37,000 and increases $3,000 per year. In how many years will Mr. and Mrs. Piant make the same salary?

____

4. Mathematical **PRACTICE** 8 **Look for a Pattern** Complete the table that shows the prime factorizations of powers of 10. Use the pattern in the table to mentally find the prime factorization of 10^9. Write using exponents.

Power of 10	Prime Factorization
10	2×5
100 or 10^2	$2^2 \times 5^2$
1,000 or 10^3	
10,000 or 10^4	

Prime factorization of 10^9 = ________.

Review the Strategies

Use any strategy to solve each problem.
- Make a table.
- Use the four-step plan.

A card shop recorded how many packs of trading cards it sold each day. Use the table to solve Exercises 5–7.

Trading Cards Sold			
Day	**Week 1**	**Week 2**	**Week 3**
Mon.	28	48	25
Tue.	32	43	37
Wed.	38	45	42
Thur.	44	41	35
Fri.	36	39	41

My Work!

5. In which week did they sell the most packs of cards?

6. In which week did they sell the least amount?

7. How many more packs did they sell in Week 2 than in Week 3?

8. A putt-putt course offers a deal that when you purchase 10 rounds of golf, you get 1 round for free. If you played a total of 35 rounds, how many rounds did you purchase?

9. Mathematical PRACTICE 1 **Plan Your Solution** Saketa is saving money to buy a new ferret cage. In the first week, she saved $24. Each week after the first, she saves $6. How much money will Saketa have saved in six weeks?

MY Homework

Lesson 5
Problem Solving: Make a Table

Homework Helper

Need help? connectED.mcgraw-hill.com

Every 10 minutes, 160 passengers can ride the Black Stallion roller coaster. How many people can ride the Black Stallion in 60 minutes?

1 Understand

What facts do you know?

There are 160 passengers that ride every 10 minutes.

What do you need to find?

the number of people that can ride the coaster in 60 minutes

2 Plan

I can make a table to solve the problem.

Solve

Minutes	10	20	30	40	50	60
Passengers	160	320	480	640	800	960

+160 +160 +160 +160 +160

So, 960 passengers can ride the Black Stallion in 60 minutes.

Check

Is my answer reasonable? Explain.

Estimate. Round 160 to the nearest hundred.

200 + 200 + 200 + 200 + 200 = 1,000

Problem Solving

Solve each problem by making a table.

1. Mathematical PRACTICE 8 **Look for a Pattern** In a recent year, one United States dollar was equal to about 82 Japanese yen. How many Japanese yen are equal to $100? $1,000? $10,000?

2. A local restaurant offers a deal if you purchase 3 medium pizzas, you get 2 side dishes for free. If you get a total of 8 side dishes, how many pizzas did you buy?

3. A recipe for potato salad calls for one teaspoon of vinegar for every 2 teaspoons of mayonnaise. How many teaspoons of vinegar are needed for 16 teaspoons of mayonnaise?

4. A package of 4 mechanical pencils comes with 2 free erasers. If you get a total of 12 free erasers, how many packages of pencils did you buy?

5. Veronica is saving money to buy a saddle for her horse that costs $175. She plans to save $10 the first month and then increase the amount she saves by $5 each month after the first month. How many months will it take her to save $175?

Check My Progress

Vocabulary Check

Draw a line to match each word to its definition.

1. Numbers such as 1, 2, 4, 5, 10, and 20. • **cubed**
2. A number raised to the second power. • **prime factorization**
3. A number raised to the third power. • **factors**
4. A way of expressing a composite number as a product of its prime factors. • **powers of ten**
5. Numbers like 10, 100, 1,000 and so on. • **squared**

Concept Check

Find the prime factorization of each number using exponents.

6. 36 = ______________ **7.** 45 = ______________

Write each product using an exponent.

8. $13 \times 13 \times 13 \times 13 =$ ________ **9.** $7 \times 7 \times 7 =$ ________

Write each power as a product of the same factor. Then find the value.

10. 10^4 = ______________

11. 14^1 = ______________

Find each product mentally.

12. 23×100 = ________ **13.** 150×400 = ________

Problem Solving

14. The amount of space in the dog carrier shown can be found by multiplying its width, length, and height. Write the amount of space, in cubic units, using an exponent. Then evaluate.

____

15. There are 10^2 fifth-grade students attending a class field trip to the city's art museum. If each admission ticket costs $6, what is the total cost of admission for the 10^2 students?

____

Test Practice

16. Dakota read 130 pages over the weekend. Dajuan read 165 pages over the weekend. Which prime number do the prime factorizations of 130 and 165 have in common?

Ⓐ 2 Ⓒ 5

Ⓑ 3 Ⓓ 7

Name

Hands On
Use Partial Products and the Distributive Property

Lesson 6

ESSENTIAL QUESTION
What strategies can be used to multiply whole numbers?

Draw It

Darnel and his four friends went to an ice skating rink and bought sandwiches. They divided the total cost and found that each person needs to pay $17. What was the total cost for ice skating and sandwiches?

Find 5 × 17 using an area model.

Label the model to find the partial products.

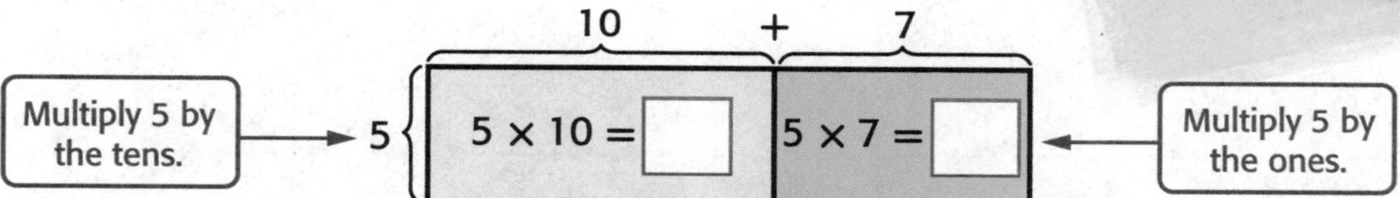

2 Add the partial products.

_______ + _______ = _______

So, the total cost for ice skating and sandwiches was _______.

When you use partial products, you are using a property. A **property** is a rule in mathematics that can be applied to all numbers. The property that you applied above is called the **Distributive Property.** You will learn more about this property in the next lesson.

Try It

Find 7 × 56 using an area model.

1 Label the model to find partial products.

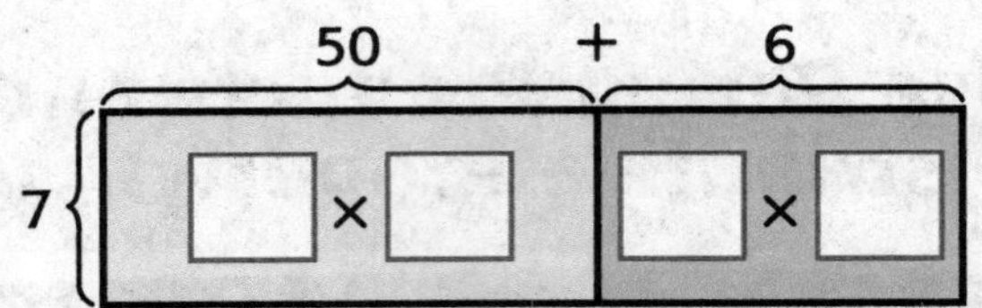

2 Multiply. Then add.

7 × 56 = (7 × 50) + (7 × ______)

= ______ + ______

= ______

So, 7 × 56 = ______.

Talk About It

1. **Mathematical PRACTICE 3 Draw a Conclusion** How do area models show partial products?

______

______

______

2. In finding 7 × 56 above, why does the area model separate 56 into 50 and 6?

______

______

3. Find 6 × 36 using partial products. Label the model to help you solve. Show your work.

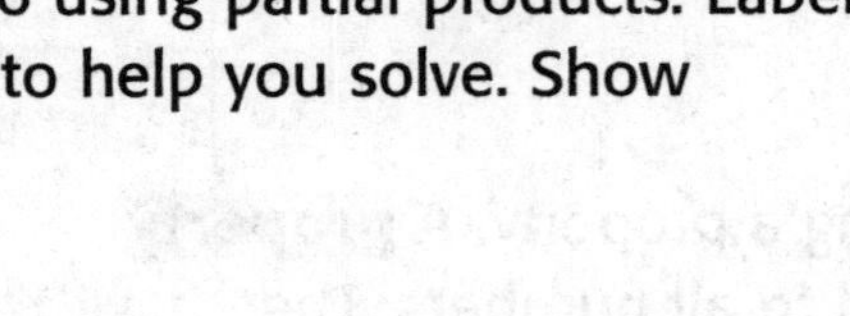

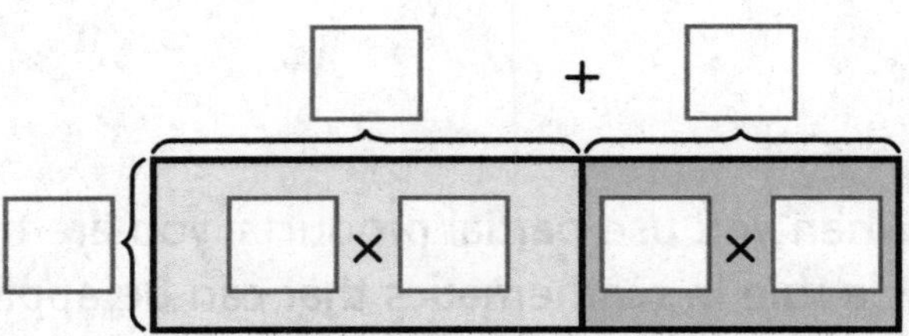

Name ....

Practice It

Multiply using the area model. Label each model.

4. 4 × 16 = ______

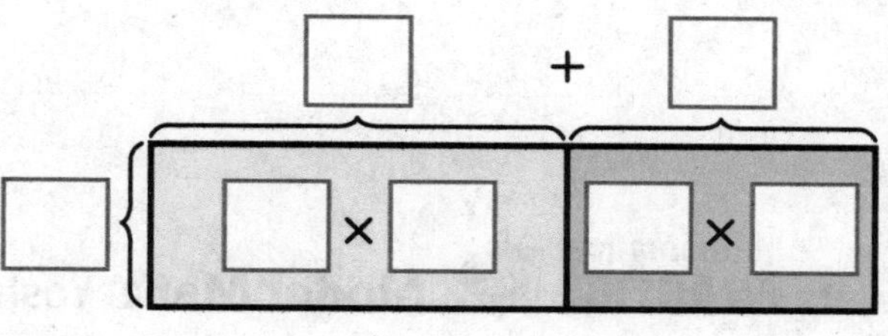

5. 6 × 81 = ______

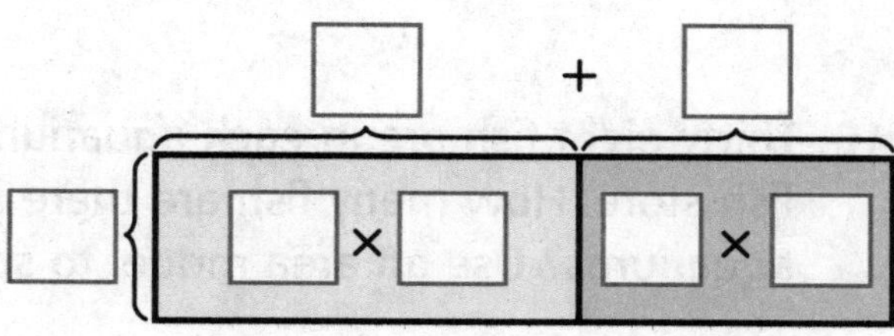

6. 7 × 29 = ______

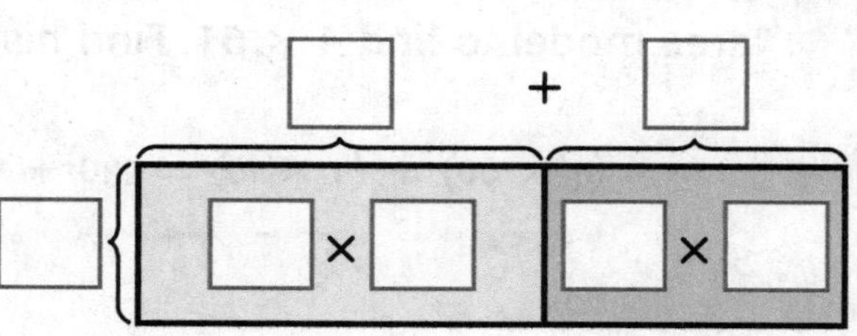

7. 5 × 39 = ______

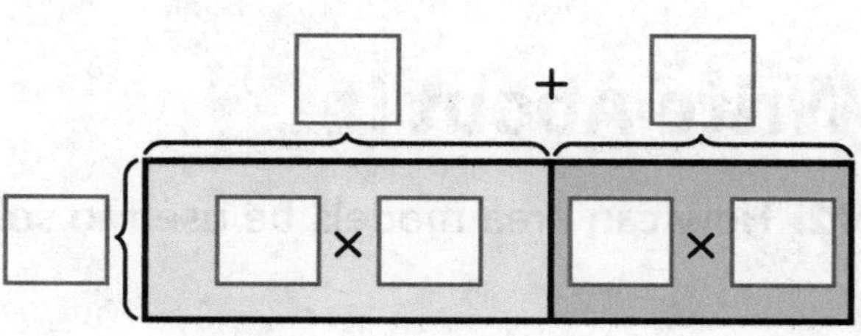

Apply It

8. Ronnie swims 4 laps each day at the pool. How many laps does he swim in 28 days? Use an area model to solve.

My Work!

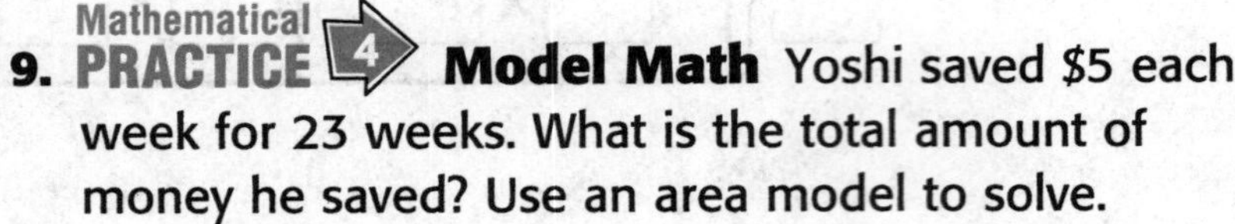

9. Mathematical PRACTICE 4 **Model Math** Yoshi saved \$5 each week for 23 weeks. What is the total amount of money he saved? Use an area model to solve.

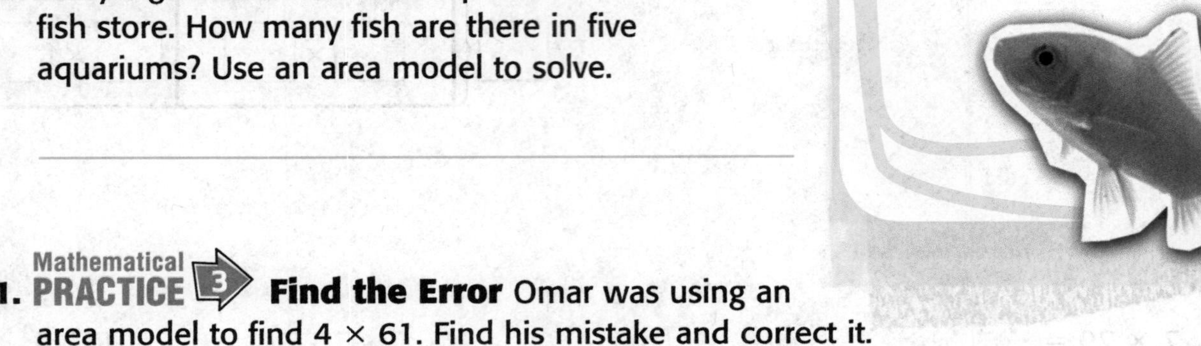

10. Thirty-eight fish are in each aquarium at a fish store. How many fish are there in five aquariums? Use an area model to solve.

11. Mathematical PRACTICE 3 **Find the Error** Omar was using an area model to find 4×61. Find his mistake and correct it.

$(4 \times 60) + (4 \times 10) = 240 + 40 = 280$

Write About It

12. How can area models be used to solve multiplication problems?

Name ....

MY Homework

Lesson 6
Hands On: Use Partial Products and the Distributive Property

Homework Helper

Need help? connectED.mcgraw-hill.com

Consuelo and her two siblings decided to buy a trampoline. They divided the total cost and found that each person needs to pay $48. What was the total cost for the trampoline?

Find 3 × 48 using an area model.

1 The model is labeled to show the partial products.

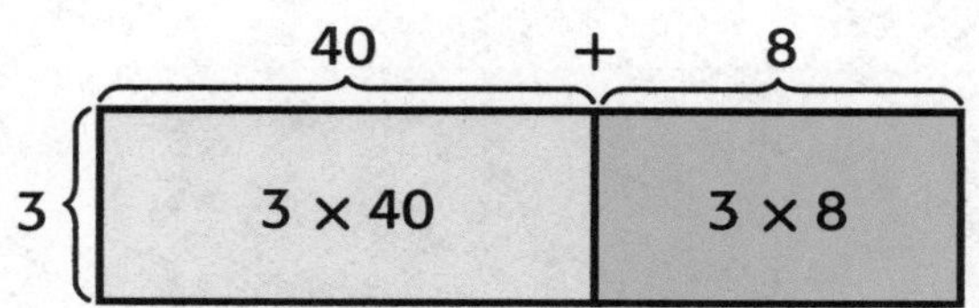

2 Multiply. Then add.

$3 \times 48 = (3 \times 40) + (3 \times 8)$

$= 120 + 24$

$= 144$

So, the total cost of the trampoline was $144.

Practice

1. Multiply 2 × 27 using the area model. Label the model.

2 × 27 = ______

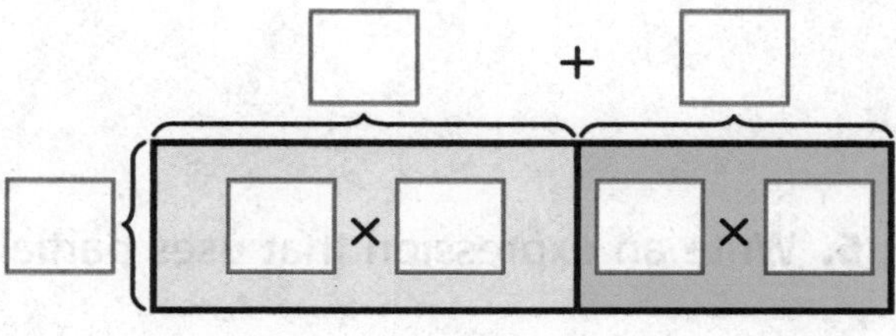

2. Multiply 9 × 43 using the area model. Label the model.

9 × 43 = ______

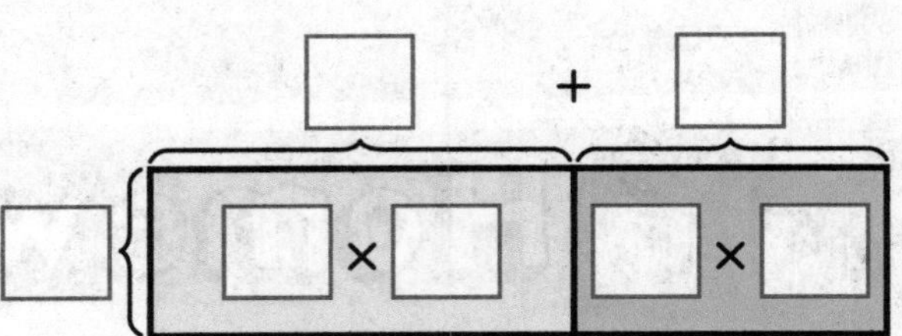

Problem Solving

3. A pet store has eighteen hamsters in each cage. How many hamsters are there in six cages? Use an area model to solve.

______

4. **Mathematical PRACTICE 4 Model Math** Brandon packs his lunch 4 days each week. How many days does he pack his lunch for 36 weeks? Use an area model to solve.

______

5. Ricarda earns a $25 allowance each week. How much allowance will she earn after 7 weeks? Use an area model to solve.

______

6. Write an expression that uses partial products to multiply 8 × 64.

______

Name ....

The Distributive Property

Lesson 7

ESSENTIAL QUESTION

What strategies can be used to multiply whole numbers?

The **Distributive Property** allows you to multiply a sum by a number. To do so, multiply each addend by the number. Then add.

$$3 \times (5 + 2) = (3 \times 5) + (3 \times 2)$$

Math in My World

Example 1

For a field trip, 42 students each paid $3 for transportation. Use mental math and the Distributive Property to find how much money was collected altogether.

Find 3×42.

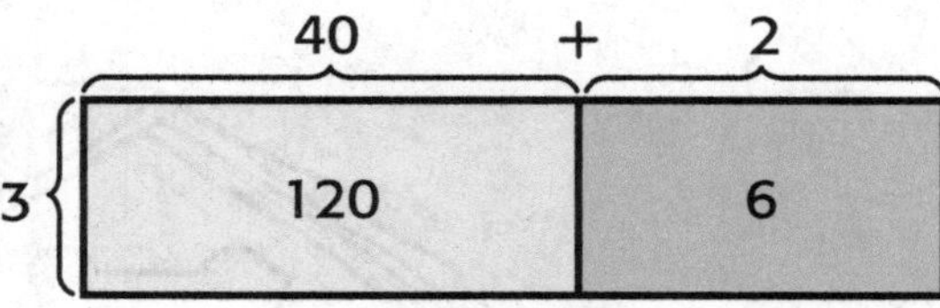

 Write 42 as 40 + 2. $3 \times 42 = 3 \times (\_\_\_\_ + \_\_\_\_)$

 Apply the Distributive Property. $= (3 \times \_\_\_\_) + (3 \times \_\_\_\_)$

 Multiply. $= \_\_\_\_ + \_\_\_\_$

 Add. $= \_\_\_\_$

So, $ ______ was collected for the field trip.

Example 2 

Find 7 × 26 mentally using the Distributive Property. Show the steps that you used.

 Write 26 as 20 + 6. $7 \times 26 = 7 \times (\_\_\_\_ + \_\_\_\_)$

 Apply the Distributive Property. $= (7 \times \_\_\_\_) + (7 \times \_\_\_\_)$

3 Multiply. $= \_\_\_\_ + \_\_\_\_$

4 Add. $= \_\_\_\_$

So, 7 × 26 = ______.

Guided Practice 

Find each product mentally using the Distributive Property. Show the steps that you used.

1. 5 × 18

$5 \times 18 = 5 \times (\_\_\_\_ + 8)$

$= (5 \times \_\_\_\_) + (5 \times \_\_\_\_)$

$= \_\_\_\_ + \_\_\_\_$

$= \_\_\_\_$

So, 5 × 18 = ______.

2. 2 × 21

$2 \times 21 = 2 \times (\_\_\_\_ + \_\_\_\_)$

$= (2 \times \_\_\_\_) + (2 \times \_\_\_\_)$

$= \_\_\_\_ + \_\_\_\_$

$= \_\_\_\_$

So, 2 × 21 = ______.

Name

Independent Practice

Find each product mentally using the Distributive Property. Show the steps that you used.

3. $6 \times 13 =$ ______

4. $3 \times 52 =$ ______

5. $5 \times 26 =$ ______

6. $4 \times 69 =$ ______

7. $2 \times 49 =$ ______

8. $7 \times 23 =$ ______

9. $26 \times 6 =$ ______

10. $55 \times 4 =$ ______

Problem Solving

11. A horse is 17 *hands* tall. If a *hand* equals 4 inches, how tall is the horse in inches? Use the Distributive Property to solve.

12. Mr. Collins is buying 5 train tickets for $36 each. What is the total cost of the tickets? Use the Distributive Property to solve.

13. Melanie runs 23 miles each week . How many miles does she run in 9 weeks? Use the Distributive Property to solve.

HOT Problems

14. Mathematical PRACTICE 3 **Find the Error** Dylan is using the Distributive Property to simplify $6 \times (9 + 4)$. Find his mistake and correct it.

$$6 \times (9 + 4) = 54 + 4$$
$$= 58$$

15. **Building on the Essential Question** How can the Distributive Property be used to multiply numbers? Explain.

Name ....

MY Homework

Lesson 7

The Distributive Property

Homework Helper

Need help? connectED.mcgraw-hill.com

At a petting zoo, 29 students each paid $2 to pet the animals. Use mental math and the Distributive Property to find how much money was paid altogether.

Write 29 as 20 + 9. ⟶ $2 \times 29 = 2 \times (20 + 9)$

Apply the Distributive Property. ⟶ $= (2 \times 20) + (2 \times 9)$

Multiply. ⟶ $= 40 + 18$

4 Add. ⟶ $= 58$

So, the students spent $58 at the petting zoo.

Practice

Find each product mentally using the Distributive Property. Show the steps that you used.

1. $4 \times 48 =$ ______

2. $3 \times 67 =$ ______

3. $6 \times 18 =$ ______

4. $8 \times 74 =$ ______

Problem Solving

5. A bookshelf in Deirdre's living room has 4 shelves. There are 12 books on each shelf. How many books are on the bookshelf altogether?

____

6. Jorge is collecting baseball cards. He has 29 stacks of cards with 4 in each stack. How many cards does he have altogether?

____

7. Mathematical PRACTICE 3 **Justify Conclusions** The Distributive Property also combines subtraction and multiplication. For example, $3 \times (5 - 2) = (3 \times 5) - (3 \times 2)$. Explain how you could use the Distributive Property and mental math to find 5×198.

____

____

____

Vocabulary Check

Fill in each blank with the correct term or number to complete the sentence.

8. The Distributive Property combines ____________ and ____________ to make multiplying whole numbers simpler.

Test Practice

9. Susana collected 5 cents at the recycling plant for each of her 78 cans. How much money did she collect altogether?

Ⓐ \$0.39 Ⓒ \$3.90

Ⓑ \$3.50 Ⓓ \$39.00

Name ....

Estimate Products

Lesson 8

ESSENTIAL QUESTION
What strategies can be used to multiply whole numbers?

When a problem asks for *about* how many, you can use estimation, rounding, and/or compatible numbers. **Compatible numbers** are numbers in a problem that are easy to compute mentally.

Math in My World

Example 1

A pet store has 12 gecko lizards for sale. Each gecko lizard costs $92. About how much money would the store make if it sells all 12 gecko lizards?

$92

Estimate the product of 92 and 12.

One Way **Round one factor.**

THINK It is easier to compute 92 × 10 than 90 × 12.

92 ⟶ 9 2

× 12 ⟶ × 1 ☐ Round 12 to the nearest ten.

 Find 92 × 10 mentally.

By rounding one factor, the estimate is ________.

Another Way **Round both factors.**

92 ⟶ 9 ☐ Round 92 to the nearest ten.

× 12 ⟶ × 1 ☐ Round 12 to the nearest ten.

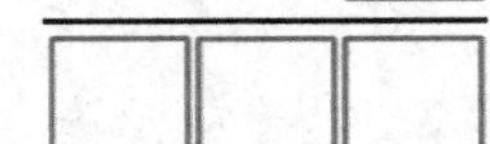 Find 90 × 10 mentally.

By rounding both factors, the estimate is ________.

Another Way **Use compatible numbers.**

92 → ☐☐☐

× 12 → × 1☐

☐,☐☐☐

Use numbers that are easy to multiply mentally such as 100 and 10.

Find 100 × 10 mentally.

By using compatible numbers, the estimate is ________.

Use a calculator to multiply 92 × 12. What do you get? ________

Circle whether the estimates were all less than or greater than the actual product.

less than　　　greater than

Guided Practice

Estimate by rounding or using compatible numbers. Show how you estimated.

1. Round both factors.

32 → 3☐

× 18 → × 2☐

☐☐☐

The product is about ______.

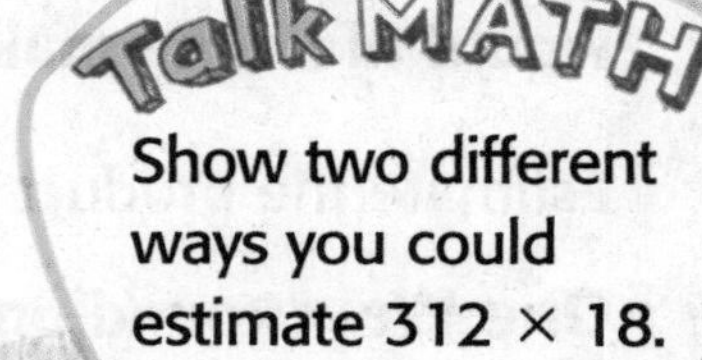

2. Use compatible numbers.

98 → ☐☐☐

× 83 → × ☐☐

☐,☐☐☐

The product is about ______.

Name

Independent Practice

Estimate by rounding. Show how you estimated.

3. $\begin{array}{r} 218 \\ \times \quad 6 \\ \hline \end{array}$

4. $\begin{array}{r} 68 \\ \times \quad 7 \\ \hline \end{array}$

5. $\begin{array}{r} 131 \\ \times \quad 29 \\ \hline \end{array}$

6. 61 × 68 ≈ ______

7. 79 × 56 ≈ ______

8. 392 × 46 ≈ ______

Estimate by using compatible numbers. Show how you estimated.

9. $\begin{array}{r} 106 \\ \times \quad 52 \\ \hline \end{array}$

10. $\begin{array}{r} 33 \\ \times \quad 6 \\ \hline \end{array}$

11. $\begin{array}{r} 127 \\ \times \quad 8 \\ \hline \end{array}$

12. 33 × 84 ≈ ______

13. 450 × 21 ≈ ______

14. 729 × 42 ≈ ______

Problem Solving

15. The table shows the number of pounds of apples that were harvested each day. Estimate how many pounds of apples were harvested altogether by rounding. Show how you estimated.

Day	Pounds of Apples
1	514
2	487
3	349
4	421
5	392

My Work!

16. In one week, a campground rented 18 cabins at $225 each. About how much did they collect in rent altogether? Show how you estimated.

17. The average weight for a guinea pig is 2 pounds. The average weight for a pot belly pig is 45 times greater than the guinea pig. About how much does an average pot belly pig weigh? Show how you estimated.

HOT Problems

18. Mathematical PRACTICE 2 **Use Number Sense** Use the digits 1, 3, 4, and 7 to create two whole numbers whose product is estimated to be about 600.

19. **Building on the Essential Question** When is the estimation of products a useful tool?

Name ....

MY Homework

Lesson 8
Estimate Products

Homework Helper

Need help? connectED.mcgraw-hill.com

Mountain View Elementary is sending 21 boxes of magazines to a school in Uruguay. There are 154 magazines in each box. About how many magazines are they sending?

South America
Uruguay

Estimate the product of 21 and 154.

One Way **Round each factor to the nearest ten.**

154 →	150	Round 154 to the nearest ten.
× 21 →	× 20	Round 21 to the nearest ten.
	3,000	Find 150 × 20 mentally.

By rounding both factors to the nearest ten, the estimate is about

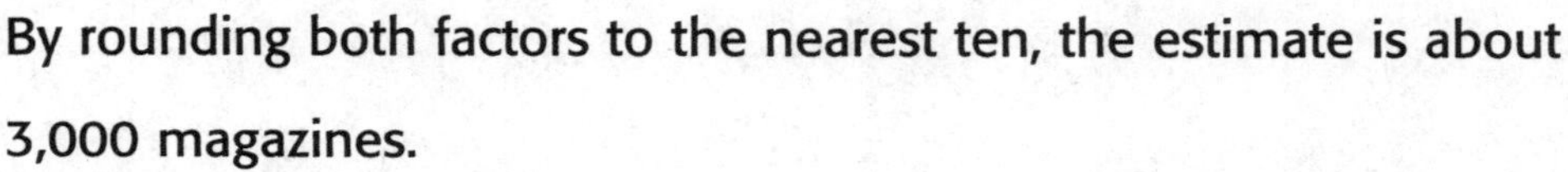

3,000 magazines.

Another Way **Use compatible numbers.**

154 →	200	Use numbers that are easy to multiply mentally such as 200 and 20.
× 21 →	× 20	
	4,000	Find 200 × 20 mentally.

By using compatible numbers, the estimate is about

4,000 magazines.

Practice

Estimate by rounding or using compatible numbers. Show how you estimated.

1. $\begin{array}{r} 4 \\ \times\ 24 \\ \hline \end{array}$

2. $\begin{array}{r} 76 \\ \times\ 78 \\ \hline \end{array}$

3. $\begin{array}{r} 508 \\ \times\ 27 \\ \hline \end{array}$

Real World Problem Solving

4. For a school assembly, students sit in chairs that are arranged in 53 rows. There are 12 chairs in each row. About how many students can be seated? Show how you estimated.

____

5. Klara bought a dozen bags of bird food for $27. Use compatible numbers to find the approximate cost of six dozen bags of bird food. Show how you estimated.

____

6. Mathematical PRACTICE 3 **Find the Error** Rico is estimating 139×18. Find his mistake and correct it.

$100 \times 10 = 1{,}000$

____

____

Vocabulary Check

Fill in the blank with the correct term or number to complete the sentence.

7. Compatible numbers are numbers in a problem that are easy to compute _______________.

Test Practice

8. On a cross-country vacation, Maria filled her 14-gallon gas tank eleven times. Which is the best estimate of how many gallons of gas she put in the tank altogether?

Ⓐ 75 gallons
Ⓑ 150 gallons
Ⓒ 200 gallons
Ⓓ 225 gallons

Name ....

Multiply by One-Digit Numbers

Lesson 9

ESSENTIAL QUESTION

What strategies can be used to multiply whole numbers?

Math in My World

Example 1

Grace and her three friends each paid $38 for an admission ticket to an amusement park. The total paid can be found by multiplying 4 and 38.

Find 38 × 4.

1 Multiply the ones.

8 ones × 4 = 32 ones

Regroup 32 ones as 3 tens and 2 ones.

2 Multiply the tens.

3 tens × 4 = 12 tens

Add any new tens.

12 tens + 3 tens = 15 tens

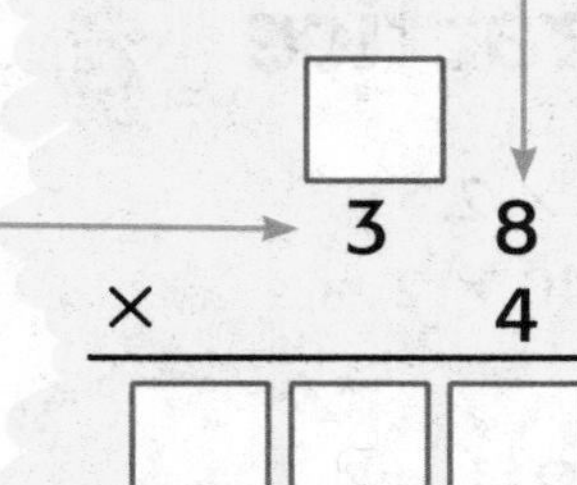

So, the total amount paid for admission to an amusement park is ______.

Check You can use an area model to check your answer.

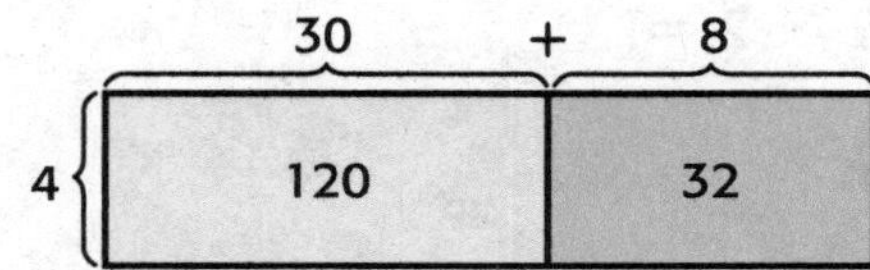

Example 2 

Find 317 × 5.

Estimate 300 × 5 = ______

1 **Multiply the ones.**

7 ones × 5 = 35 ones

Regroup 35 ones as 3 tens and 5 ones.

 **Multiply the tens.**

1 ten × 5 = 5 tens

Add any new tens.

5 tens + 3 tens = 8 tens

 **Multiply the hundreds.**

3 hundreds × 5 = 15 hundreds

$$\begin{array}{r} \square \\ 3\ 1\ 7 \\ \times \quad\ \ 5 \\ \hline \square,\square\square\square \end{array}$$

So, 317 × 5 = ______.

Check Compare to the estimate. ______ ≈ ______

Guided Practice

1. Multiply 42 × 2.
Estimate 40 × 2 = ______

$$\begin{array}{r} 4\ 2 \\ \times \quad 2 \\ \hline \square\square \end{array}$$

So, 42 × 2 = ______.

Check

______ ≈ ______

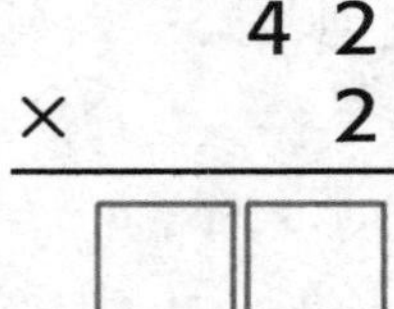

Name ....

Independent Practice

Estimate. Then multiply. Use your estimate to check your answer.

2. $\begin{array}{r} 21 \\ \times\ 3 \\ \hline \end{array}$

3. $\begin{array}{r} 32 \\ \times\ 6 \\ \hline \end{array}$

4. $\begin{array}{r} 52 \\ \times\ 9 \\ \hline \end{array}$

5. $\begin{array}{r} 401 \\ \times\ \ 7 \\ \hline \end{array}$

6. $\begin{array}{r} 712 \\ \times\ \ 3 \\ \hline \end{array}$

7. $\begin{array}{r} 143 \\ \times\ \ 9 \\ \hline \end{array}$

8. 31 × 5 = ________

9. 208 × 3 = ________

10. 47 × 6 = ________

11. 211 × 7 = ________

12. 182 × 6 = ________

13. 806 × 7 = ________

Problem Solving

14. One 747-airplane can carry 420 passengers. How many total passengers can three planes carry?

15. Mathematical PRACTICE 5 **Use Math Tools** In an auditorium, there are 9 rows of seats with 18 seats in each row. There are also 6 rows of seats with 24 seats in each row. How many seats are there in the auditorium? Estimate first. Then check for reasonableness.

16. Malia bought 14 cans of cat food. Each can weighs 8 ounces. How many total ounces of cat food did she buy?

HOT Problems

17. Mathematical PRACTICE 2 **Use Number Sense** Catalina multiplied 842 and 3 and got 3,326. How can she check to see if her answer is reasonable?

18. **Building on the Essential Question** What is the procedure for multiplying by a one-digit number?

Name

MY Homework

Lesson 9
Multiply by One-Digit Numbers

Homework Helper

eHelp

Need help? connectED.mcgraw-hill.com

The world's largest cactus is 5 times as tall as the cactus shown. How tall is the world's largest cactus?

Find 15 × 5.

Estimate 20 × 5 = 100

1 Multiply the ones.

5 ones × 5 = 25 ones
Regroup 25 ones as 2 tens and 5 ones.

2 Multiply the tens.

1 ten × 5 = 5 tens
Add any new tens.
5 tens + 2 tens = 7 tens

$$\begin{array}{r} {}^{2}\\ 15 \\ \times\ 5 \\ \hline 75 \end{array}$$

So, the world's largest cactus is 75 feet tall.

Check Compare to the estimate. 75 ≈ 100

Practice

Estimate. Then multiply. Use your estimate to check your answer.

1. $\begin{array}{r} 18 \\ \times\ 8 \\ \hline \end{array}$

2. $\begin{array}{r} 72 \\ \times\ 4 \\ \hline \end{array}$

3. $\begin{array}{r} 341 \\ \times\ \ 4 \\ \hline \end{array}$

Real World Problem Solving

4. **Mathematical PRACTICE 3** **Check for Reasonableness** Malcolm ran the 440-yard dash and the 220-yard dash at a track meet. There are 3 feet in one yard. How many total feet did Malcolm run? Estimate first. Then check for reasonableness.

5. Each student in Mrs. Henderson's science class brought in 3 books for the book donation. If there are 25 students in the class, how many total books did they collect?

6. Karen and Anthony are setting up rows for the piano recital. They set up 24 rows with 6 chairs in each row. How many total people will the rows seat?

7. Veronica brought her turtle out of its aquarium for 15 minutes every night for 7 days. How many total minutes did she bring her turtle out of its aquarium?

Test Practice

8. A restaurant has 36 tables. If each table can sit five people, how many people can be seated at the restaurant?

   Ⓐ 216 people

   Ⓑ 180 people

   Ⓒ 150 people

   Ⓓ 41 people

Name ....

Multiply by Two-Digit Numbers

Lesson 10

ESSENTIAL QUESTION
What strategies can be used to multiply whole numbers?

Math in My World

Example 1

Domestic cats can run up to 44 feet per second on land. At this rate, how many feet could a cat run in 12 seconds?

Find 44×12.

Estimate $44 \times 10 =$ ______

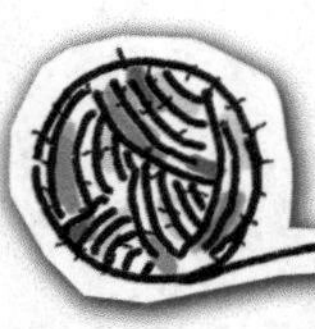

Multiply the ones.

$44 \times 2 = 88$

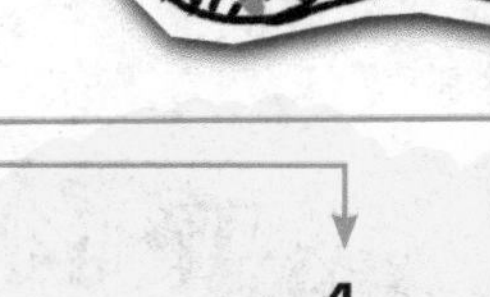

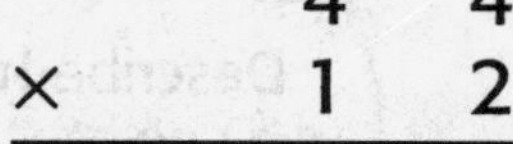

Multiply the tens.

$44 \times 10 = 440$

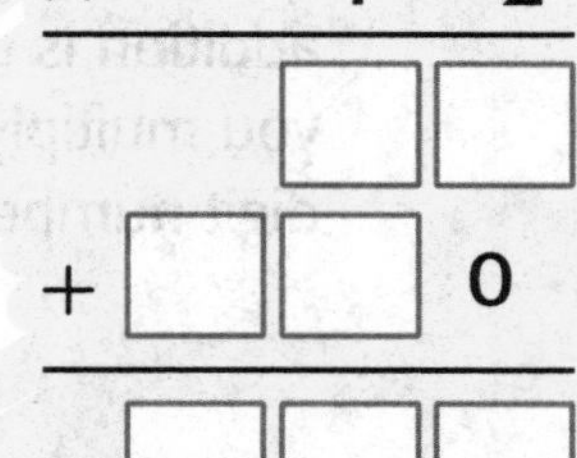

$$\begin{array}{r} 4\ 4 \\ \times\ 1\ 2 \\ \hline \square\ \square \\ +\ \square\ \square\ 0 \\ \hline \square\ \square\ \square \end{array}$$

Add.

$88 + 440 = 528$

Helpful Hint
By estimating first, you can determine if your answer is reasonable.

So, a domestic cat can run ______ feet in 12 seconds.

Check Compare to the estimate. ______ ≈ ______

Example 2  

Find 165 × 31.

Estimate 200 × 30 = ______

 **Multiply the ones.**

165 × 1 = 165

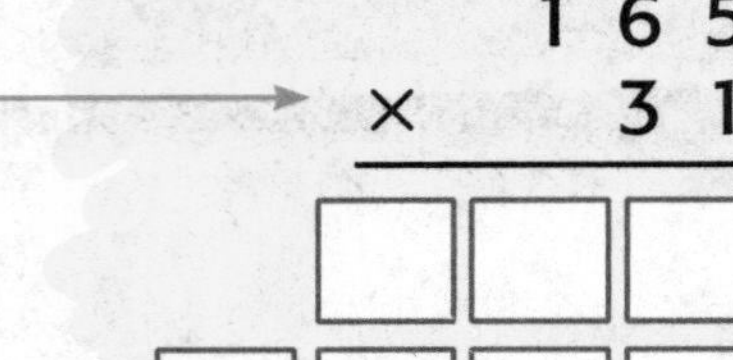

 **Multiply the tens.**

165 × 30 = 4,950

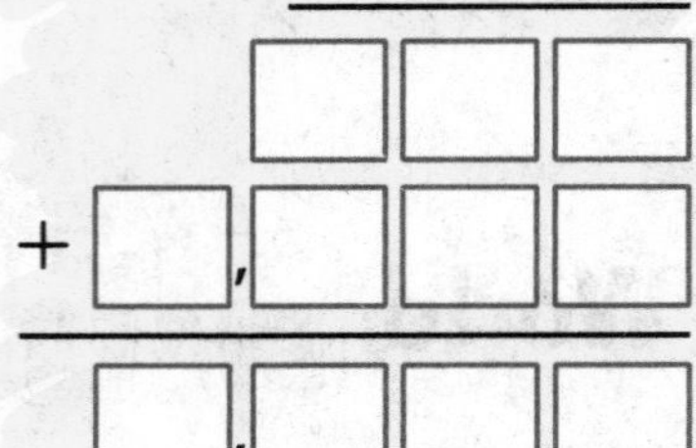

 **Add.**

165 + 4,950 = 5,115

So, 165 × 31 = ______.

Check Compare to the estimate. ______ ≈ ______

Guided Practice 

Describe how addition is used when you multiply by two-digit numbers.

1. Multiply 32 × 13.

Estimate 30 × 10 = ______

```
    3 2
×   1 3
-------
  [ ][ ]
+[ ][ ] 0
-------
[ ][ ][ ]
```

So, 32 × 13 = ______.

Check for Reasonableness

______ ≈ ______

Name

Independent Practice

Estimate. Then multiply. Use your estimate to check your answer.

2. $\begin{array}{r} 102 \\ \times \ 12 \\ \hline \end{array}$

3. $\begin{array}{r} 102 \\ \times \ 56 \\ \hline \end{array}$

4. $\begin{array}{r} 24 \\ \times \ 21 \\ \hline \end{array}$

5. $\begin{array}{r} 39 \\ \times \ 34 \\ \hline \end{array}$

6. $\begin{array}{r} 13 \\ \times \ 54 \\ \hline \end{array}$

7. $\begin{array}{r} 51 \\ \times \ 82 \\ \hline \end{array}$

8. 21 × 42 = ______

9. 69 × 14 = ______

10. 83 × 367 = ______

11. 534 × 67 = ______

12. 141 × 25 = ______

13. 229 × 31 = ______

Problem Solving

14. Marshall's mother buys 2 boxes of granola bars each week. Each box contains 8 granola bars. If she continues buying 2 boxes each week, how many granola bars will she buy in a year (1 year = 52 weeks)?

15. Mathematical PRACTICE 5 **Use Math Tools** A delivery truck travels 278 miles each day. How far does it travel in 25 days?

16. A cow can eat 25 pounds of hay a day. At that rate, how many pounds of hay can a cow eat in 31 days?

HOT Problems

17. Mathematical PRACTICE 2 **Use Number Sense** Find 235 × 124. Use the same strategy that you used for multiplying by a two-digit number except include multiplying by the hundreds place.

18. **Building on the Essential Question** How do you multiply by two-digit numbers?

Name

MY Homework

Lesson 10
Multiply by Two-Digit Numbers

Homework Helper

Need help? connectED.mcgraw-hill.com

Alicia lives in Nashville, Tennessee. Last year her family drove to Atlanta, Georgia, each month to visit her grandmother. Find the total distance they drove to visit her grandmother for the year.

Destination City From Nashville	Round-Trip Distance (mi)
Atlanta	498
Raleigh	1,080

Find 498 × 12.

Estimate 500 × 10 = 5,000

1. **Multiply the ones.**
   498 × 2 = 996

2. **Multiply the tens.**
   498 × 10 = 4,980

3. **Add.**
   996 + 4,980 = 5,976

$$\begin{array}{r} 498 \\ \times \quad 12 \\ \hline 996 \\ +\ 4,980 \\ \hline 5,976 \end{array}$$

So, they drove a total of 5,976 miles for the year.

Check Compare to the estimate. 5,976 ≈ 5,000

Practice

Estimate. Then multiply. Use your estimate to check your answer.

1. $\begin{array}{r} 19 \\ \times\ 15 \\ \hline \end{array}$

2. $\begin{array}{r} 43 \\ \times\ 65 \\ \hline \end{array}$

3. 470 × 56 = ______

Real World Problem Solving

4. Ms. Jenkins was arranging chairs for a school awards assembly. Each row contained 15 chairs. If there were 21 rows, how many chairs had to be arranged?

5. Leon earns $14 an hour. How much does he earn in 4 weeks if he works 12 hours each week?

6. Mathematical PRACTICE 5 **Use Math Tools** Without actually calculating, how much greater is the product of 98 × 50 than the product of 97 × 50?

7. The table shows Katrina's prices for dog walking. If she walks 5 medium-sized dogs and 8 large-sized dogs for 12 weeks, how much will she earn?

Dog Type	Cost Per Week ($)
Small	10
Medium	12
Large	14

Test Practice

8. Each day there are 12 tours at the glass factory. Twenty-eight people can go on a tour. How many people can tour the glass factory each day?

Ⓐ 236 people
Ⓑ 280 people
Ⓒ 336 people
Ⓓ 436 people

Name

Mathematical PRACTICE 6

Multiply.

1. 17×6

2. 24×7

3. 31×3

4. 68×2

5. 41×8

6. 92×5

7. 19×4

8. 67×7

9. 32×4

10. 90×6

11. 83×2

12. 62×5

13. 18×9

14. 38×5

15. 26×6

16. 74×8

17. 87×5

18. 53×7

19. 49×3

20. 71×4

Name

Fluency Practice

Multiply.

1. $\begin{array}{r} 11 \\ \times\ 23 \\ \hline \end{array}$

2. $\begin{array}{r} 54 \\ \times\ 41 \\ \hline \end{array}$

3. $\begin{array}{r} 76 \\ \times\ 15 \\ \hline \end{array}$

4. $\begin{array}{r} 35 \\ \times\ 64 \\ \hline \end{array}$

5. $\begin{array}{r} 27 \\ \times\ 10 \\ \hline \end{array}$

6. $\begin{array}{r} 89 \\ \times\ 33 \\ \hline \end{array}$

7. $\begin{array}{r} 41 \\ \times\ 48 \\ \hline \end{array}$

8. $\begin{array}{r} 92 \\ \times\ 13 \\ \hline \end{array}$

9. $\begin{array}{r} 63 \\ \times\ 25 \\ \hline \end{array}$

10. $\begin{array}{r} 39 \\ \times\ 67 \\ \hline \end{array}$

11. $\begin{array}{r} 89 \\ \times\ 40 \\ \hline \end{array}$

12. $\begin{array}{r} 19 \\ \times\ 84 \\ \hline \end{array}$

13. $\begin{array}{r} 218 \\ \times\ 13 \\ \hline \end{array}$

14. $\begin{array}{r} 104 \\ \times\ 37 \\ \hline \end{array}$

15. $\begin{array}{r} 921 \\ \times\ 26 \\ \hline \end{array}$

16. $\begin{array}{r} 585 \\ \times\ 48 \\ \hline \end{array}$

17. $\begin{array}{r} 732 \\ \times\ 55 \\ \hline \end{array}$

18. $\begin{array}{r} 337 \\ \times\ 79 \\ \hline \end{array}$

19. $\begin{array}{r} 376 \\ \times\ 80 \\ \hline \end{array}$

20. $\begin{array}{r} 890 \\ \times\ 14 \\ \hline \end{array}$

Review

Chapter 2

Multiply Whole Numbers

Vocabulary Check

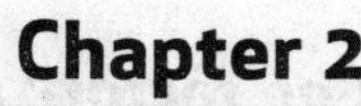

Match each word to its definition. Write your answers on the lines provided.

1. **compatible numbers** ________
2. **product** ________
3. **power of 10** ________
4. **factor** ________
5. **prime factorization** ________
6. **Distributive Property** ________
7. **exponent** ________
8. **base** ________
9. **power** ________

A. This is a number obtained by raising a base to an exponent.

B. This property states that to multiply a number by a sum, you can multiply each addend by the number and add the products.

C. This is a number that divides into a whole number evenly. It is also a number that is multiplied by another number.

D. This is a way of expressing a composite number as a product of its prime factors.

E. This is a number like 10, 100, 1,000, and so on. It is the result of using only 10 as a factor.

F. These are numbers that are easy to multiply mentally.

G. This is the answer to a multiplication problem.

H. In a power, this number represents the number of times the base is used as a factor.

I. In a power, this is the number used as the repeated factor.

Concept Check

Find the prime factorization of each number.

10. 12 = ________ **11.** 42 = ________

Write each product using an exponent.

12. 10 × 10 = ________ **13.** 5 × 5 × 5 × 5 = ________

Find each product mentally.

14. 73×10^2 = ________ **15.** 60 × 40 = ________

Estimate. Then multiply. Use your estimate to check your answer.

16. $\begin{array}{r} 72 \\ \times\ 36 \\ \hline \end{array}$

17. $\begin{array}{r} 23 \\ \times\ 84 \\ \hline \end{array}$

18. $\begin{array}{r} 321 \\ \times\ 64 \\ \hline \end{array}$

Find each product mentally using the Distributive Property. Show the steps that you used.

19. 8 × 71 = ________ **20.** 6 × 83 = ________

Name

Problem Solving

For Exercises 21–23, use the following information. Then estimate to find the distance sound travels through each material in each given time.

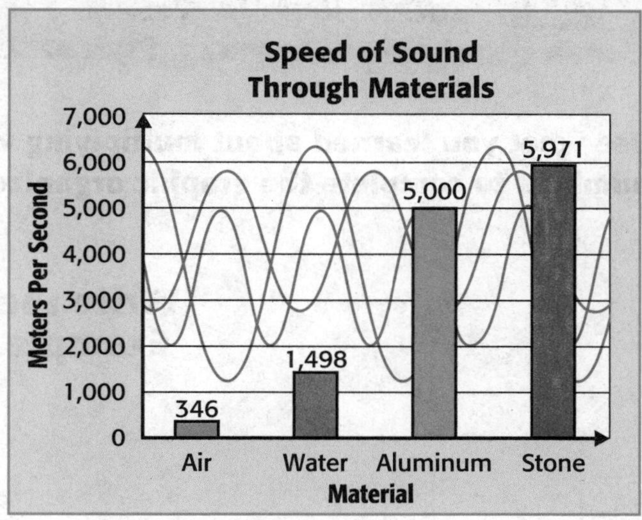

Sound travels through different materials at different speeds. For example, the graph shows that in one second, sound travels 5,971 meters through stone. However, it travels only 346 meters through air in one second.

21. air, 20 seconds

22. stone, 12 seconds

23. Estimate how much farther sound travels through stone in 17 seconds than through aluminum in the same time.

24. Sylvia is saving to buy a new terrarium for her iguana. She saves \$2 the first week, \$4 the second week, \$8 the third week, and so on. How much total money will she save in 5 weeks? Solve by completing the table.

Week	1	2	3	4	5
Amount Saved ($)	2	4	8		

I want a new home.

Test Practice

25. Colin bought 7 flats of flowers. Each flat contains 24 flowers. How many flowers did he buy?

Ⓐ 140 flowers Ⓒ 168 flowers

Ⓑ 154 flowers Ⓓ 200 flowers

Reflect

Chapter 2

Answering the ESSENTIAL QUESTION

Use what you learned about multiplying whole numbers to complete the graphic organizer.

Write the Example	Real-World Example
Vocabulary	Model

ESSENTIAL QUESTION

What strategies can be used to multiply whole numbers?

Now reflect on the ESSENTIAL QUESTION  **Write your answer below.**

Chapter

3 Divide by a One-Digit Divisor

ESSENTIAL QUESTION

What strategies can be used to divide whole numbers?

Let's Help Others!

Watch

Watch a video!

MY Common Core State Standards

Number and Operations in Base Ten

CCSS

5.NBT.6 Find whole-number quotients of whole numbers with up to four-digit dividends and two-digit divisors, using strategies based on place value, the properties of operations, and/or the relationship between multiplication and division. Illustrate and explain the calculation by using equations, rectangular arrays, and/or area models.

Ok, this'll be good to know!

Standards for Mathematical PRACTICE

1. Make sense of problems and persevere in solving them.
2. Reason abstractly and quantitatively.
3. Construct viable arguments and critique the reasoning of others.
4. Model with mathematics.
5. Use appropriate tools strategically.
6. Attend to precision.
7. Look for and make use of structure.
8. Look for and express regularity in repeated reasoning.

= focused on in this chapter

Name ....

Am I Ready?

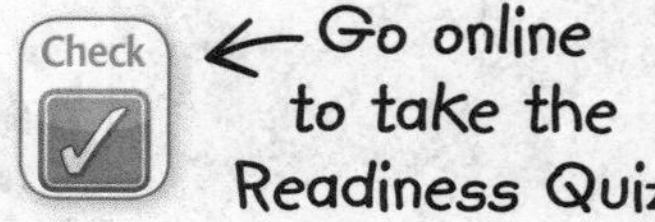

Multiply.

1. 12 × 7 = ________ **2.** 42 × 8 = ________ **3.** 51 × 9 = ________

4. 7 × 18 = ________ **5.** 3 × $75 = ________ **6.** 3 × $89 = ________

7. Turner's bookshelf has 6 shelves. Each shelf has 17 books. How many books are on the bookshelf?

____

Round each number to its greatest place value.

8. 36 ________ **9.** $451 ________ **10.** 7,499 ________

11. $33,103 ________ **12.** $271 ________ **13.** $5,001 ________

14. There are 7,209 students at the amusement park. To the nearest thousand, how many students are at the park?

____

Divide.

15. 8 ÷ 2 = ________ **16.** 15 ÷ 3 = ________ **17.** 27 ÷ 3 = ________

18. 28 ÷ 4 = ________ **19.** 48 ÷ 6 = ________ **20.** 54 ÷ 9 = ________

21. Three people spent a total of $24 for lunch. If they divided the total cost equally, how much did each person pay?

Item	Cost
Pizza	$12
Salads	$6
Drinks	$6

____

How Did I Do?

Shade the boxes to show the problems you answered correctly.

1	2	3	4	5	6	7	8	9	10	11	12	13	14	15	16	17	18	19	20	21

Name ....

MY Math Words

Vocab

Review Vocabulary

compatible numbers multiples place value product

Making Connections

Use the review vocabulary to complete the concept wheel.

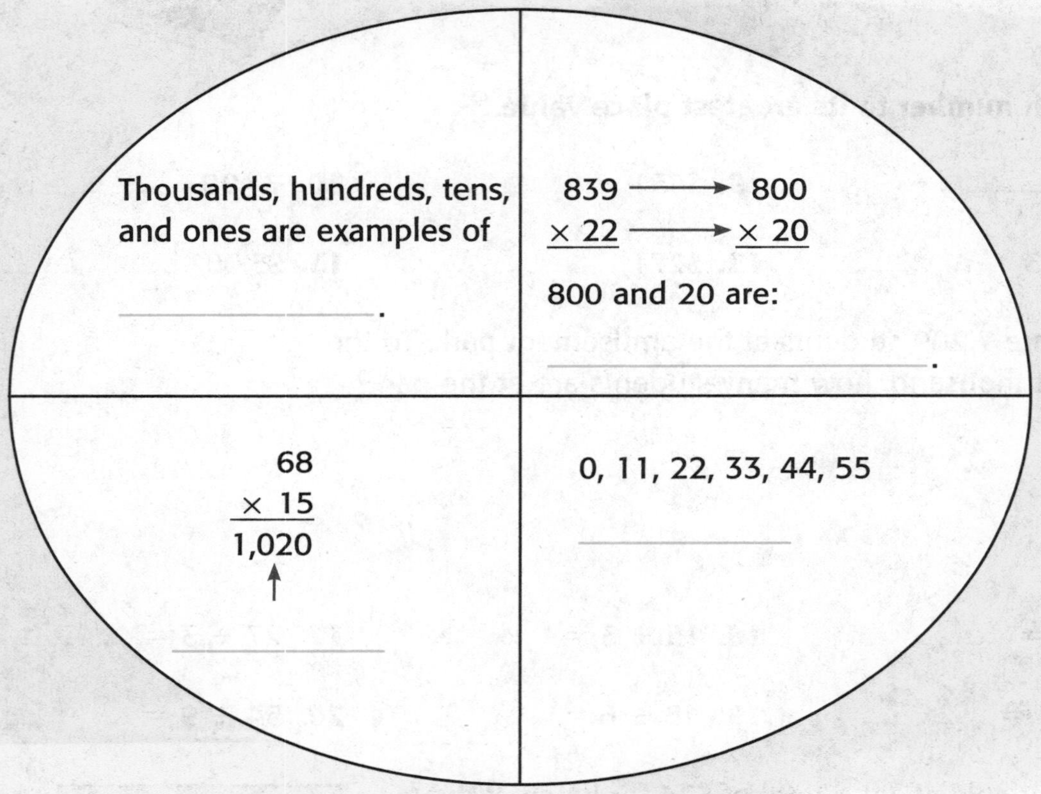

How can these vocabulary words help you solve multiplication problems?

____

____

MY Vocabulary Cards

Mathematical PRACTICE

Lesson 3–3

dividend

$76 \div 4 = 19$

Lesson 3–3

divisor

$76 \div 4 = 19$

Lesson 3–1

fact family

28
×, ÷
4 7

$4 \times 7 = 28$
$7 \times 4 = 28$
$28 \div 7 = 4$
$28 \div 4 = 7$

Lesson 3–7

partial quotients

```
8)536
 −480    60
   56   + 7
 − 56    67
    0
```

Lesson 3–3

quotient

$76 \div 4 = 19$

Lesson 3–3

remainder

```
   79 R3
4)319
 −28
   39
  −36
    3
```

Lesson 3–1

unknown

$42 \div 6 = ■$

$6 \times 7 = 42$

$42 \div 6 = 7$

Lesson 3–1

variable

$4 \times k = 32$

Ideas for Use

- Work with a partner to name the part of speech of each word. Consult a dictionary to check your answers.
- Draw or write examples for each card. Be sure your examples are different from what is shown on each card.

The number that divides the dividend.

Write 3 division problems that have 2-digit dividends and 1-digit divisors. Circle each divisor.

A number that is being divided.

A division problem has a divisor of 12 and a quotient of 7. What is the dividend?

A dividing method in which the dividend is separated into addends that are easy to divide.

How can you use place value when using partial quotients?

A group of related facts that use the same numbers.

Write a set of numbers that are a fact family. Explain why they are a fact family.

The number that is left after one whole number is divided by another.

Write the letter used to represent *remainder*.

The result of a division problem.

Which review word is also the result of an operation on numbers?

A letter or symbol used to represent an unknown quantity.

The Latin root *var* means "different." Explain how this helps you understand the meaning of *variable*.

A missing value.

50 divided by an unknown number is equal to 5. Find the unknown number.

FOLDABLES® Follow the steps on the back to make your Foldable.

My Division Strategies

Quotients with Zeros

Example:

Two-Digit Dividends

Example:

Division Models

Example:

Interpret the Remainder

Example:

Estimate Quotients

Example:

Place the First Digit

Example:

Name ....

Relate Division to Multiplication

Lesson 1

ESSENTIAL QUESTION
What strategies can be used to divide whole numbers?

A **fact family** is a group of related facts that use the same numbers. You can use fact families to relate multiplication and division.

Helping others is a SLAM DUNK!

Math in My World

Example 1

Sheryl is helping to put away 18 basketballs after practice. She places them on a rack that has 3 shelves. How many basketballs can she put on each self?

Use a fact family.

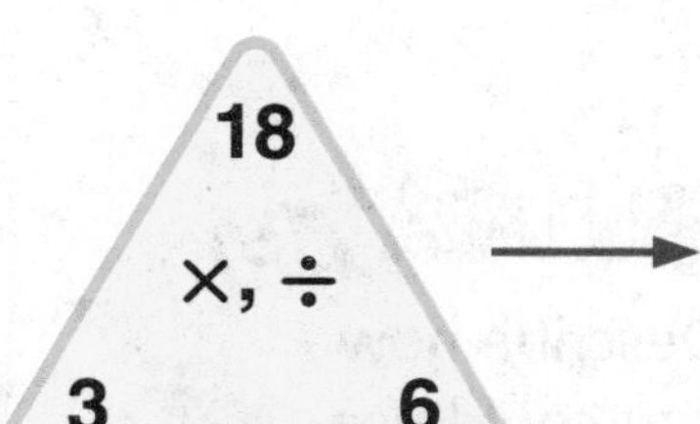

3 × ____ = 18

6 × ____ = 18

18 ÷ 6 = ____

18 ÷ 3 = ____

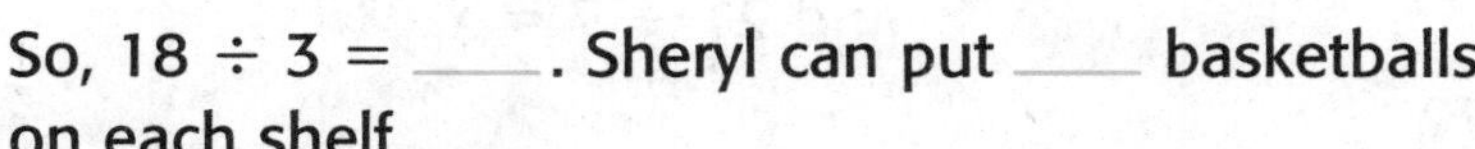

So, 18 ÷ 3 = ____. Sheryl can put ____ basketballs on each shelf.

Check Draw an equal amount of basketballs on each shelf.

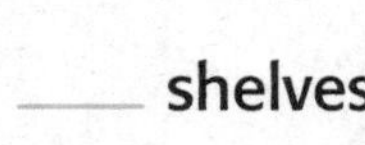

____ shelves

____ × ____ = 18

____ basketballs on each shelf

An equation is a number sentence that contains an equals sign (=). You can use related facts to help find the **unknown,** or missing value, in an equation. You can use a **variable,** or a letter, to represent the unknown number.

Example 2

Ellie is creating gift bags for her party guests. She wants to divide 56 pencils equally among the 7 gift bags. How many pencils will go in each bag?

Let p represent the number of pencils in each bag.

$\_\_\_\_ \div \_\_\_\_ = p$

Think: What number times 7 is 56?

Write a related multiplication fact.

$7 \times \_\_\_\_ = 56$

So, $56 \div 7 = \_\_\_\_$. Since $p = \_\_\_\_$, Ellie will put ____ pencils in each bag.

Guided Practice

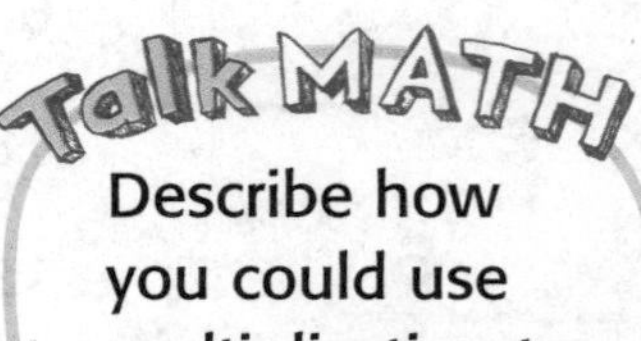

Describe how you could use multiplication to find $21 \div 7 = x$.

1. Complete the fact family for 8, 9, 72.

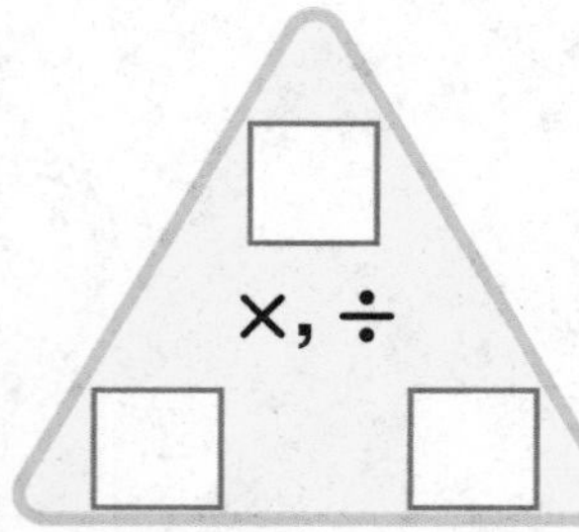

$8 \times \_\_\_\_ = 72$

$9 \times \_\_\_\_ = 72$

$\_\_\_\_ \div 8 = \_\_\_\_$

$\_\_\_\_ \div 9 = \_\_\_\_$

Divide. Use a related multiplication fact.

2. $48 \div \_\_\_\_ = 6$

Think: $\_\_\_\_ \times 6 = 48$

3. $40 \div 5 = \_\_\_\_$

Think: $5 \times \_\_\_\_ = 40$

Name

Independent Practice

Write a fact family for each set.

4.

5. 

6. 4, 9, 36

7. 5, 7, 35

8. 3, 8, 24

Divide. Write the related multiplication fact.

9. $64 \div 8 =$ ____

____ $\times 8 = 64$

10. $45 \div 9 =$ ____

____ $\times 9 = 45$

11. ____ $\div 9 = 9$

$9 \times 9 =$ ____

12. ____ $\div 8 = 4$

$4 \times 8 =$ ____

13. $40 \div$ ____ $= 8$

$8 \times$ ____ $= 40$

14. $63 \div$ ____ $= 7$

$7 \times$ ____ $= 63$

Algebra Find the unknown number in each equation. Use a related division fact.

15. $2 \times m = 12$

$m =$ ____

16. $8 \times y = 24$

$y =$ ____

17. $9 \times g = 72$

$g =$ ____

Real World Problem Solving

Algebra For Exercises 18–20, use the information below.

Orange blossoms have 5 petals and are some of the most fragrant flowers.

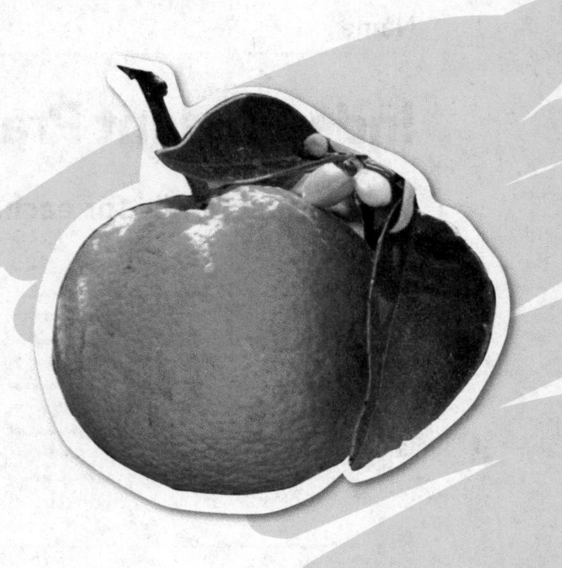

18. How many petals would there be in a group of 7 flowers?

19. How many petals p would there be in a group of 11 flowers? Write an equation to find the unknown. Then find the unknown.

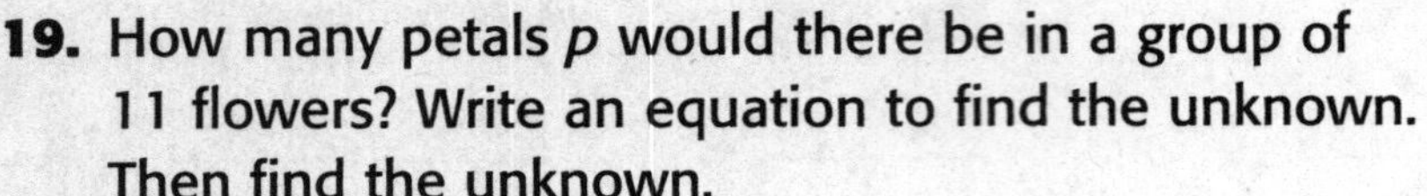

20. A group of f flowers has 40 petals in all. Write an equation to find the unknown. Then find the unknown.

HOT Problems

21. Mathematical PRACTICE 2 **Reason** Can the number 12 be part of more than one fact family? Explain.

22. Mathematical PRACTICE 3 **Which One Doesn't Belong?** Circle the equation that does not belong with the other three. Explain why it does not belong.

$54 \div 9 = 6$ $\quad$ $54 \div 6 = 9$ $\quad$ $9 \times 3 = 27$ $\quad$ $6 \times 9 = 54$

23. **Building on the Essential Question** How do multiplication facts help me divide?

Name ....

MY Homework

Lesson 1

Relate Division to Multiplication

Homework Helper

eHelp

Need help? connectED.mcgraw-hill.com

There are 20 students in the after-school program. There are 4 students in each group. How many groups are there?

Use a fact family.

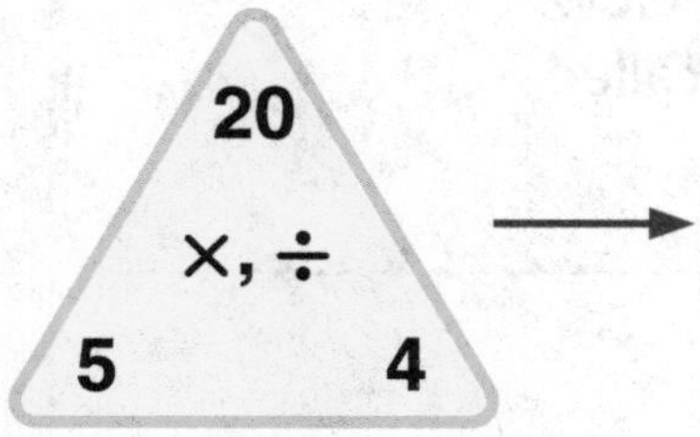

$4 \times 5 = 20$

$5 \times 4 = 20$

$20 \div 4 = 5$

$20 \div 5 = 4$

So, $20 \div 4 = 5$. There are 5 groups in the after-school program.

Check Use a drawing.

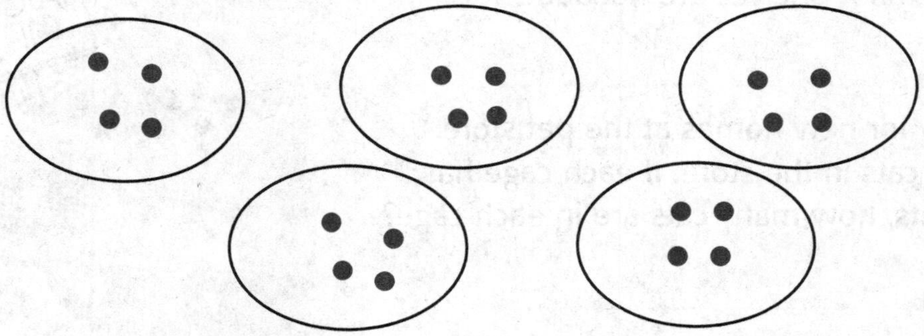

Practice

Write a fact family for each set of numbers.

1. 2, 10, 20

2. 8, 7, 56

3. 7, 9, 63

Divide. Write the related multiplication fact.

4. $36 \div$ ______ $= 6$

$6 \times$ ______ $= 36$

5. $77 \div 11 =$ ______

______ $\times 11 = 77$

6. ______ $\div 8 = 12$

$12 \times 8 =$ ______

Mathematical PRACTICE 2 **Use Algebra Find the unknown number in each equation. Use a related division fact.**

7. $4 \times b = 12$

$b =$ ______

8. $3 \times n = 45$

$n =$ ______

9. $6 \times k = 18$

$k =$ ______

Vocabulary Check

10. Choose the correct word(s) to complete the sentence below.
A group of related facts that use the same numbers is called

a(n) ______________.

Problem Solving

11. Julian places 8 books on each shelf of a bookcase. If he shelves 32 books, how many shelves are needed?

____

12. There are 15 cats ready for new homes at the pet store. There are 5 cages with cats in the store. If each cage has the same number of cats, how many cats are in each cage?

____

13. The leaves of a poison ivy plant are found in clusters of 3. One poison ivy plant has a total of 21 leaves. How many leaf clusters does this plant have?

____

Test Practice

14. A local pet store has a total of 72 fish in 9 tanks. Each tank holds the same number of fish. How many fish are in each tank?

Ⓐ 6 fish

Ⓑ 7 fish

Ⓒ 8 fish

Ⓓ 9 fish

Name

Hands On
Division Models

Lesson 2

ESSENTIAL QUESTION
What strategies can be used to divide whole numbers?

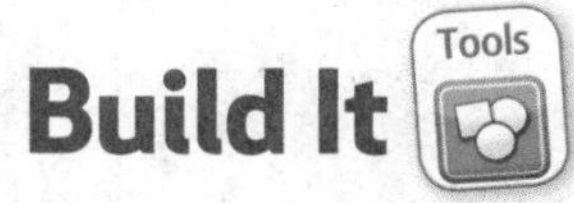

Build It

In art class, three students share 48 markers evenly. How many markers will each student have?

1 Model 48 using base-ten blocks.

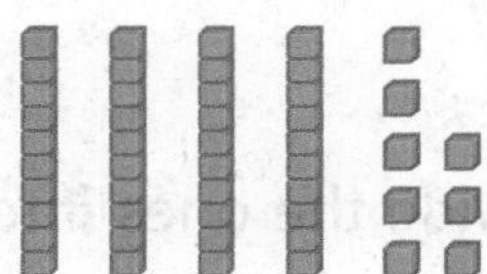

2 Divide the tens into 3 equal groups.
Circle 3 groups of tens. Draw the equal groups.

3 Regroup the remaining tens blocks into 10 ones.
How many ones are there altogether? ______ ones

4 Divide the ones into 3 equal groups. Draw an equal amount of ones in each group. Each group has ______ ones.

Each group has ______ ten and ______ ones.

So, 48 ÷ 3 = ______.

Each student will have ______ markers.

Try It

Find 56 ÷ 5.

Model 56 using base-ten blocks.

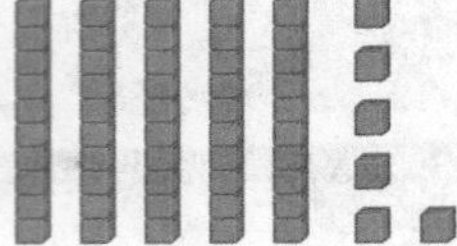

2 Divide the tens into 5 equal groups. Circle the groups of tens.

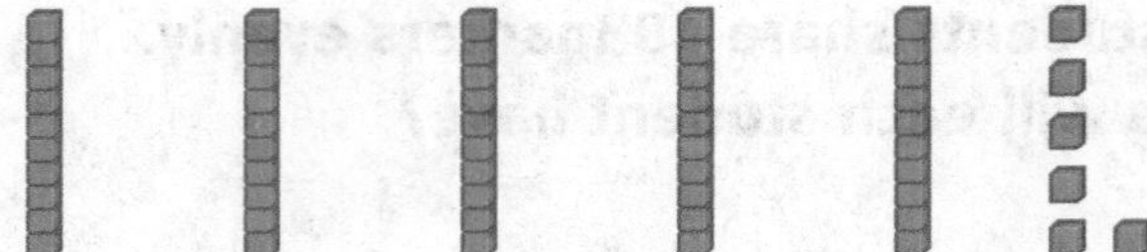

3 Divide the ones into 5 equal groups. Draw an equal amount of tens and ones in each group.

How many ones are left over? ______

Each group has ______ ten and ______ one.

So, when you divide 56 into 5 groups, there are ______ in each group with one left over.

Talk About It

1. Mathematical PRACTICE 5 **Use Math Tools** Model 32 ÷ 3 using base-ten blocks. Will there be any left over? Explain.

____

Name

CCSS

Practice It

Divide. Use base-ten blocks. Draw the equal groups. State if there are any left over.

2. 44 ÷ 4
How many are in each group? ______
Are there any left over? If so, state how many. ______

3. 39 ÷ 3
How many are in each group? ______
Are there any left over? If so, state how many. ______

4. 32 ÷ 5
How many are in each group? ______
Are there any left over? If so, state how many. ______

5. 57 ÷ 8
How many are in each group? ______ Are there any left over? If so, state how many. ______

6. Use base-ten blocks to divide 64 ÷ 6. How many are left over?

______

Apply It

Solve. Use base-ten blocks.

7. Mathematical PRACTICE 5 **Use Math Tools** Kendrick has 42 craft sticks to make 3 identical crafts. How many sticks will he use for each craft?

8. Allison uses 41 stickers to decorate 3 picture frames. Each picture frame has the same number of stickers. How many stickers are on each picture frame? How many stickers are left over?

9. Kara and six of her friends are playing miniature golf. If it costs $42 for all of them to play a round of golf, how much does each round cost per person?

10. Justin has 71 newspapers to deliver in 3 days. He delivers the same number of newspapers each day. How many newspapers does Justin deliver each day? How many newspapers are left over?

11. Mathematical PRACTICE 2 **Use Number Sense** Write the division sentence that is shown by the model.

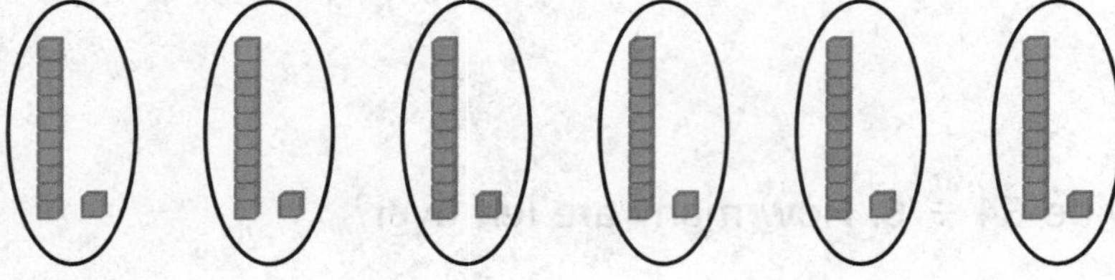

Write About It

12. How does place value help model division?

Name

MY Homework

Lesson 2
Hands On:
Division Models

Homework Helper

Need help? connectED.mcgraw-hill.com

Find 71 ÷ 6.

1 Model 71 using base-ten blocks.

2 Divide the tens into 6 equal groups. Circle 6 groups of ten.

3 Regroup the remaining tens blocks into 10 ones.
You now have a total of 11 ones.

4 Divide the ones into 6 equal groups. Draw an equal amount of tens and ones in each group.

There are 5 ones left over.
So, when you divide 71 into 6 groups, there are 11 in each group with five left over.

Practice

Divide. Use base-ten blocks. Draw the equal groups. State if there are any left over.

1. $42 \div 3$

How many are in each group? ______

Are there any left over? ______

2. $87 \div 4$

How many are in each group? ______

Are there any left over? ______

Problem Solving

3. Jamil has 3 pet lizards. The pet store owner said that Jamil will need to buy a total of 36 crickets to feed his lizards. If each lizard eats the same number of crickets, how many crickets does each lizard eat?

______

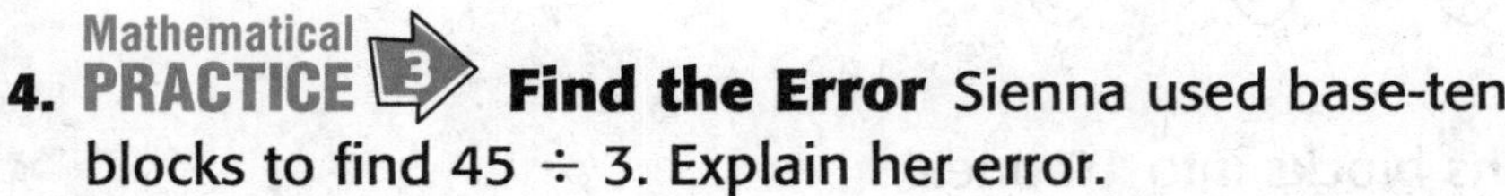

4. **Find the Error** Sienna used base-ten blocks to find $45 \div 3$. Explain her error.

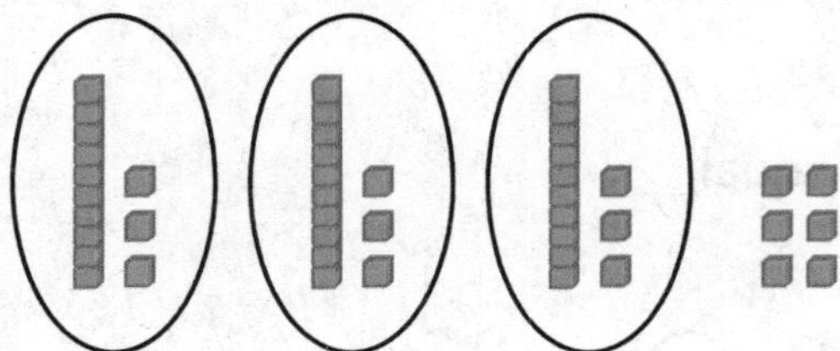

______

______

5. An airplane that can hold 63 passengers is separated into 3 sections. Each section holds the same number of passengers. Write a division sentence to correctly describe the situation.

______

Name ....

Two-Digit Dividends

Lesson 3

ESSENTIAL QUESTION
What strategies can be used to divide whole numbers?

The **dividend** is the number that is being divided.
The **divisor** tells you how many groups.

dividend: $36 \div 3$ ← divisor → $3\overline{)36}$

The result of the division is called the **quotient.**

Speeding to the TOY DRIVE!

Math in My World

Example 1

Eli donates his toys to 5 different charities. He has a total 75 toys to donate. Eli donates the same number of toys to each charity. How many toys does each charity receive?

Let *t* represent the number of toys each charity receives.

______ ÷ ______ = *t*

Find 75 ÷ 5.

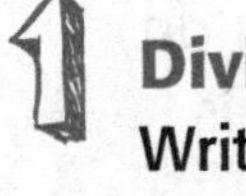

1. **Divide the tens.** 7 ÷ 5
Write 1 in the quotient over the tens place.

2. **Multiply.** 5 × 1
Subtract. 7 − 5

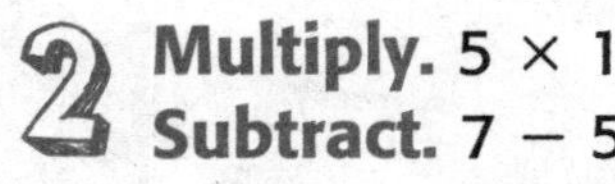

3. **Bring down the ones.**

```
   [ ][ ]
 5)7  5
  −5  ↓
   2  5
  −2  5
      0
```

4. **Divide the ones.**
25 ÷ 5
Write 5 in the quotient over the ones place.

5. **Multiply.** 5 × 5
Subtract. 25 − 25

The model shows 5 groups of fifteen.

So, 75 ÷ 5 = ______.

Each charity receives ______ toys.

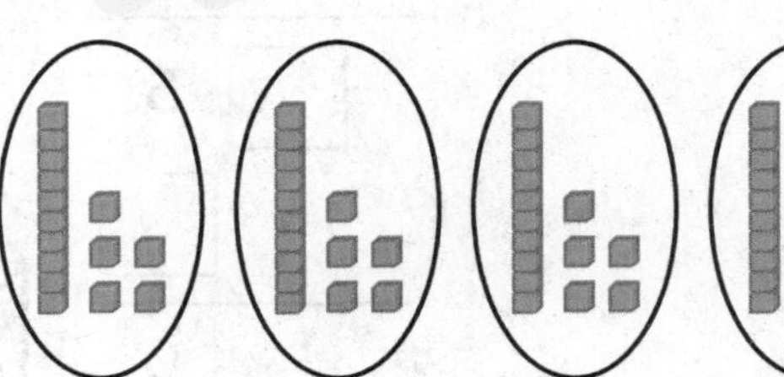

A **remainder** is the number, or part left, after you divide. Use R to represent the remainder.

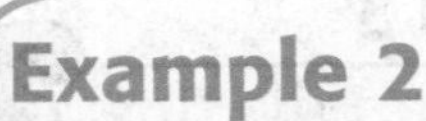

Example 2

Caleb is putting his baseball cards in an album. He has 57 cards and can put 4 cards on each page. How many full pages will Caleb have? Will there be any cards left?

Find 57 ÷ 4.

Divide the tens.

5 ÷ 4
Write 1 in the quotient over the tens place.

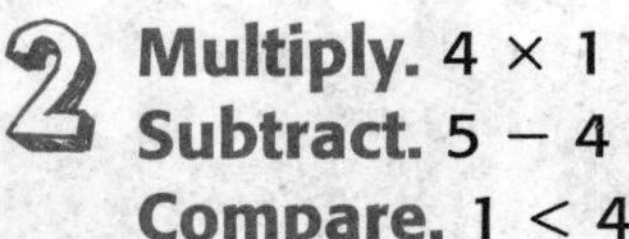

Multiply. 4 × 1
Subtract. 5 − 4
Compare. 1 < 4

Bring down the ones.

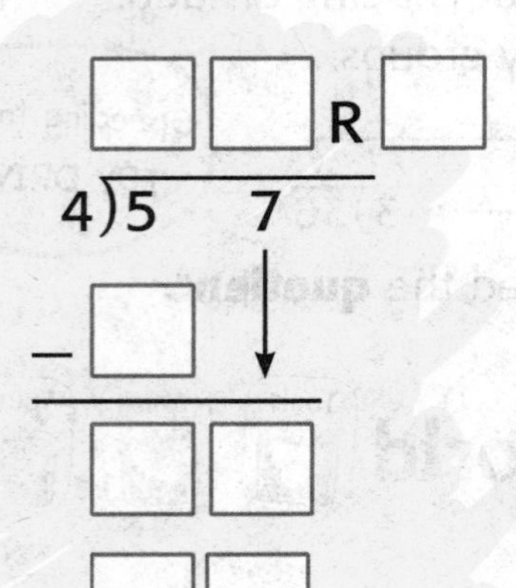

Divide the ones.

17 ÷ 4
Write 4 in the quotient over the ones place.

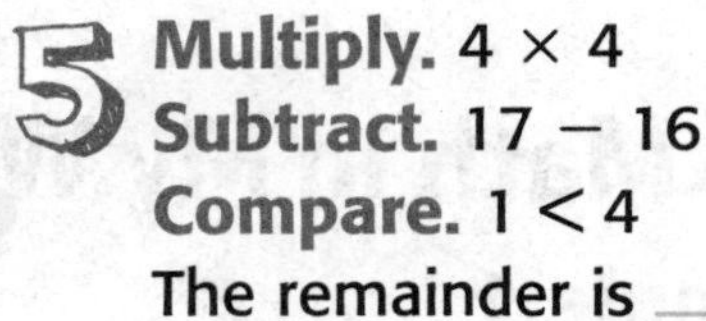

Multiply. 4 × 4
Subtract. 17 − 16
Compare. 1 < 4
The remainder is ______.

So, there will be ______ full pages and ______ card will be left over.

Talk MATH

What should you do if the remainder is greater than or equal to the divisor?

Guided Practice

1.
```
   □□
5)6 5
 − 5
   □5
 − □□
     0
```

2.
```
   □ 5
3)4 5
 − □
   □5
 − 1□
     0
```

Name ....

Independent Practice

Divide.

3.

$$\begin{array}{r} \square\,\square \\ 3\overline{)4\;\;2} \\ -\;3 \\ \hline \square\;\;2 \\ -\;\square\,\square \\ \hline 0 \end{array}$$

4.

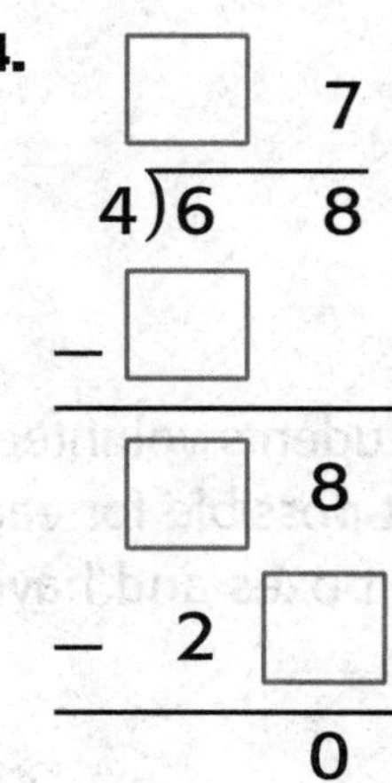

5.

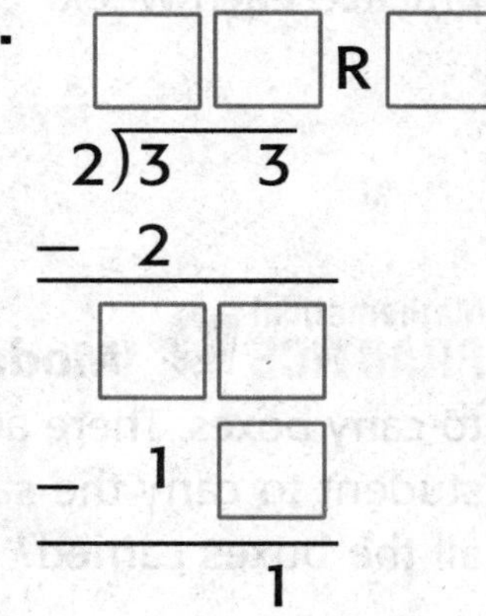

6. $2\overline{)28}$

7. $6\overline{)74}$

8. $7\overline{)85}$

9. $60 \div 4 =$ ________

10. $64 \div 5 =$ ________

11. $70 \div 6 =$ ________

Algebra Divide to find the unknown number in each equation.

12. $48 \div 3 = h$

$h =$ ________

13. $44 \div 2 = b$

$b =$ ________

14. $72 \div 4 = w$

$w =$ ________

Problem Solving

15. Maranda practiced a total of 52 hours in 4 weeks to prepare for a piano recital. If she practiced the same number of hours each week, how many hours did she practice each week?

16. Mathematical PRACTICE 4 **Model Math** Five students volunteered to carry boxes. There are 62 boxes. Is it possible for each student to carry the same number of boxes and have all the boxes carried? Explain.

HOT Problems

17. Mathematical PRACTICE 2 **Reason** The following equations show the relationship between multiplication and division.

$18 \div 3 = 6$ $\quad$ $3 \times 6 = 18$

$18 \div 0 = ?$ $\quad$ $0 \times ? = 18$

Explain why it is not possible to divide by zero.

18. **Building on the Essential Question** How does place value help me divide?

Name ....

MY Homework

Lesson 3
Two-Digit Dividends

Homework Helper

Need help? connectED.mcgraw-hill.com

Find 87 ÷ 6.

1 Divide the tens.
8 ÷ 6
Write 1 in the quotient over the tens place.

2 Multiply. 6 × 1
Subtract. 8 − 6
Compare. 2 < 6

3 Bring down the ones.

$$\begin{array}{r} 14\text{ R3} \\ 6\overline{)87} \\ -6\downarrow \\ \hline 27 \\ -24 \\ \hline 3 \end{array}$$

4 Divide the ones.
27 ÷ 6
Write 4 in the quotient over the ones place.

5 Multiply. 6 × 4
Subtract. 27 − 24
Compare. 3 < 6
The remainder is 3.

Practice

Divide.

1. $3\overline{)63}$

2. $7\overline{)96}$

3. $5\overline{)68}$

Algebra Divide to find the unknown number in each equation.

4. $72 \div 6 = n$

$n =$ ______

5. $45 \div 3 = p$

$p =$ ______

6. $52 \div 2 = k$

$k =$ ______

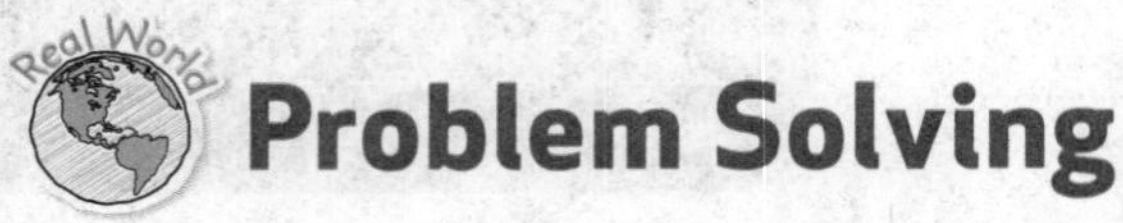

Problem Solving

7. A book has 5 chapters and a total of 90 pages. If each chapter has the same number of pages, how many pages are in each chapter?

8. Mathematical PRACTICE 2 **Reason** Caitlin divides a bag of fruit snacks with 4 friends. She divides the 89 snacks equally. How many fruit snacks does each person receive? How many fruit snacks will be left over?

9. Chet is camping with his after-school club. They have 9 tents and 72 people. The same number of campers will be in each tent. How many campers will be in each tent?

10. Isaiah helped pick 72 bananas on the weekend. There were a total of 6 people picking bananas. If they each picked an equal number of bananas, how many bananas did each person pick?

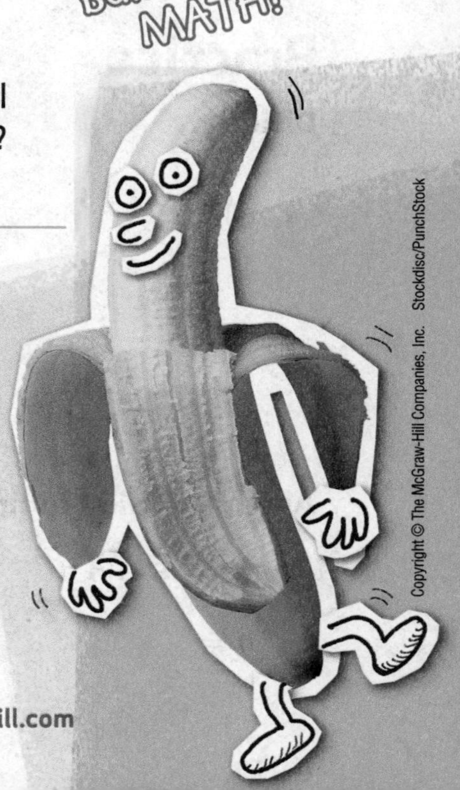

Test Practice

11. A box of granola bars has 26 bars. If 7 friends split the bars equally, how many bars will be left?

Ⓐ 2 bars Ⓒ 4 bars
Ⓑ 3 bars Ⓓ 5 bars

Name ....

Division Patterns

Lesson 4

ESSENTIAL QUESTION
What strategies can be used to divide whole numbers?

You can use basic facts and patterns to divide multiples of 10.

Math in My World

Example 1

A monarch butterfly can fly 240 miles in 3 days. Suppose it flies the same distance each day. How many miles can it fly each day?

Find 240 ÷ 3.

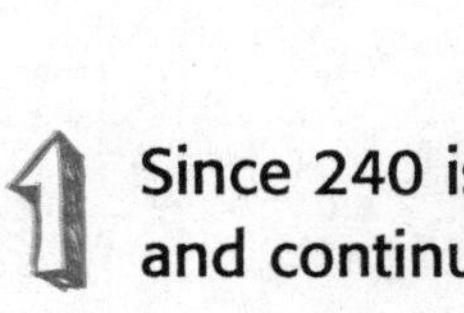

1 Since 240 is a multiple of 10, use the basic fact and continue the pattern.

2 4	÷ 3	basic fact
2 4 ☐	÷ 3	24 tens divided by 3 equals 8 tens.
2, 4 ☐ ☐	÷ 3	24 hundreds divided by 3 equals 8 hundreds.
2 4, ☐ ☐ ☐	÷ 3	24 thousands divided by 3 equals 8 thousands.

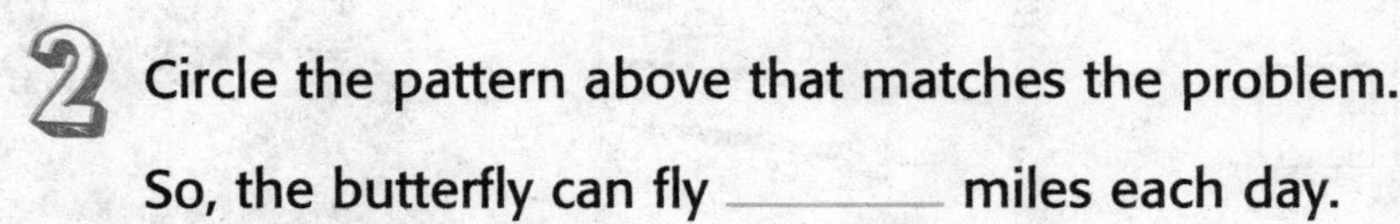

2 Circle the pattern above that matches the problem.

So, the butterfly can fly ________ miles each day.

Example 2

A cow eats 900 pounds of hay over a period of 30 days. How many pounds of hay would the cow eat each day at this rate?

Let h represent the number of pounds of hay.

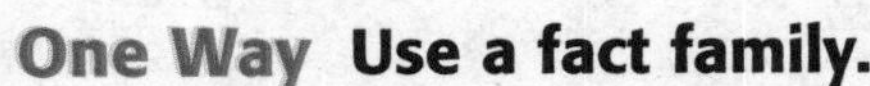

_______ ÷ _______ = h

Hay.

One Way Use a fact family.

Use the fact family of 3, 3, and 9 to help represent the problem.

$3 \times 3 = 9 \longleftrightarrow 9 \div 3 = 3$

$3\square \times 3 = 90 \longleftrightarrow 9\square \div 3\square = 3$

$3\square \times 3\square = 900 \longleftrightarrow 9\square\square \div 3\square = 3\square$

Another Way Use a pattern in the number of zeros.

You can cross out the same number of zeros in the dividend and the divisor to make the division easier.

$900 \div 30$ Cross out the same number of zeros in both the dividend and divisor.

$90 \div 3 = 30$ Divide. THINK: 9 tens ÷ 3 = 3 tens.

So, 900 ÷ 30 = _______. Since h = _______, the cow eats _______ pounds each day.

Talk MATH

Is the quotient 48 ÷ 6 equal to the quotient 480 ÷ 60? Explain.

Guided Practice

1. Find 500 ÷ 5 mentally.

500 ÷ 5 = _______

$5 \div 5 = 1$

$5\square \div 5 = 1\square$

$5\square\square \div 5 = 1\square\square$

Name ______________________________

Independent Practice

Divide mentally.

2. 800 ÷ 2 = ________

3. 900 ÷ 3 = ________

4. 150 ÷ 5 = ________

5. 140 ÷ 7 = ________

6. 450 ÷ 9 = ________

7. 280 ÷ 4 = ________

8. 180 ÷ 60 = ________

9. 240 ÷ 30 = ________

10. 420 ÷ 70 = ________

Algebra Mentally find each unknown.

11. 1,800 ÷ 30 = *k*

k = ________

12. 2,000 ÷ 400 = *z*

z = ________

13. 2,400 ÷ 300 = *s*

s = ________

Problem Solving

14. **Mathematical PRACTICE 1 Make Sense of Problems** A group of 10 people bought tickets to a reptile exhibit and paid a total of $130. What was the price of one ticket?

15. The fastest team in a wheelbarrow race traveled 100 meters in about 20 seconds. On average, how many meters did the team travel each second?

16. A video store took in $450 in DVD rentals during one day. If DVDs rent for $9 each, how many DVDs were rented?

17. Daniela has a 160-ounce bag of potting soil. She puts an equal amount of soil in 4 pots. How much soil will she put in each pot?

HOT Problems

18. **Mathematical PRACTICE 3 Find the Error** Sonia is finding 5,400 ÷ 90 mentally. Find her mistake and correct it.

5,4~~00~~ ÷ 9~~0~~
↓
54 ÷ 9 = 6

19. **Building on the Essential Question** How can patterns help me divide multiples of 10?

Name ....

MY Homework

Lesson 4
Division Patterns

Homework Helper

Need help? connectED.mcgraw-hill.com

Find 630 ÷ 7.

1 Since 630 is a multiple of 10, use the basic fact and continue the pattern.

63 ÷ 7	basic fact
630 ÷ 7	63 tens divided by 7 equals 9 tens.
6,300 ÷ 7	63 hundreds divided by 7 equals 9 hundreds.
63,000 ÷ 7	63 thousands divided by 7 equals 9 thousands.

2 Circle the pattern above that matches the problem.

So, 630 ÷ 7 = 90.

Practice

Divide mentally.

1. 270 ÷ 3 = ______ **2.** 3,200 ÷ 80 = ______ **3.** 320 ÷ 8 = ______

Algebra Mentally find each unknown.

4. 2,000 ÷ 10 = m **5.** 8,100 ÷ 90 = b **6.** 450 ÷ 9 = r

m = ______ b = ______ r = ______

Problem Solving

7. Mathematical PRACTICE 1 **Make a Plan** Peyton has collected 120 aluminum cans for recycling. If 20 cans will fit in each blue plastic bag, how many bags will she need to carry all the cans?

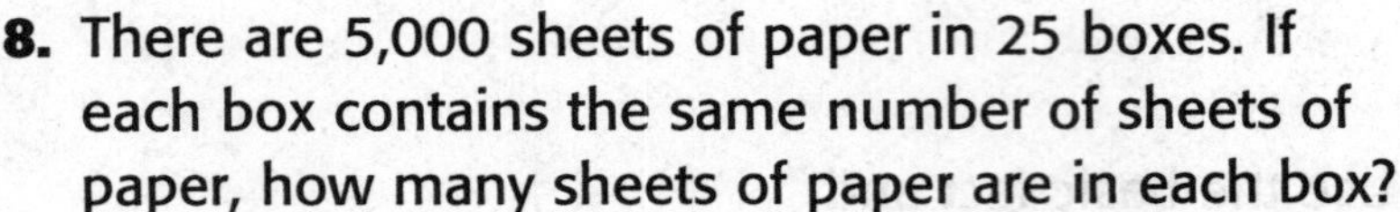

8. There are 5,000 sheets of paper in 25 boxes. If each box contains the same number of sheets of paper, how many sheets of paper are in each box?

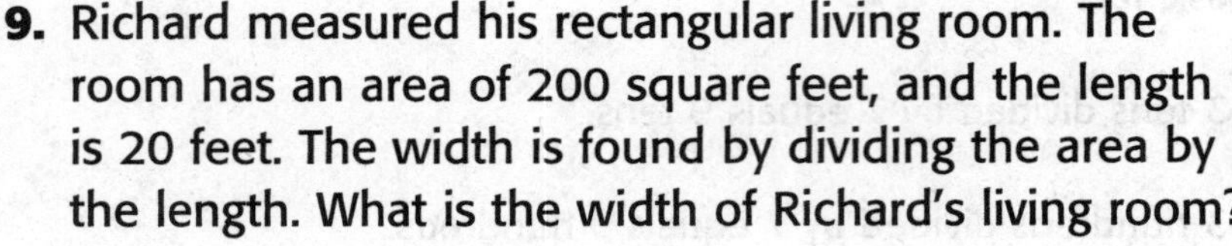

9. Richard measured his rectangular living room. The room has an area of 200 square feet, and the length is 20 feet. The width is found by dividing the area by the length. What is the width of Richard's living room?

10. The average polar bear weighs 1,200 pounds. The average grizzly bear weighs 800 pounds. The average black bear weighs 400 pounds. The average polar bear weighs how many times more than the average black bear?

The average grizzly bear weighs how many times more than the average black bear?

Test Practice

11. An elementary school has 320 students. All of the students are going on a field trip. If 40 students can ride a bus, how many buses are needed?

Ⓐ 6 buses
Ⓑ 7 buses
Ⓒ 8 buses
Ⓓ 9 buses

Check My Progress

Vocabulary Check

Draw lines to match each definition to each vocabulary word.

1. the number that divides the dividend	• **fact family**
2. a group of related facts using the same numbers	• **variable**
3. a missing value in a number sentence or equation	• **divisor**
4. a number that is being divided	• **quotient**
5. the result of a division problem	• **dividend**
6. a letter or symbol used to represent an unknown quantity	• **remainder**
7. the number that is left after one whole number is divided by another	• **unknown**

Concept Check

Divide. Write the related multiplication fact.

8. $54 \div 9 =$ ______

______ $\times 9 = 54$

9. ______ $\div 9 = 8$

$8 \times 9 =$ ______

Algebra **Divide to find the unknown number in each equation.**

10. $95 \div 5 = n$

$n =$ ______

11. $96 \div 8 = b$

$b =$ ______

Divide.

12. $2\overline{)48}$

13. $7\overline{)81}$

Divide mentally.

14. $3{,}500 \div 5 =$ ______

15. $420 \div 60 =$ ______

Problem Solving

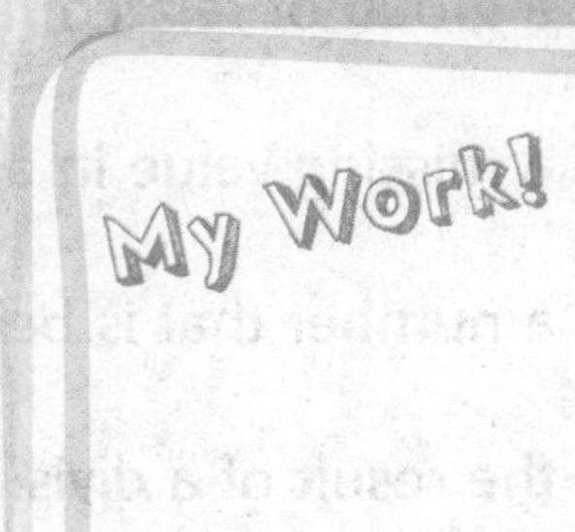

16. Suki received $87 for working 3 days. If she made the same amount each day, how much did Suki earn each day?

______

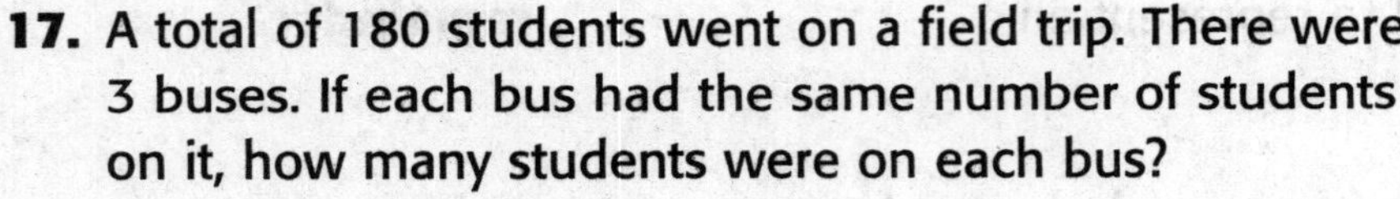

17. A total of 180 students went on a field trip. There were 3 buses. If each bus had the same number of students on it, how many students were on each bus?

______

18. Marc is helping out with the school bake sale. He has 50 cookies to place in bags. He places 3 cookies in each bag. How many bags will he use? How many cookies will be left over?

______

Test Practice

19. A train traveled 300 miles in 5 hours. How far did the train travel each hour, on average?

Ⓐ 60 miles　　Ⓒ 600 miles

Ⓑ 150 miles　　Ⓓ 1,500 miles

Name ....

Estimate Quotients

Lesson 5

ESSENTIAL QUESTION
What strategies can be used to divide whole numbers?

To estimate a quotient, you can use compatible numbers, or numbers that are easy to divide mentally. Look for numbers that are part of fact families.

Math in My World

Example 1

A dog's heart beats 365 times in 3 minutes. About how many times does a dog's heart beat in 1 minute?

Estimate 365 ÷ 3.

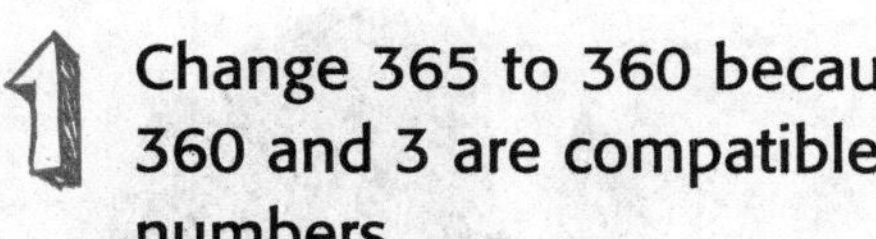

1. Change 365 to 360 because 360 and 3 are compatible numbers.

   365 ÷ 3
   ↓
   ☐☐☐ ÷ 3 = ☐☐☐

2. Divide mentally.

So, a dog's heart beats about ______ times a minute.

Check Multiply to check your answer.

______ × 3 = ______

You can use both rounding and compatible numbers to help estimate.

Example 2

Tutor

Estimate 208 ÷ 8.

1. Round the dividend to the nearest hundred.
2. Change the divisor to a number that is compatible with the rounded dividend.
3. Divide mentally.

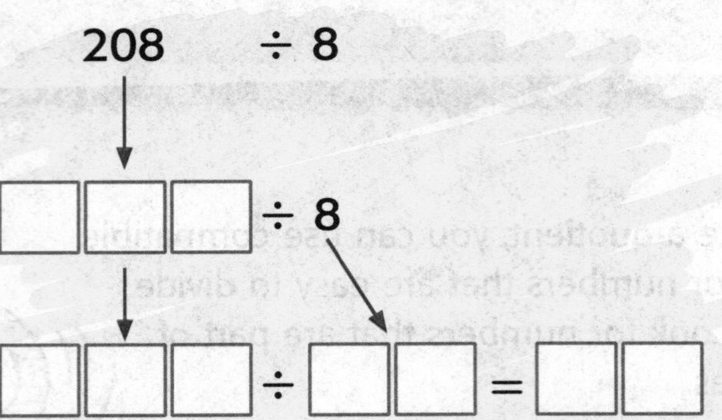

So, 208 ÷ 8 is about ______.

Guided Practice

Check

1. Estimate 850 ÷ 9.

Change 850 to 900 because 900 and 9 are compatible numbers.

850 ÷ 9

______ ÷ 9 = ______

Divide mentally.

So, 850 ÷ 9 is about ______.

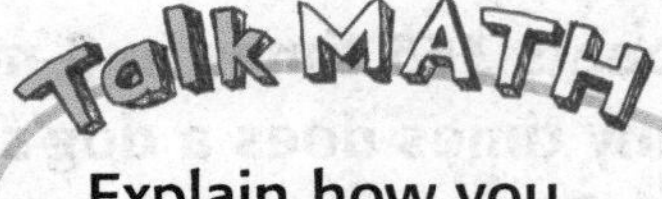

Name ________________________________

Independent Practice

Estimate. Show how you estimated.

2. 635 ÷ 8

↓

_____ ÷ 8 = _____

3. 545 ÷ 5

↓

_____ ÷ 5 = _____

4. 431 ÷ 2

↓

_____ ÷ 2 = _____

5. 374 ÷ 9

____

6. 395 ÷ 4

____

7. 660 ÷ 7

____

8. 289 ÷ 9

____

9. 477 ÷ 9

____

10. 230 ÷ 7

____

11. 639 ÷ 7

____

12. 350 ÷ 8

____

13. 584 ÷ 6

____

CCSS

Problem Solving

14. Mathematical PRACTICE 5 **Use Math Tools** A grocery store employee puts 8 bagels in each bag. If she has 385 bagels, about how many bags does she need?

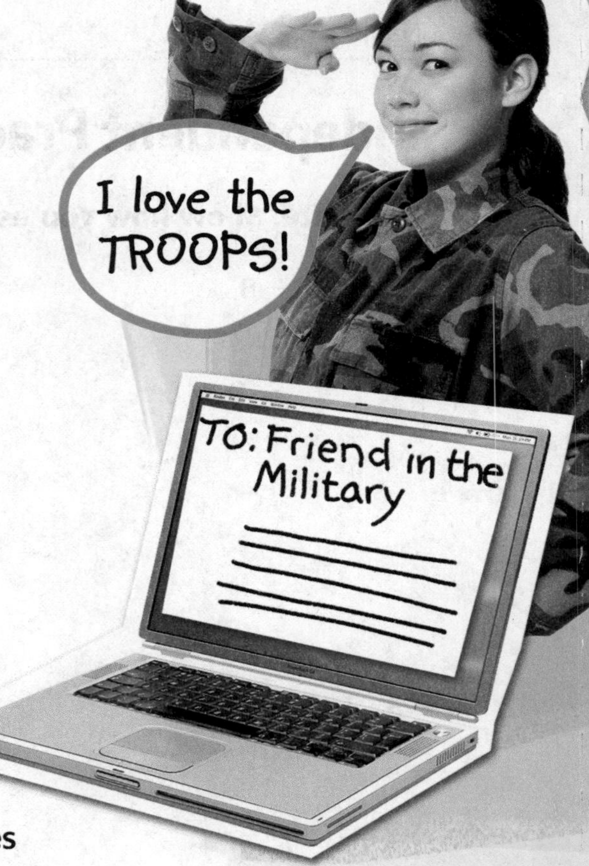

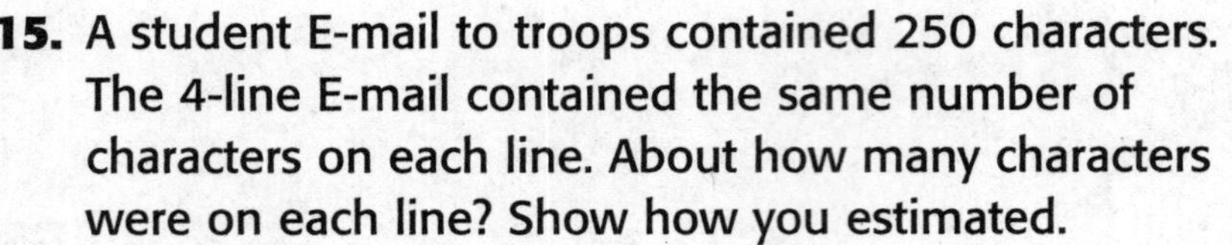

15. A student E-mail to troops contained 250 characters. The 4-line E-mail contained the same number of characters on each line. About how many characters were on each line? Show how you estimated.

16. Jani drives 290 miles in 5 hours. About how many miles does she drive each hour?

HOT Problems

17. Mathematical PRACTICE 2 **Use Number Sense** Write a division problem. Show two different ways that you can estimate the quotient using compatible numbers.

18. **Building on the Essential Question** Why is it important to know how to estimate quotients?

Name ....

MY Homework

Lesson 5
Estimate Quotients

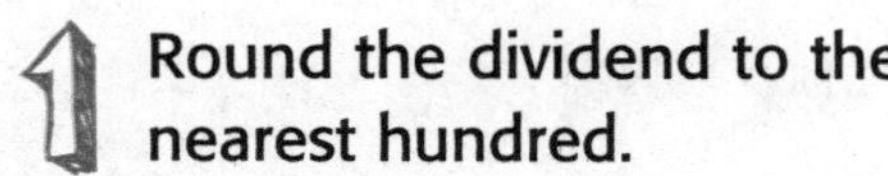

Homework Helper

eHelp

Need help? connectED.mcgraw-hill.com

Estimate 488 ÷ 9.

1 Round the dividend to the nearest hundred.

$$488 \div 9$$
$$\downarrow$$
$$500 \div 9$$

2 Change the divisor to a number that is compatible with the rounded dividend.

$$\downarrow \quad \downarrow$$
$$500 \div 10 = 50$$

3 Divide mentally.

So, 488 ÷ 9 is about 50.

Practice

Estimate. Show how you estimated.

1. 115 ÷ 2 ____________

2. 791 ÷ 2 ____________

3. 151 ÷ 3 ____________

4. 460 ÷ 9 ____________

5. 477 ÷ 7 ____________

6. 392 ÷ 9 ____________

Problem Solving

7. The table shows how much each fifth grade room earned from a bake sale. The money is going to be given to 6 different charities. If each charity is given an equal amount, about how much will each charity receive? Show how you estimated.

Bake Sale

Room	Earnings($)
110	327
112	425
114	550
116	486

My Work!

8. There were 317 marbles divided equally among 8 bowls. About how many marbles were in each bowl?

9. Mathematical PRACTICE 5 **Use Math Tools** Emilio has 5 bags of birdseed. Each bag has about 28 ounces of birdseed. If he divides the birdseed equally into 3 containers, about how much birdseed will he put in each container?

Test Practice

10. Which of the following is the most reasonable estimate for the number of Calories in one serving of milk?

Ⓐ between 8 and 9

Ⓑ less than 80

Ⓒ between 80 and 90

Ⓓ more than 90

Servings of Milk	Calories
5	430

Name ____________________

Hands On
Division Models with Greater Numbers

Lesson 6

ESSENTIAL QUESTION
What strategies can be used to divide whole numbers?

Build It 

At the fair, you need tickets to ride the rides. Three friends share 336 tickets equally. How many tickets will each friend receive?

Find 336 ÷ 3.

1 Model 336 using base-ten blocks.

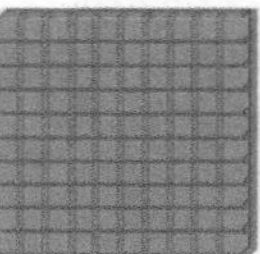   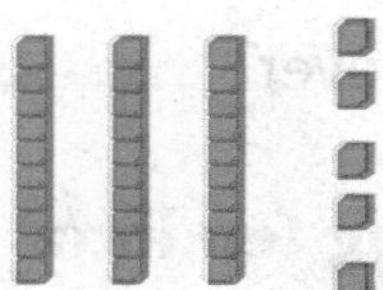

2 Divide the hundreds into 3 groups. Circle the equal groups above.

How many hundreds are in each group? ________

3 Divide the tens into 3 groups. Circle the equal groups above.

How many tens are in each group? ________

4 Divide the ones into 3 groups. Circle the equal groups above.

How many ones are in each group? ________

1	1	2
hundred	ten	ones

________ + ________ + ________ = ________

So, each friend will receive ________ tickets.

Check Use multiplication to check your answer.

________ × 3 = 336

Try It

Find 319 ÷ 2.

The base-ten blocks show 319.

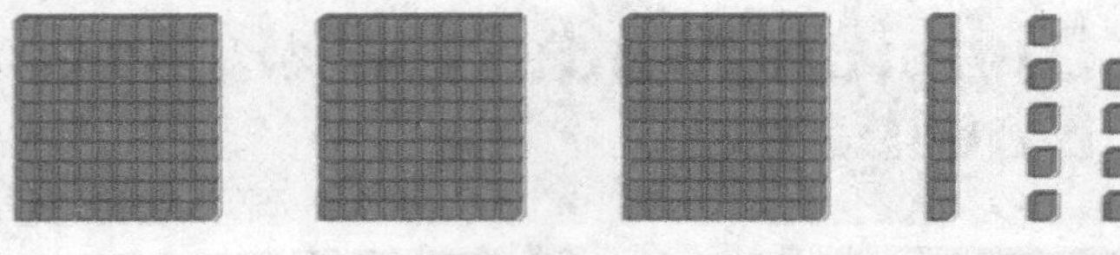

1 Divide the hundreds into 2 groups.

Draw the equal groups.

How many hundreds are left over? ______

My Drawing!

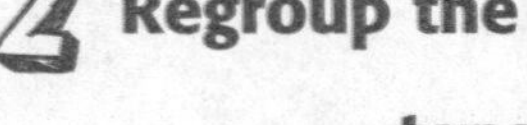

2 Regroup the remaining hundred into ______ tens.

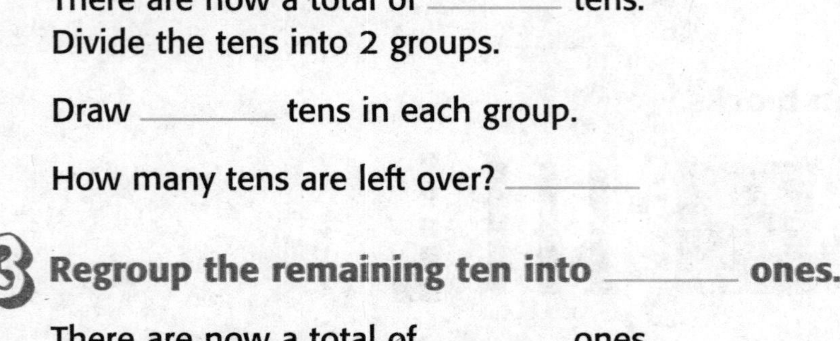

There are now a total of ______ tens.
Divide the tens into 2 groups.

Draw ______ tens in each group.

How many tens are left over? ______

3 Regroup the remaining ten into ______ ones.

There are now a total of ______ ones.
Divide the ones into 2 groups.

Draw ______ ones in each group.

How many ones are left over? ______

Each group has ______ hundred, ______ tens, and ______ ones.

There is ______ one left over. So, 319 ÷ 2 = ______.

Check Use multiplication to check your answer.

______ × 2 = 318   318 + ______ = 319

Talk About It

1. Mathematical PRACTICE 1 **Plan Your Solution** How can you divide 3 hundreds into 2 equal groups?

______

______

Name

Practice It

Use models to find each quotient. Draw the equal groups.

2. 344 ÷ 2 = ________

3. 468 ÷ 4 = ________

4. 383 ÷ 3 = ________

5. 257 ÷ 2 = ________

Apply It

For Exercises 6-7 and 9, use models to help you find each quotient.

My Work!

6. Sarah ordered 366 school newspapers to be printed. The newspapers arrived in 3 boxes. How many newspapers were in each box?

7. A monkey's heart beats about 576 times in 3 minutes. How many times does a monkey's heart beat in 1 minute?

8. Mathematical PRACTICE 6 **Explain to a Friend** If you are dividing a three-digit even number by two, will you ever have a remainder? Explain to a friend.

Time to HELP!

9. Gretchen spent 236 hours helping out the neighbors on their farm over the past 2 months. She helped out the same number of hours each month. How many hours did she help each month?

Write About It

10. How can I use models to help me divide?

Name ....

MY Homework

Lesson 6

Hands On: Division Models with Greater Numbers

Homework Helper

Need help? connectED.mcgraw-hill.com

At the arcade, Jamie and her friends won 476 tickets playing games. The four friends share 476 tickets equally. How many tickets will each friend receive? Find 476 ÷ 4.

1 Model 476 using base-ten blocks.

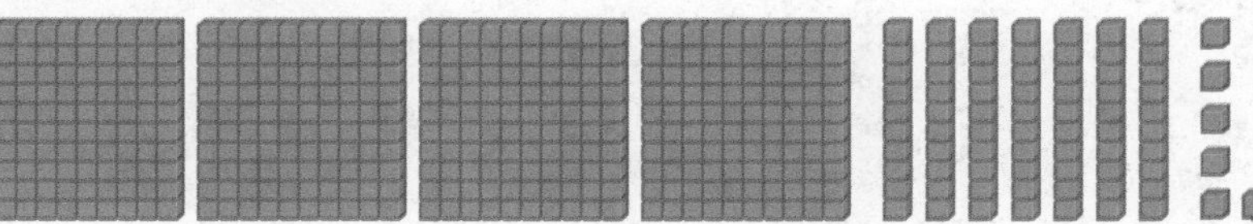

2 Divide the hundreds into 4 groups.

How many hundreds are in each group? 1

Group 1

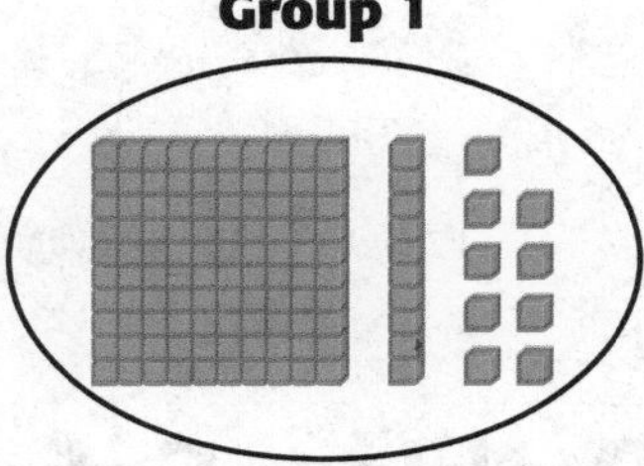

3 There are a total of 7 tens. Divide the tens into 4 groups.

How many tens are in each group? 1

How many tens are left over? 3

Group 2

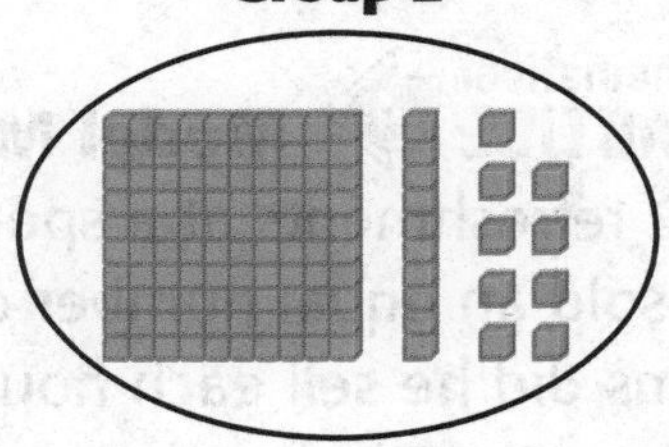

4 Regroup the remaining tens into 30 ones.

There are a total of 36 ones.
Divide the ones into 4 groups.

How many ones are in each group? 9

Group 3

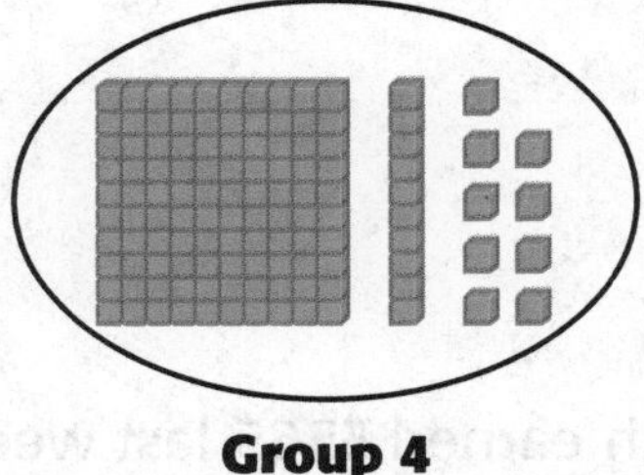

Group 4

Each group has 1 hundred, 1 ten, and 9 ones.

So, each friend will receive 119 tickets.

Check Use multiplication to check your answer.

$119 \times 4 = 476$

Practice

Use models to find each quotient. Draw the equal groups.

1. $328 \div 2 =$ ________

2. $443 \div 3 =$ ________

Problem Solving

3. Mathematical PRACTICE 4 **Model Math** Denny volunteered to sell 372 refreshments at a sports game. He helped for 3 hours. If he sold an equal number of items each hour, how many items did he sell each hour? Draw models to help you find the quotient.

____

4. Faith earned \$565 last week. She earned the same amount each day. If she worked 5 days, how much did she earn each day? Draw models to help you find the quotient.

____

Name ....

Hands On
Distributive Property and Partial Quotients

Lesson 7

ESSENTIAL QUESTION
What strategies can be used to divide whole numbers?

The Distributive Property allows you to divide each place-value position by the same factor.

Draw It

Jesse has 369 beads to be split evenly among 3 necklaces. How many beads can Jesse put on each necklace?

Find 369 ÷ 3.

 **Model 369 as (300 + ______ + ______).**

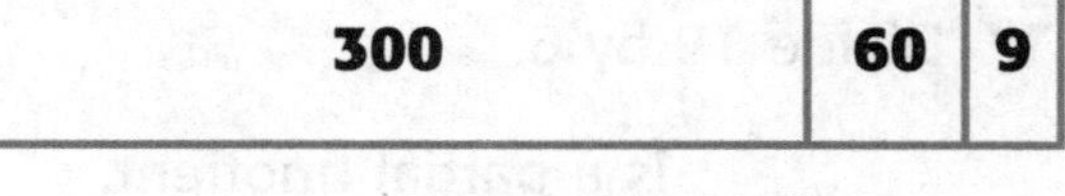

2 Divide each section by 3.

Write each quotient above the bar.

300 ÷ 3 = ______

60 ÷ 3 = ______

9 ÷ 3 = ______

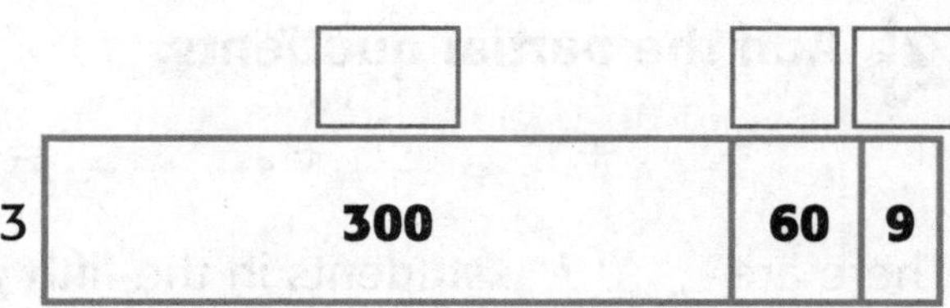

3 Add the quotients.

______ + ______ + ______ = ______

So, 369 ÷ 3 = ______.

Jesse can put ______ beads on each necklace.

Check Multiply to check your answer.

______ × 3 = 369

Partial quotients is a method of dividing where you break the dividend into addends that are easy to divide.

Try It

There are 738 students in Manuel's school. There are 6 grade levels in his school, with each grade having the same number of students. How many students are in the fifth grade?

Find 738 ÷ 6.

Partial Quotients

$$\begin{array}{r|l} 6\overline{)738} & \\ -\ 600 & \_\_\_\_ \\ \hline 138 & \\ -\ 120 & \_\_\_\_ \\ \hline 18 & \\ -\ 18 & \_\_\_\_ \\ \hline 0 & \end{array}$$

1. **Divide the hundreds.**
   600 is close to 738 and is compatible with 6.
   Divide 600 by 6.

   ______ is a partial quotient.
   Subtract 600 from 738.

2. **Divide the tens.**
   120 is close to 138 and is compatible with 6.
   Divide 120 by 6.

   ______ is a partial quotient.
   Subtract 120 from 138.

3. **Divide the ones.**
   Divide 18 by 6.

   ______ is a partial quotient.

4. **Add the partial quotients.**

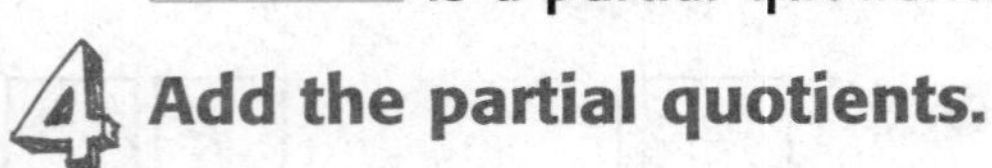

______ + ______ + ______ = ______

There are ______ students in the fifth grade.

Check Multiply to check your answer.

______ × 6 = 738

Talk About It

1. Mathematical PRACTICE 3 **Draw a Conclusion** How would you use the Distributive Property to find 242 ÷ 2?

____

____

Name ....

Practice It

Divide. Use the Distributive Property to draw bar diagrams.

2. 248 ÷ 2 = ________

3. 963 ÷ 3 = ________

4. 585 ÷ 5 = ________

5. 488 ÷ 4 = ________

Divide. Use partial quotients.

6. 654 ÷ 3 = ________

7. 675 ÷ 5 = ________

8. 351 ÷ 3 = ________

Apply It

9. Joshua is saving his money to donate to his favorite charity. He has saved $432 so far. If he has been saving for 3 years, how much did he save each year?

____

10. Mr. Keaton planted 918 cornstalks. There are 9 cornstalks in each row. How many rows of corn did he plant?

____

11. Mathematical PRACTICE 2 **Use Number Sense** Colton wants to buy a trampoline in 2 months. The trampoline costs $228. If he saves the same amount each month, how much will he have to save each month?

____

12. Mathematical PRACTICE 2 **Reason** Suppose you are finding 296 ÷ 4 using partial quotients. Is 50 or 70 a more reasonable partial quotient? Explain.

____

____

Write About It

13. How can properties help me divide?

____

____

____

Name ....

MY Homework

Lesson 7
Hands On: Distributive Property and Partial Quotients

Homework Helper

Need help? connectED.mcgraw-hill.com

The Statue of Liberty has a total of 354 steps. Janice decides to climb the steps in 2 equal sections. How many steps will she climb in each section?

Find 354 ÷ 2.

Model 354 as (300 + 50 + 4).

300	50	4

Divide each section by 2.

Write each quotient above the bar.

300 ÷ 2 = 150
50 ÷ 2 = 25
4 ÷ 2 = 2

	150	25	2
2	300	50	4

Add the quotients.

150 + 25 + 2 = 177

So, 354 ÷ 2 = 177. Janice will climb 177 steps in each section.

Check Multiply to check your answer. 177 × 2 = 354

Practice

1. Divide 844 ÷ 4. Use the Distributive Property to draw a bar diagram.

____

Problem Solving

2. The fourth, fifth and sixth grade classes of 396 students take a field trip to the historical museum over 3 days. How many students can go each day?

____

3. Colton wants to buy a video game system in 3 months. The video game system costs $396. If he saves the same amount each month, how much will he have to save each month?

____

4. Paul's MP3 player has a total of 852 songs. He has 4 different groups of songs. If he puts an equal number of songs in each group, how many songs will be in each group?

____

My Work!

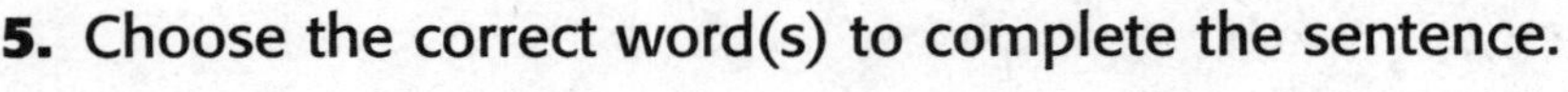

Vocabulary Check

Vocab

5. Choose the correct word(s) to complete the sentence.

Using partial quotients is a method of dividing where you break the ______________ into sections that are easy to divide.

6. Mathematical PRACTICE 5 **Use Math Tools** Stanley is using the bar diagram to help him find 482 ÷ 2. What is the quotient?

2	400	80	2

____

Name ______________________

Divide Three- and Four-Digit Dividends

Lesson 8

ESSENTIAL QUESTION
What strategies can be used to divide whole numbers?

To divide a greater dividend, use the same process as dividing a two-digit dividend.

Math in My World

Example 1

In a 4-hour period, 852 people rode an amusement park ride. If the same number of people rode the ride each hour, how many people rode the ride in the first hour?

Let p represent the number of people.

_______ $\div$ _______ $= p$

Find 852 ÷ 4.

Estimate $900 \div 4 =$ _______

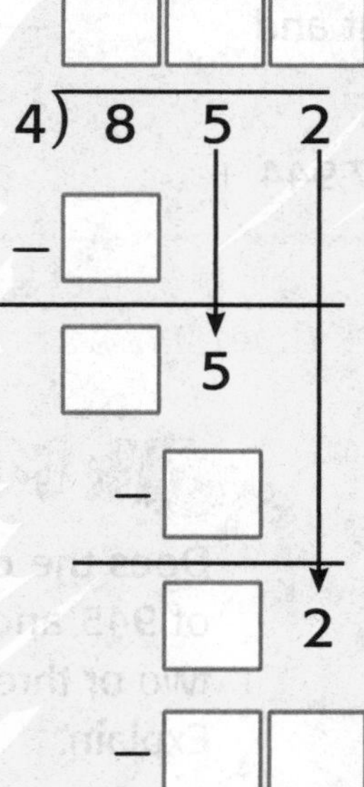

Divide the hundreds.
Divide. $8 \div 4$
Multiply. 2×4
Subtract. $8 - 8$
Compare. $0 < 4$

2 **Divide the tens.**
Divide. $5 \div 4$
Multiply. 1×4
Subtract. $5 - 4$
Compare. $1 < 4$

3 **Divide the ones.**
Divide. $12 \div 4$
Multiply. 3×4
Subtract. $12 - 12$
Compare. $0 < 4$

So, _______ people rode the ride in the first hour.

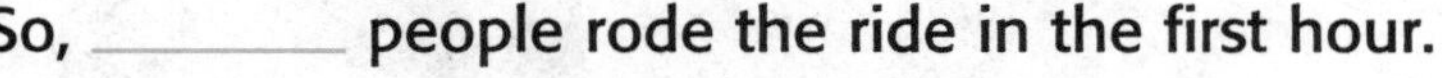

Check Multiply to check your answer.

_______ $\times 4 = 852$

Example 2

Find $6\overline{)7{,}946}$.

1 Divide the thousands.
Divide. 7 ÷ 6
Multiply. 1 × 6
Subtract. 7 − 6
Compare. 1 < 6

2 Divide the hundreds.
Divide. 19 ÷ 6
Multiply. 3 × 6
Subtract. 19 − 18
Compare. 1 < 6

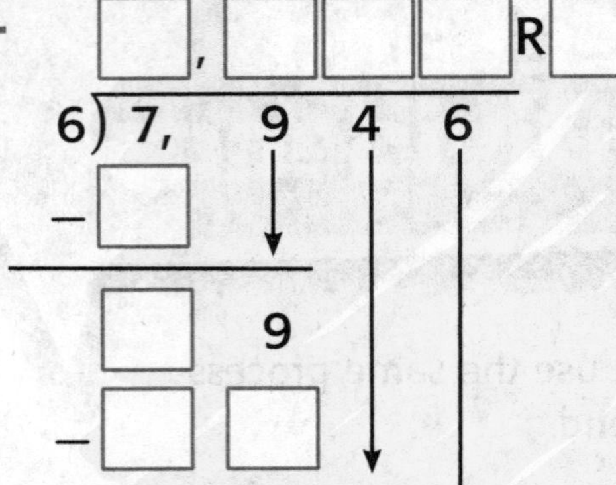
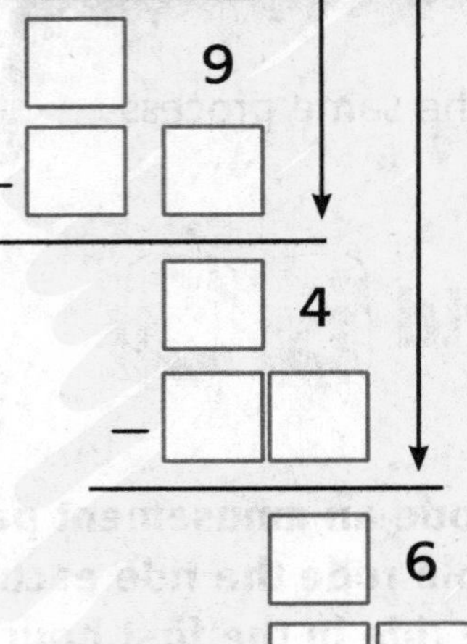

3 Divide the tens.
Divide. 14 ÷ 6
Multiply. 2 × 6
Subtract. 14 − 12
Compare. 2 < 6

4 Divide the ones.
Divide. 26 ÷ 6
Multiply. 4 × 6
Subtract. 26 − 24
Compare. 2 < 6

So, 7,946 ÷ 6 = ________.

Check To check division with a remainder, first multiply the quotient and the divisor. Then add the remainder.

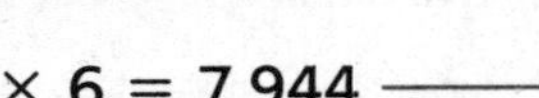

________ × 6 = 7,944 ⟶ 7,944 + ________ = 7,946

Guided Practice

1. Divide.

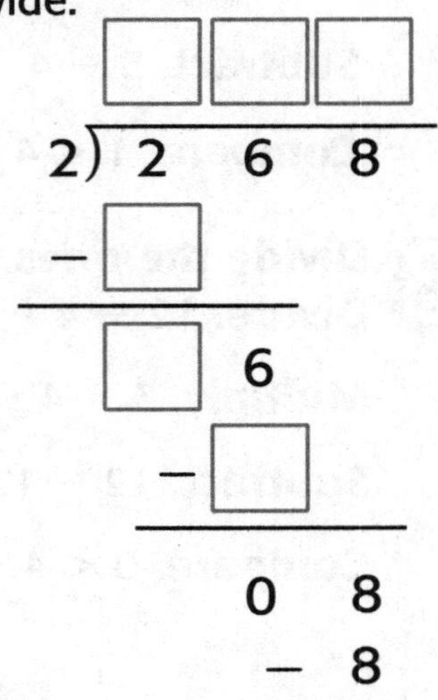

Name ____________________

Independent Practice

Divide.

2. $5\overline{)595}$

3. $4\overline{)625}$

4. $5\overline{)5{,}815}$

5. 516 ÷ 3 = ______

6. 6,418 ÷ 3 = ______

7. 9,345 ÷ 7 = ______

8. $5\overline{)755}$

9. $4\overline{)8{,}468}$

10. $2\overline{)2{,}349}$

Algebra Divide to find the unknown number in each equation.

11. 414 ÷ 3 = *c*

c = ______

12. 5,120 ÷ 4 = *m*

m = ______

13. 1,535 ÷ 5 = *x*

x = ______

Problem Solving

14. Three new video game systems cost $645. If all the game systems cost the same, what is the cost of each game system?

15. Mathematical PRACTICE 2 **Use Algebra** A state park has cable cars that travel about 864 yards in 4 minutes. The cars travel the same amount of yards each minute. How many yards do the cars travel per minute? Use the bar diagram to write an equation to find the unknown. Then find the unknown.

|------------ 864 yards ------------|

1 min	1 min	1 min	1 min

|--? yd--|

16. Three adult kangaroos weigh 435 pounds. If each adult weighs the same, how much would one adult kangaroo weigh?

HOT Problems

17. Mathematical PRACTICE 2 **Use Number Sense** Place the digits 2, 4, 7, and 8 in ■ ■ ■ ÷ ■ to create a division problem with the greatest quotient.

18. **Building on the Essential Question** How can I divide larger dividends?

Name ....

MY Homework

Lesson 8
Divide Three- and Four-Digit Dividends

Homework Helper

Need help? connectED.mcgraw-hill.com

Find 630 ÷ 5.

Estimate 600 ÷ 5 = 120

1 Divide the hundreds.
Divide. 6 ÷ 5
Multiply. 1 × 5
Subtract. 6 − 5
Compare. 1 < 5

```
   126
5)630
 − 5
   13
 − 10
    30
  − 30
     0
```

2 Divide the tens.
Divide. 13 ÷ 5
Multiply. 2 × 5
Subtract. 13 − 10
Compare. 3 < 5

3 Divide the ones.
Divide. 30 ÷ 5
Multiply. 6 × 5
Subtract. 30 − 30
Compare. 0 < 5

So, 630 ÷ 5 = 126.

Check Multiply to check your answer. 126 × 5 = 630

Practice

Divide.

1. 3)945

2. 3)493

3. 6,315 ÷ 5 = ________

Problem Solving

4. Mathematical PRACTICE 2 **Use Number Sense** Mr. Peters has 80 sheets of colored paper. Seven of his students need the paper for a project. How many sheets does each student get? How many sheets are left over?

My Work!

5. Tina has earned a total of 9,644 frequent flyer miles by traveling between Twin Falls and Preston. She has made this trip 4 times. How many miles is one trip between these two cities?

6. A family of 4 spent $104 for tickets to a concert. All of the tickets were the same price. What was the cost of each ticket?

7. On Monday, a concession stand manager ordered 985 popcorn bags. She splits the bags evenly among 5 concession stands. How many popcorn bags will each concession stand receive?

Test Practice

8. A great white shark weighed 4,302 pounds. This weight was 3 times the weight of a blue marlin fish. What was the weight of the blue marlin?

   Ⓐ 1,402 pounds

   Ⓑ 1,424 pounds

   Ⓒ 1,434 pounds

   Ⓓ 1,502 pounds

Check My Progress

Vocabulary Check

1. Circle the method that correctly uses **partial quotients**.

$4\overline{)644}$	
− 640	140
4	
− 4	4
0	

$3\overline{)513}$	
− 300	100
210	
− 150	50
60	
− 60	20
0	

$4\overline{)528}$	
− 400	100
128	
− 120	30
8	
− 8	2
0	

Concept Check

Estimate. Show how you estimated.

2. 244 ÷ 8 ______

3. 700 ÷ 6 ______

4. 890 ÷ 4 ______

Divide.

5. $5\overline{)630}$

6. 1,766 ÷ 6 = ______

7. $2\overline{)87}$

Real World Problem Solving

8. Each of the 9 parking lots at an automobile plant holds the same number of new cars. The lots are full. If there are 431 cars in the lots, about how many cars are in each lot? Show how you estimated.

____

9. A total of 176 valves were used for 8 cars as they were being assembled. The same number of valves were used for each car. How many valves were used for each car?

____

10. A construction company estimates that it will take 852 hours to complete a remodeling project. If there are 6 employees that each work an equal number of hours, how many hours will each employee have to work?

____

11. Rachel has 145 mugs displayed at a craft show. She displays them in 8 rows with the same number of mugs in each row. How many mugs are in each row? Explain how you interpreted the remainder.

____

My Work!

Test Practice

12. Valley Schools have a student population of 1,608 students. If there are an equal number of students in the 6 grade levels, how many students are there in each grade?

Ⓐ 28 students Ⓒ 248 students

Ⓑ 208 students Ⓓ 268 students

Name ____________

Place the First Digit

Lesson 9

ESSENTIAL QUESTION
What strategies can be used to divide whole numbers?

A three-digit dividend may not have enough hundreds to divide. If so, the quotient should start at the next place-value position.

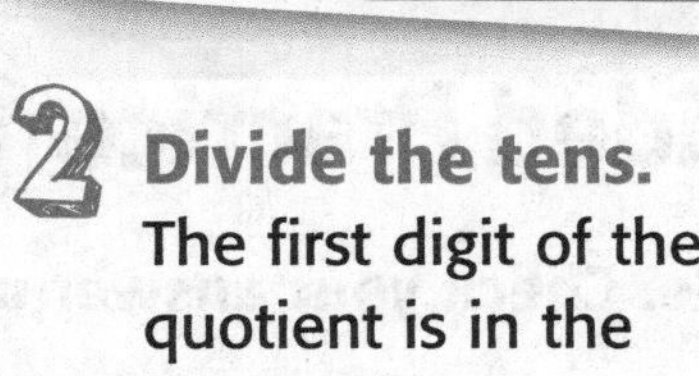

Math in My World

Example 1

Raven received 135 E-mails over 3 weeks. If she received the same number of E-mails each week, how many E-mails did she receive in the first week?

Find 135 ÷ 3.

Estimate 150 ÷ 3 = ______

Divide the hundreds. There are not enough hundreds to divide into 3 groups. So, regroup 1 hundred and 3 tens as 13 tens.

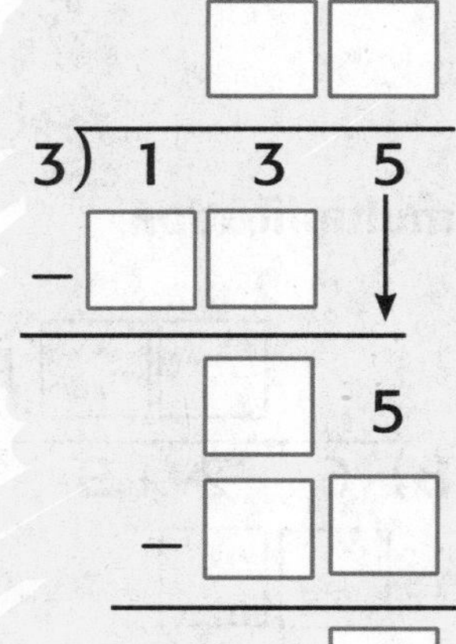

2 **Divide the tens.** The first digit of the quotient is in the tens place.

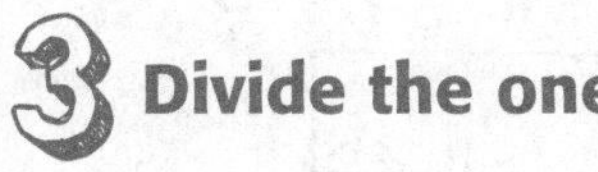

So, Raven received ______ E-mails in the first week.

The symbol ≈ *means about* or *almost equal to.*

Check for Reasonableness Compare to the estimate.

______ ≈ 50

Example 2

Find $7\overline{)6{,}784}$.

Estimate 7,000 ÷ 7 = ______

Divide the hundreds. There are not enough thousands to divide into 7 groups. So, regroup 6 thousands and 7 hundreds as 67 hundreds.

```
      [ ][ ][ ]R[ ]
 7) 6,  7  8  4
   -[ ][ ]
   ------  ↓
      [ ]  8
      -[ ][ ]
      -------
         [ ]  4
         -[ ][ ]
         -------
             [ ]
```

2 Divide the tens. The first digit of the quotient is in the hundreds place.

3 Divide the ones.

So, 6,784 ÷ 7 = ______.

Check for Reasonableness Compare to the estimate.

______ ≈ ______

Talk MATH

You want to find 510 ÷ 6. Tell how you know where to place the quotient's first digit.

Guided Practice

Divide. Check your answer using multiplication.

1.

```
      [ ][ ]
 5) 4  3  5
  -[ ][ ]
  -------
      [ ]  5
     -[ ][ ]
     -------
         [ ]
```

2.

```
      [ ][ ]R[ ]
 8) 6  2  9
  -[ ][ ]
  -------
      [ ]  9
     -[ ][ ]
     -------
         [ ]
```

Name ....

CCSS

Independent Practice

Divide.

3. $6\overline{)576}$

4. $5\overline{)3{,}085}$

5. $4\overline{)256}$

6. $6\overline{)4{,}527}$

7. $4\overline{)217}$

8. $4\overline{)274}$

9. $2{,}181 \div 3 =$ ______

10. $108 \div 9 =$ ______

11. $3{,}417 \div 4 =$ ______

Algebra **Find the unknown number in each equation.**

12. $232 \div 8 = q$

$q =$ ______

13. $324 \div 9 = s$

$s =$ ______

14. $192 \div 4 = y$

$y =$ ______

Problem Solving

My Work!

15. There are 624 envelopes to be sorted into 8 different mail bags. If the same number of envelopes will be in each bag, how many envelopes will be in one bag?

16. Mathematical PRACTICE 2 **Use Symbols** There are 594 people standing in line to see a movie premiere. The movie is playing in 6 theaters. If the same number of people will see the movie in each theater, how many people will be in each theater? Write an equation to find the unknown. Then find the unknown.

17. The Environmental Club is having a trash pickup day. There are 130 people signed up to help. For the trash pickup day, they will work in groups of 4 people. No more than 4 people can join a group. How many groups are there? Explain how you interpreted the remainder.

HOT Problems

18. Mathematical PRACTICE 2 **Use Number Sense** Can you determine the number of digits in the quotient of 637 ÷ 7 without dividing? Explain.

19. **Building on the Essential Question** How can I know where to place the first digit of a quotient?

Name

MY Homework

Lesson 9
Place the First Digit

Homework Helper

eHelp

Need help? connectED.mcgraw-hill.com

Find 498 ÷ 6.

Estimate 500 ÷ 5 = 100

Divide the hundreds.
There are not enough hundreds to divide into 6 groups. So, regroup 4 hundreds and 9 tens as 49 tens.

$$\begin{array}{r} 83 \\ 6\overline{)498} \\ -48 \\ \hline 18 \\ -18 \\ \hline 0 \end{array}$$

2 Divide the tens.
The first digit of the quotient is in the tens place.

3 Divide the ones.

So, 498 ÷ 6 = 83.

Check Multiply to check your answer. 83 × 6 = 498

Practice

Divide.

1. $6\overline{)486}$

2. $6\overline{)392}$

3. 5,920 ÷ 6 = ______

Problem Solving

4. The phone company needs 420 poles to repair the telephone lines. Each truck holds 6 poles. How many trucks will they need?

5. The festival committee has $1,544 to spend on pies for the pie-eating contest. If each pie costs $8, how many pies can the committee purchase?

6. **Algebra** A group of 273 people take a canoe trip. Each canoe holds 3 people. How many canoes will the group need? Write an equation to find the unknown. Then find the unknown.

HOT Problems

7. Mathematical PRACTICE 3 **Which One Doesn't Belong?** Circle the expression that does not have a two-digit quotient.

519 ÷ 6	915 ÷ 7	439 ÷ 7	812 ÷ 9

Test Practice

8. A glass company ships 470 glass ornaments. Each box holds 5 ornaments. How many boxes will the company need?

Ⓐ 84 boxes

Ⓑ 92 boxes

Ⓒ 93 boxes

Ⓓ 94 boxes

Name

Quotients with Zeros

Lesson 10

ESSENTIAL QUESTION

What strategies can be used to divide whole numbers?

Math in My World

Example 1

Maya is saving to buy a television. The television costs $327. She plans to save money for 3 months. How much does Maya need to save each month to buy the television?

Let m represent the amount Maya needs to save each month. Find the unknown in the equation $\$327 \div 3 = m$.

Estimate $\$300 \div 3 =$ ______

 **Divide the hundreds.**

 **Divide the tens.**

There are not enough tens to divide.

Place a 0 in the quotient.

 Regroup the two tens as twenty ones.

There are now 27 ones.

 **Divide the ones.**

```
     □ □ □
 3 ) 3  2  7
   - □  ↓  |
     □  2  |
      - □  ↓
        □  7
      - □  □
           □
```

So, $\$327 \div 3 =$ ______. Since $m =$ ______, Maya needs to save ______ each month.

Check for Reasonableness Compare to the estimate. ______ ≈ ______

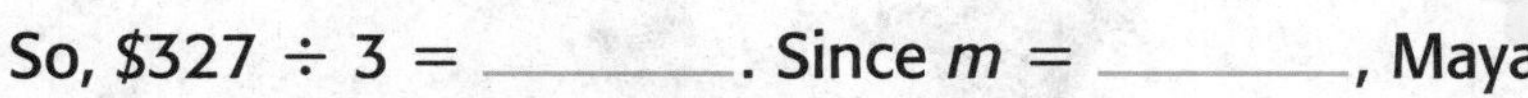

Example 2

Find 5,231 ÷ 4.

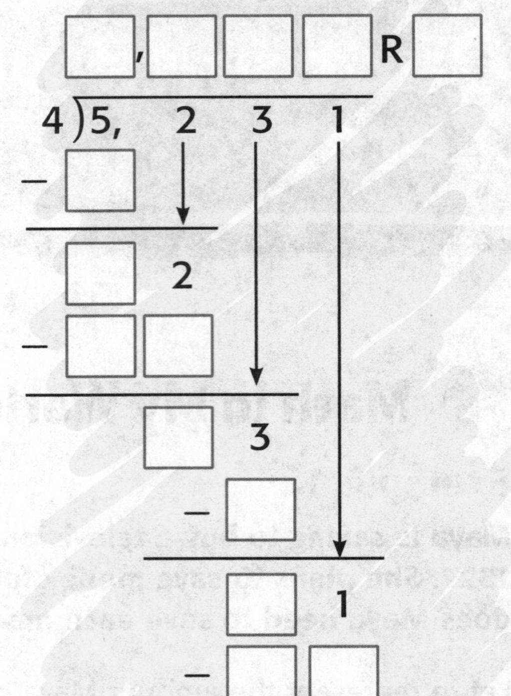

1 Divide the thousands.

2 Divide the hundreds.

3 Divide the tens.
There are not enough tens to divide. Place a 0 in the tens place.

4 Divide the ones.

The remainder is ______.

So, 5,231 ÷ 4 = ____________.

Check Use multiplication to check your answer.

________ × 4 = ________ and 5,228 + ________ = 5,231

Guided Practice

1. Divide.

4)8 4 2

Name ....

Independent Practice

Divide.

2. $2\overline{)418}$

3. $2\overline{)210}$

4. $4\overline{)4{,}324}$

5. $6\overline{)782}$

6. $2\overline{)6{,}213}$

7. $3\overline{)6{,}192}$

8. 840 ÷ 7 = ________

9. 627 ÷ 3 = ________

10. 5,330 ÷ 5 = ________

11. 8,017 ÷ 9 = ________

12. 413 ÷ 4 = ________

13. 9,163 ÷ 3 = ________

Problem Solving

14. **Mathematical PRACTICE 2 Use Algebra** There are 312 fish at the aquarium in 3 different fish tanks. Each tank has the same number of fish. How many fish are in each tank? Write an equation to find the unknown. Then find the unknown.

15. There are 1,620 minutes of music to be put on 9 CDs. If the same number of minutes fits on each CD, how many minutes of music fit on each CD?

16. Gina spent 120 minutes helping her neighbors rake leaves in the last 4 days. She helped the same amount of minutes each day. How many minutes did she rake leaves each day?

HOT Problems

17. **Mathematical PRACTICE 1 Keep Trying** Write two division problems that have zeros in the quotient. One of the problems should have a remainder and the other should not.

18. **Building on the Essential Question** How can I know when to place a zero in the quotient?

Name ....

MY Homework

Lesson 10

Quotients with Zeros

Homework Helper

Need help? connectED.mcgraw-hill.com

Find 815 ÷ 2.

1 Divide the hundreds.

2 Divide the tens.
There are not enough tens to divide.
Place a 0 in the tens place.

3 Divide the ones.

The remainder is 1.

```
   407R1
 2)815
  - 8↓
    01
   - 0↓
    15
   -14
     1
```

So, 815 ÷ 2 = 407 R1.

Check 407 × 2 = 814 and 814 + 1 = 815

Practice

Divide.

1. $8\overline{)856}$

2. $3\overline{)2{,}926}$

3. 841 ÷ 4 = ________

Real World Problem Solving

4. Monica wants to join the swim team. She practices 812 minutes in 4 weeks. She practices the same number of minutes each week. How many minutes does Monica practice each week?

____

5. **Algebra** The art teacher asks her students to cut out apples from construction paper. Five apples can be cut from each sheet of paper. If she needs 1,045 apples, how many sheets of paper does she need? Write an equation to find the unknown. Then find the unknown.

____

6. Mathematical PRACTICE 3 **Which One Doesn't Belong?** Circle the division problem that does not belong with the other three. Explain.

$621 \div 6$	$384 \div 3$	$719 \div 7$	$514 \div 5$

____

Test Practice

7. Abby and her family are going to Yellowstone National Park this summer. They drive 1,212 miles from their home to the park. If they drive the same number of miles each day for 4 days, how many miles will they drive each day?

Ⓐ 303 miles

Ⓑ 330 miles

Ⓒ 403 miles

Ⓓ 3,030 miles

Name

Hands On
Use Models to Interpret the Remainder

Lesson 11

ESSENTIAL QUESTION
What strategies can be used to divide whole numbers?

Build It

Tools

A group of fifth graders collected 46 cans of food to donate to 3 food banks. If each food bank is to get an equal number of cans, how many cans do they each receive?

1. Use _______ connecting cubes to represent the cans of food.

   Use _______ paper plates to represent the food banks.

   Divide the cubes equally among the _______ plates.

   How many cubes are on each plate? _______

   How many cubes are left over? _______

2. Interpret the remainder.
   Since each food bank is to get the same number of cans of food, they will each receive

   _______ cans.

   There is _______ can left over.

Try It

A total of 35 students are going on a field trip to NASA's Johnson Space Center in Houston, Texas. If there needs to be one adult for every 8 students, how many adults are needed?

1. Use ______ connecting cubes to represent the students.

   Use paper plates to represent the adults.

   How many cubes are on each plate? ______

   How many cubes are left over? ______

2. Interpret the remainder.

   There are ______ groups of ______ students. They will each need an adult.

   There are ______ students who are not enough for a full group of 8. They will also need an adult.

   So, ______ + ______, or ______ adults are needed.

Talk About It

1. Mathematical PRACTICE 3 **Justify Conclusions** In Activity 1, the remainder was dropped. Explain why.

______

______

______

2. In Activity 2, the quotient was "rounded up" to 5. Explain why.

______

______

Name

Practice It

Solve using models. Explain how to interpret the remainder. Draw your models.

3. Each picnic table at a park seats 6 people. How many tables will 83 people at a family reunion need?

My Drawing!

4. Mrs. Malone has $75 to buy volleyballs for Lincoln Middle School. How many can she buy at $9 each?

My Drawing!

5. Darcy has 63 flowers to use to make centerpieces. Each centerpiece uses 8 flowers. How many centerpieces can she make?

My Drawing!

Apply It

For Exercises 6–8, solve using models. Explain how to interpret the remainder.

6. A teacher received 30 new calculators. Only 8 calculators fit in each carrier. How many carriers does the teacher need?

7. Mathematical PRACTICE 5 **Use Math Tools** Joel has 48 oranges. He puts 7 oranges in a bag. How many bags can he fill?

8. Anita is helping to make gift boxes for a community center. She has a total of 34 toys. She places 3 toys in each box. How many boxes will she need?

My Work!

9. Mathematical PRACTICE 2 **Reason** Suppose 2 friends want to share 5 cookies evenly. Interpret the remainder in two different ways.

Write About It

10. How can I use models to interpret the remainder?

Name

MY Homework

Lesson 11
Hands On: Use Models to Interpret the Remainder

Homework Helper

Need help? connectED.mcgraw-hill.com

A total of 47 students signed up to play soccer. If there are 4 teams, how many students are on each team?

1. Use 47 connecting cubes to represent the students.
   Use 4 paper plates to represent the teams.
   How many cubes are on each plate? 11
   How many cubes are left over? 3

2. Interpret the remainder.
   There are 4 teams of 11 students. There are 3 students who are left over and will be placed on teams.

So, 3 teams will have 12 students and 1 team will have 11 students.

Practice

1. Jarrod's sports card holder can hold 9 cards on each page. How many pages will Jarrod need if he has 75 sports cards? Solve using models. Explain how to interpret the remainder. Draw your models.

My Drawing!

Solve using models. Explain how to interpret the remainder. Draw your models.

2. A group of 4 students are selling candy bars to raise money for a field trip. The group needs to sell 53 candy bars. If each student sells an equal amount, how many candy bars do they each sell? Solve using models. Explain how to interpret the remainder. Draw your models.

Problem Solving

For Exercises 3–5, solve using models. Explain how you interpreted the remainder. Draw your models.

3. Mathematical PRACTICE 4 **Model Math** Joel has 59 songs on his MP3 player. He equally divides them in 7 groups. How many songs will be in each group?

4. Lauren's shoe organizer can hold 5 pairs of shoes of each row. How many rows will Lauren need if she has 24 pairs of shoes?

5. Mr. Staley has $62 to buy binders for his mathematics class. How many binders can he buy at $3 each?

Interpret the Remainder

Lesson 12

ESSENTIAL QUESTION
What strategies can be used to divide whole numbers?

Let's put down roots!

Math in My World

Example 1

A state park has 257 evergreens to plant equally in 9 areas. How many evergreens are planted in each area? What does the remainder represent?

Divide 257 ÷ 9.

Place the first digit.

$25 \div 9 \approx 2$

Put 2 in the tens place of the quotient.

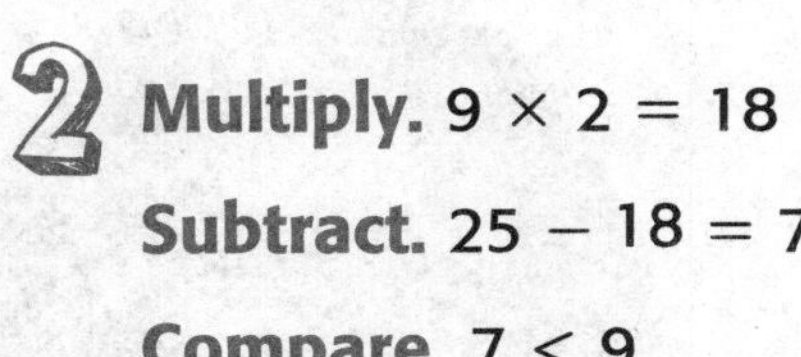

Multiply. $9 \times 2 = 18$

Subtract. $25 - 18 = 7$

Compare. $7 < 9$

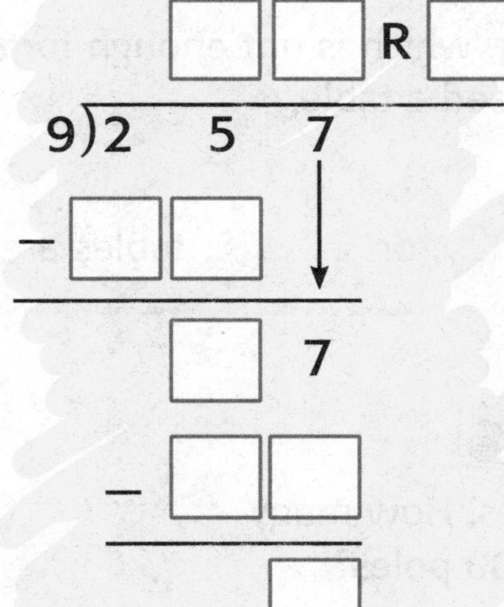

Divide the ones.

$77 \div 9 \approx 8$

$9 \times 8 = 72$

$77 - 72 = 5$

$5 < 9$

Write the remainder.

The remainder is 5.

Interpret the remainder, 5.

The remainder, 5, means there are 5 evergreens left over.

So, the park plants ______ evergreens in each area and ______ evergreens are left.

Example 2

There are 174 guests invited to a dinner. Each table seats 8 guests. How many tables are needed?

Divide 174 ÷ 8.

Place the first digit.

$17 \div 8 \approx 2$

Put 2 in the tens place of the quotient.

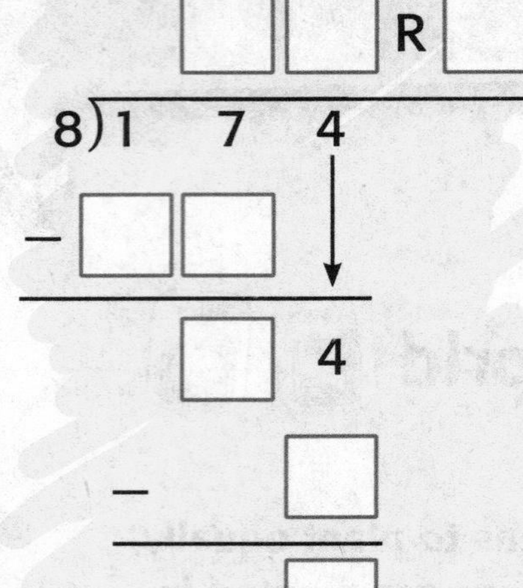

Multiply. $8 \times 2 = 16$

Subtract. $17 - 16 = 1$

Compare. $1 < 8$

3 Divide the ones.

$14 \div 8 \approx 1$

$1 \times 8 = 8$

$14 - 8 = 6$

$6 < 8$

4 Write the remainder.

The remainder is 6.

Interpret the remainder, 6.

There are 6 guests left over, which is not enough for a full table of 8. But, they also need a table.

So, a total of ______ + ______, or ______ tables are needed.

Talk MATH Discuss the different ways you can interpret the remainder.

Guided Practice

1. A tent is put up with 7 poles. How many tents can be put up with 200 poles?

Divide 200 ÷ 7.

Interpret the remainder, ______.

There are ______ poles left over, which is not enough for another tent.

So, a total of ______ tents can be put up.

Name

Independent Practice

Solve. Explain how you interpreted the remainder.

2. There are 50 students traveling in vans on a field trip. Each van seats 8 students. How many vans are needed?

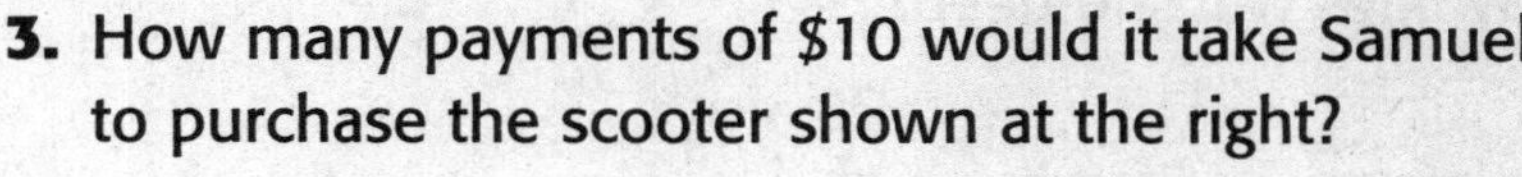

3. How many payments of $10 would it take Samuel to purchase the scooter shown at the right?

4. Mrs. Hodges made 144 muffins for a bake sale. She puts them into tins of 5 muffins each. How many tins of muffins can she make?

5. How many 8-foot sections of fencing are needed for 189 feet of fence?

Problem Solving

Solve Exercises 6–8. Explain how you interpreted the remainder.

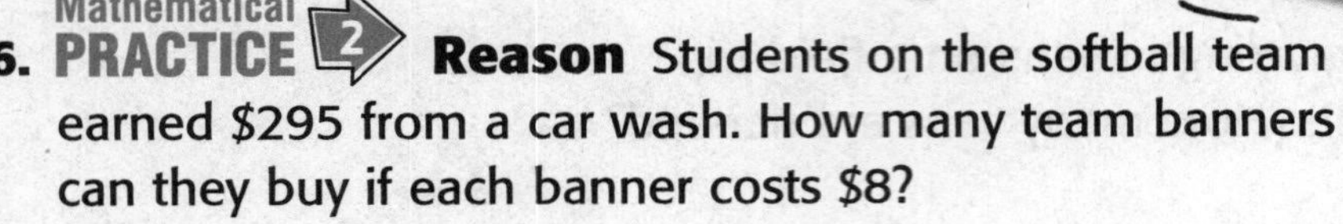

6. **Mathematical PRACTICE 2 Reason** Students on the softball team earned $295 from a car wash. How many team banners can they buy if each banner costs $8?

7. Valerie has 20 stuffed animals. She wants to store them in plastic bags. She estimates she can fit three stuffed animals in each bag. How many bags will she need?

8. How many 6-ounce cups can be filled from 4 gallons of juice? (*Hint:* 1 gallon = 128 ounces)

HOT Problems

9. **Mathematical PRACTICE 1 Make Sense of Problems** If the divisor is 30, what is the least three-digit dividend that would give a remainder of 8? Explain.

10. **Building on the Essential Question** How can I interpret the remainder?

MY Homework

Lesson 12
Interpret the Remainder

Homework Helper

Need help? connectED.mcgraw-hill.com

There are 46 students volunteering at a senior citizen community. There is a maximum of 6 students in each group. How many groups are needed?

Divide $46 \div 6$.

Place the first digit.

$46 \div 6 \approx 7$

Put 7 in the ones digit of the quotient.

$$\begin{array}{r} 7\text{ R }4 \\ 6\overline{)46} \\ -\,42 \\ \hline 4 \end{array}$$

2 Multiply. $6 \times 7 = 42$

Subtract. $46 - 42 = 4$

Compare. $4 < 6$

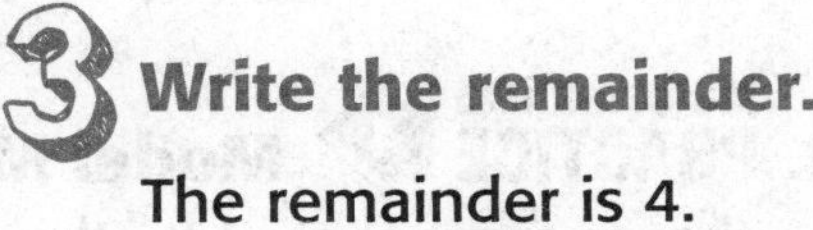

3 Write the remainder.

The remainder is 4.

Interpret the remainder, 4.

There are 4 students left over, which is not enough for another group of 6. But, they are volunteering.

So, there are a total of 7 + 1, or 8 groups of students volunteering.

Practice

1. Four employees of Papa Tony's Pizza are cleaning up at the end of a busy night. There is a list of 43 clean-up tasks that need to be completed. If each employee does the same number of tasks, how many tasks should each employee do? Solve. Explain how you interpreted the remainder.

Real World Problem Solving

Solve Exercises 2 and 3. Explain how you interpreted the remainder.

2. Mrs. Flores is buying scrapbooks for her store. Her budget is \$350. Each scrapbook costs \$9. How many scrapbooks can she buy?

3. Water stations will be placed every 400 meters of a 5-kilometer race. How many water stations are needed? (*Hint:* 1 kilometer = 1,000 meters)

4. Mathematical PRACTICE 4 **Model Math** Write a real-world problem that is represented by the division problem 38 ÷ 5 = 7 R3 in which it makes sense to round the quotient up to 8.

Test Practice

5. Three yards of fabric will be cut into pieces so that each piece is 8 inches long. How many pieces can be cut? (*Hint:* 1 yard = 36 inches)

Ⓐ 4 pieces with 4 inches left over

Ⓑ 10 pieces with 6 inches left over

Ⓒ 13 pieces with 4 inches left over

Ⓓ 16 pieces with 6 inches left over

Name ....

Problem-Solving Investigation

STRATEGY: Determine Extra or Missing Information

Lesson 13

ESSENTIAL QUESTION
What strategies can be used to divide whole numbers?

Learn the Strategy

Watch Tutor

Kayla was collecting book orders. The cost of each book is $3. There were 7 orders on Wednesday, 5 orders on Thursday, and more orders on Friday and Monday. How many book orders were collected altogether?

1 Understand

What facts do you know?

I know the cost of a book is $3 and the number of book orders on Wednesday was ________ and Thursday was ________.

What do you need to find?

I need to find the total number of ____________.

2 Plan

Determine if there is extra or missing information.

The ________ of a book is not needed. The number of book orders that were collected on Friday and Monday is missing.

3 Solve

Some information is ____________. So, I cannot solve the problem.

4 Check

Is my answer reasonable? Explain.

____

Practice the Strategy

GREAT, ...any way you slice it!

Rocco is slicing a loaf of Italian bread for dinner. The bread costs $4. He cuts the loaf into slices that are 1 inch thick. If the loaf is 18 inches long, how many pieces of bread did he cut?

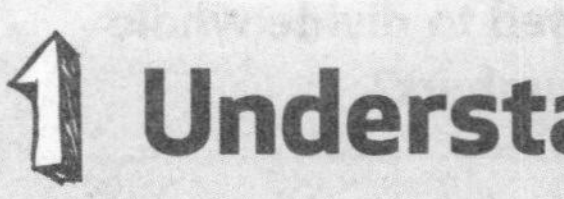

1 Understand

What facts do you know?

____

____

____

What do you need to find?

____

2 Plan

____

____

____

3 Solve

4 Check

Is my answer reasonable? Explain.

____

____

Name

Apply the Strategy

Determine if there is extra or missing information. Then solve the problem, if possible.

My Work!

1. Jayden is downloading songs onto his MP3 player. One song is 5 minutes long, another is 2 minutes long, and a third is between the lengths of the other two songs. What is the total length of all three songs?

2. Room 220 and Room 222 are having a canned food drive. How many more cans has Room 222 collected than Room 220?

3. Karly is collecting money for a bowl-a-thon. Her goal is to collect $125. Last year the bowl-a-thon raised $100. If she charges $5 for each person, how many people need to participate in the bowl-a-thon?

4. Mathematical PRACTICE 1 **Make Sense of Problems** Sari made pancake batter. She has 1 cup of batter left. How much batter did she use?

Review the Strategies

Use any strategy to solve each problem.
- Determine extra or missing information.
- Make a table.
- Use the four-step plan.

5. Mrs. Rollins raises prize chickens. Each chicken eats the same amount of food. Mrs. Rollins bought 100 pounds of chicken food last week. How much food did each chicken eat?

6. At the school bake sale, Kenji's mom bought 3 cookies, 1 brownie, and 1 cupcake. She gave the cashier $2 and received $1.05 in change. Find the cost of a cupcake and write it in the table.

Item	Price ($)
cookie	0.15
brownie	0.20
cupcake	

7. Mathematical PRACTICE 1 **Plan Your Solution** Todd has $50 to spend on a video game. The game he wants cost $30. If he buys one game, he gets a second game for half price. How much money will he have left if he purchases two games?

8. Twelve students are going roller skating. Each student pays $8 for a ticket and $4 for snacks. Find the total cost for tickets and snacks.

9. The table shows the number of miles the Wong family drove each day on their vacation. How many more miles did they drive on day 1 than on day 4?

Day	Miles
1	345
2	50
3	89
4	279

Name ______________________________

MY Homework

Lesson 13

Problem Solving: Extra or Missing Information

Homework Helper

Need help? connectED.mcgraw-hill.com

Anna collected 50 cans for a food drive. She collected 10 cans each day of the drive, mostly green beans. How many days did she collect cans?

1 Understand

What facts do you know?

- I know that Anna collected 50 cans.
- I know that each day of the drive she collected 10 cans.

What do you need to find?

- I need to find the number of days she collected cans.

2 Plan

Determine if there is extra or missing information.

The information that Anna collected mostly green beans is not needed.

I have all the necessary information to solve the problem.

3 Solve

$50 \div 10 = 5$ days

So, Anna collected cans for 5 days.

4 Check

Is my answer reasonable? Explain.

$5 \times 10 = 50$ cans

My answer is reasonable.

Practice

Determine if there is extra or missing information. Then solve the problem, if possible.

1. Mrs. Blackwell gives each of her students two pencils. How many pencils did she hand out?

____

2. If David plays 3 tennis matches every week for 9 weeks, how many matches will he play altogether?

____

3. The Alvarez family bought a car for $2,000. They made a down payment of $500. If they want to pay the balance in 5 equal payments, how much will each of these payments be?

____

4. Mathematical PRACTICE 1 **Make Sense of Problems** Melanie runs 5 miles a day. How many miles will her brother run in one week?

____

____

5. Marco does 14 extra math problems each school night. How many extra problems does he do each week? There are 5 school nights each week.

____

Review

Chapter 3

Divide by a One-Digit Divisor

Vocabulary Check

Complete the crossword puzzle using the words in the word bank.

dividend	**divisor**	**fact family**
partial quotients	**quotient**	**remainder**

Across

1. the number that divides the dividend

2. a method of dividing where you break the dividend into sections that are easy to divide

3. the result of a division problem

4. a number that is being divided

Down

5. the number that is left after one whole number is divided by another

6. a group of related facts using the same numbers

Concept Check

Check

Divide. Use a related multiplication fact.

7. $14 \div 7 =$ ______

8. $40 \div 5 =$ ______

Find the quotient using a model.

9. $48 \div 3 =$ ______

Divide.

10. $3\overline{)93}$

11. $5\overline{)78}$

12. $2\overline{)47}$

Divide mentally.

13. $300 \div 3 =$ ______

14. $160 \div 8 =$ ______

Estimate. Show how you estimated.

15. $219 \div 2$ ______

16. $720 \div 7$ ______

17. $182 \div 8$ ______

Divide.

18. $5\overline{)625}$

19. $4\overline{)8{,}642}$

20. $2\overline{)533}$

Name

Problem Solving

21. Melanie swam 918 meters in 3 days. If she swam the same distance each day, how far did Melanie swim in one day?

22. Luke has 48 books. He puts 7 books in a box. How many boxes can he fill? Explain how you interpreted the remainder.

23. In 2 hours Cara read 48 pages. If she read the same number of pages each hour, how many pages did Cara read in one hour?

24. Three plane tickets to New York cost $2,472. If each plane ticket costs the same amount, about how much does one ticket cost?

25. Carmen has 468 trading cards in 4 binders. If each binder has the same number of cards, how many cards are in each binder?

Test Practice

26. A cabinet with 4 shelves can hold 1,640 CDs. If the shelves each hold the same number of CDs, how many CDs does each shelf hold?

Ⓐ 400 CDs
Ⓑ 410 CDs
Ⓒ 420 CDs
Ⓓ 424 CDs

Reflect

Chapter 3

Answering the ESSENTIAL QUESTION

Use what you learned about dividing by a one-digit divisor to complete the graphic organizer.

ESSENTIAL QUESTION

What strategies can be used to divide whole numbers?

Place Value

Properties

Relationship between Division and Multiplication

Now reflect on the ESSENTIAL QUESTION  **Write your answer below.**

Chapter

Divide by a Two-Digit Divisor

ESSENTIAL QUESTION

What strategies can I use to divide by a two-digit divisor?

Around My School

Watch a video!

MY Common Core State Standards

Number and Operations in Base Ten

5.NBT.6 Find whole-number quotients of whole numbers with up to four-digit dividends and two-digit divisors, using strategies based on place value, the properties of operations, and/or the relationship between multiplication and division. Illustrate and explain the calculation by using equations, rectangular arrays, and/or area models.

I'll be able to get this—no problem!

Standards for Mathematical PRACTICE

1. Make sense of problems and persevere in solving them.
2. Reason abstractly and quantitatively.
3. Construct viable arguments and critique the reasoning of others.
4. Model with mathematics.
5. Use appropriate tools strategically.
6. Attend to precision.
7. Look for and make use of structure.
8. Look for and express regularity in repeated reasoning.

= focused on in this chapter

Name ______________________________

Am I Ready?

← Go online to take the Readiness Quiz

Estimate each product. Tell whether the estimate is *greater than* or *less than* the actual product.

1. $224 × 12 = ______________________

2. 372 × 36 = ______________________

3. 488 × 85 = ______________________

4. 515 × 41 = ______________________

Multiply.

5. 14 × 3 = ________ **6.** 36 × 5 = ________ **7.** 76 × 4 = ________

8. The teacher purchased 13 packs of crayons. There are 24 crayons in each pack. How many crayons are there in all?

____

9. A movie was sold out for four straight days. If 535 tickets were sold each day, how many tickets were sold in all?

____

How Did I Do? **Shade the boxes to show the problems you answered correctly.**

1	2	3	4	5	6	7	8	9

Name ....

MY Math Words

Vocab

Review Vocabulary

dividend   divisor   quotient

Making Connections

Use the review vocabulary to label the second example in the first row. Then write 2 more real-world problems for the remaining examples. Label each example.

Real-World Problem	Show ÷	Show $\overline{)\ \ }$
Thirty-two students were asked to make 8 signs for a car wash. How many groups did the students work in, if each group made 1 sign and had an equal number of students?	32 ÷ 8 = 4	$8\overline{)32}$ with quotient 4
	48 ÷ 6 = 8	$6\overline{)48}$ with quotient 8
	45 ÷ 15 = 3	$15\overline{)45}$ with quotient 3

MY Vocabulary Cards

Vocab

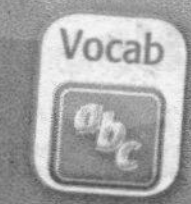

Mathematical PRACTICE

Ideas for Use

- Use the blank cards to write review vocabulary terms that relate to division of whole numbers.
- Write the name of each lesson on the front of each card. On the back, include a few study tips or important concepts to remember.

MY Foldable

FOLDABLES® Follow the steps on the back to make your Foldable.

Write the problem	Estimate	Divide	Check

FOLDABLES®
Study Organizer

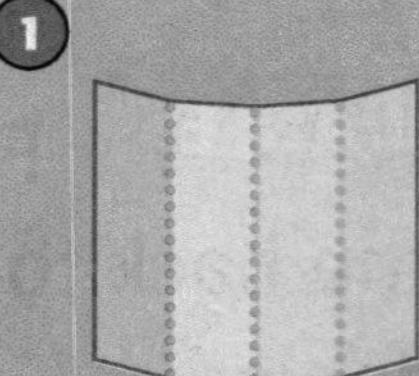

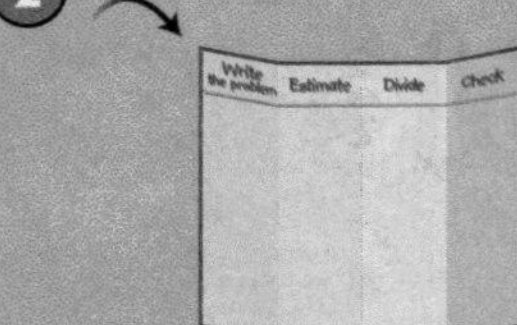
Write the problem.
Estimate
Divide
Check

Name ______________________________

Estimate Quotients

Lesson 1

ESSENTIAL QUESTION
What strategies can I use to divide by a two-digit divisor?

You can use rounding and compatible numbers to estimate quotients when dividing by two-digit divisors. By estimating first, you can determine the reasonableness of your results.

Math in My World

Example 1

The school principal has 812 fliers to pass out equally to 19 different teachers. About how many fliers would each teacher receive?

Estimate 812 ÷ 19.

1. Round the divisor to the nearest ten.

812 ÷ 19

2. Round the dividend to the nearest hundred.

812 ÷ ______

______ ÷ ______

3. Divide mentally.

______ ÷ ______ = ______

$8 \div 2 = 4$
$80 \div 20 = 4$
$800 \div 20 = 40$

So, each teacher would receive about ______ fliers.

Example 2

Estimate 234 ÷ 41.

1. Round the divisor to the nearest ten.

41 → ______

2. Change the dividend to a number that is compatible with the rounded divisor, ______.

234 → ______

3. Divide mentally.

______ ÷ ______ = ______

So, 234 ÷ 41 is about ______.

Check Use multiplication to check your answer.

______ × ______ = ______ and ______ ≈ 234

Helpful Hint

Another way to write 234 ÷ 41 is $41\overline{)234}$.

Talk MATH

Is it possible to have more than one estimate for a division problem? Explain. Give an example.

Guided Practice

1. Estimate 312 ÷ 31. Show how you estimated.

Round the divisor to the nearest ten.

31 → ______

Round the dividend to the nearest hundred.

312 → ______

Divide mentally.

______ ÷ ______ = ______

So, 312 ÷ 31 is about ______.

Name ....

Independent Practice

Estimate by rounding. Show how you estimated.

2. 121 ÷ 42

3. 400 ÷ 23

4. 642 ÷ 83

5. $28\overline{)597}$

6. $38\overline{)244}$

7. $24\overline{)943}$

Estimate using compatible numbers. Show how you estimated.

8. 653 ÷ 52

9. 208 ÷ 51

10. 300 ÷ 59

11. $32\overline{)619}$

12. $43\overline{)847}$

13. $34\overline{)272}$

Problem Solving

14. There are 598 goldfish divided equally among 23 fish tanks. About how many goldfish are in each tank?

15. The area of a rectangle is 138 square meters, and the length is 21 meters. About how many meters long is the width?

16. Mathematical PRACTICE 1 **Check for Reasonableness** A box of cereal contains 340 grams of carbohydrates. If there are 12 servings, about how many grams are there in one serving? Show how you estimated. Explain why your answer is reasonable.

HOT Problems

17. Mathematical PRACTICE 3 **Which One Doesn't Belong?** Circle the equation that is not a reasonable estimate for 533 ÷ 57.

540 ÷ 60 = 9	500 ÷ 50 = 10
550 ÷ 55 = 10	420 ÷ 60 = 7

18. **Building on the Essential Question** Explain when it would be useful to estimate.

MY Homework

Lesson 1
Estimate Quotients

Homework Helper

eHelp

Need help? connectED.mcgraw-hill.com

Estimate 304 ÷ 18.

1 Round the divisor to the nearest ten. 18 → 20

2 Change the dividend to a number that is compatible with the rounded divisor, 20. 304 → 300

3 Divide mentally. 300 ÷ 20 = 15

So, 304 ÷ 18 is about 15.

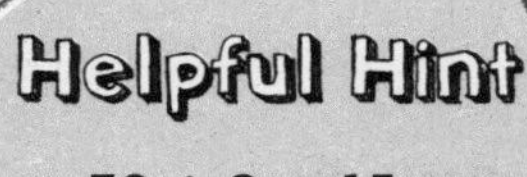

30 ÷ 2 = 15
300 ÷ 2 = 150
300 ÷ 20 = 15

Check Use multiplication to check your answer.
15 × 20 = 300 and 300 ≈ 304

Practice

Estimate. Show how you estimated.

1. 512 ÷ 52

2. 412 ÷ 97

3. 83$\overline{)237}$

4. 31$\overline{)458}$

Problem Solving

5. The Booster Club is having a bake sale. The members place 25 baked goods in each bag. There are 630 baked goods donated. About how many bags will the club have for sale?

6. A farmer has 212 acres of land for sale. He divides the land into 18 equal sections. About how many acres are in each section?

7. Alexander has 418 songs on his MP3 player. He divides the songs into 11 equal groups. About how many songs are in each group?

8. Mathematical PRACTICE 1 **Check for Reasonableness** A restaurant ordered 833 ounces of chicken. There are 16 ounces in a pound. About how many pounds of chicken did the restaurant order?

Test Practice

9. Ms. Robbins has 600 sheets of paper for a project. She has 48 students. Which is the best estimate for the number of sheets of paper she can give to each student?

Ⓐ 10 sheets

Ⓑ 12 sheets

Ⓒ 15 sheets

Ⓓ 20 sheets

Name ....

Hands On
Divide Using Base-Ten Blocks

Lesson 2

ESSENTIAL QUESTION
What strategies can I use to divide by a two-digit divisor?

Build It

Gary is saving up to buy a trumpet for band that costs $156. Suppose he saves the same amount each month for 12 months. How much money does he need to save each month?

Find 156 ÷ 12. Use base-ten blocks to find the quotient.

1 Model 156 using base-ten blocks.

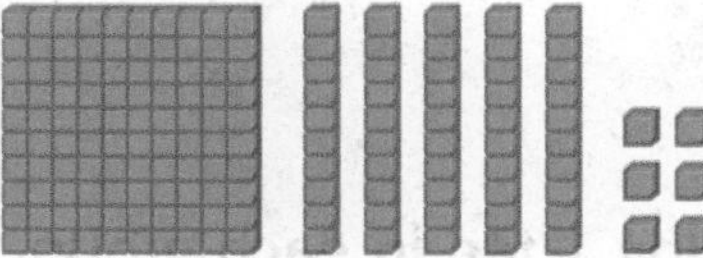

2 Since you can not separate the hundreds block into 12 groups, regroup it into tens.

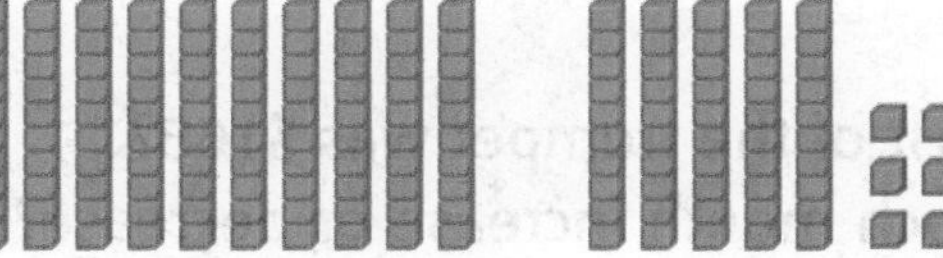

There are ________ tens total.

3 Divide the tens equally into 12 groups. Circle each group.

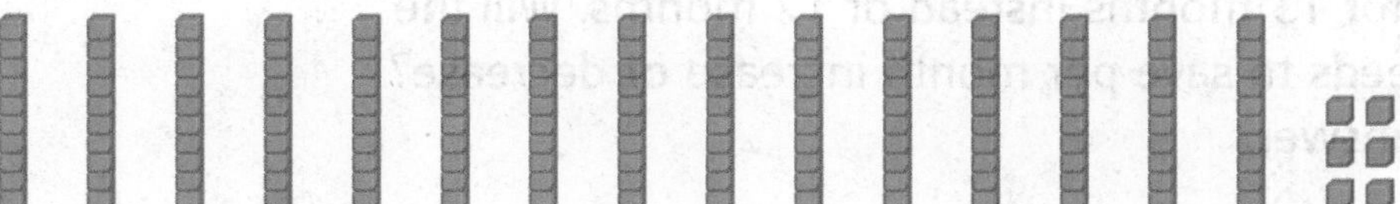

How many tens are in each group? ________

There are still ________ tens and ________ ones that need to be divided.

4 Use the remaining tens and ones and regroup them as ones. Then divide the ones equally into 12 groups. Draw the results in the work space below.

There are ________ ones.

How many ones are in each group? ________

Each group contains ________ ten and

________ ones, or ________.

My Work!

So, Gary needs to save ________ each month.

Check Use multiplication to check your answer.

________ × 12 = $156

Talk About It

1. In the activity, you started by placing 1 ten in each group. What will be the place value of the first digit of the quotient?

____

2. What would happen if the cost of the trumpet was $168? Would the amount to save each month increase or decrease?

____

3. Mathematical PRACTICE 3 **Justify Conclusions** Suppose Gary chose to save $156 for 13 months instead of 12 months. Will the amount he needs to save per month increase or decrease? Explain your answer.

____

____

____

Name

Practice It

Use models to find each quotient. Draw the equal groups.

4. 117 ÷ 13 = ______

5. 136 ÷ 17 = ______

6. 231 ÷ 11 = ______

7. 105 ÷ 15 = ______

Apply It

8. The average person eats an average of 26 pounds of bananas each year. How many years would it take a person to eat 104 pounds of bananas? Draw models to find the quotient.

9. A travel van holds 11 people. There are a total of 143 people signed up to take a trip to the zoo. How many vans are needed? Draw models to find the quotient.

10. Mathematical PRACTICE 1 **Plan Your Solution** Cheryl has a collection of sports cards and has 8 pages in the album. Each page of the album holds 14 sports cards. If she completely fills the album, how many total sports cards does she have? Draw models to find the dividend.

11. Mathematical PRACTICE 5 **Use Math Tools** The average person eats an average of 16 pounds of apples each year. How many years would it take a person to eat 144 pounds of apples? Draw models to find the quotient.

Write About It

12. How can base-ten blocks be used to divide by a two-digit divisor? Explain.

MY Homework

Lesson 2
Hands On: Divide Using Base-Ten Blocks

Homework Helper

Need help? connectED.mcgraw-hill.com

Joey has 120 stamps. He puts an equal number of stamps on each of the 10 pages in an album. How many stamps are on each page?

Find 120 ÷ 10. Use base-ten blocks to find the quotient.

1. Base-ten blocks are used to model 120.

2. Since you can not separate the hundreds block into 10 groups, it was regrouped into tens. There are 12 tens total.

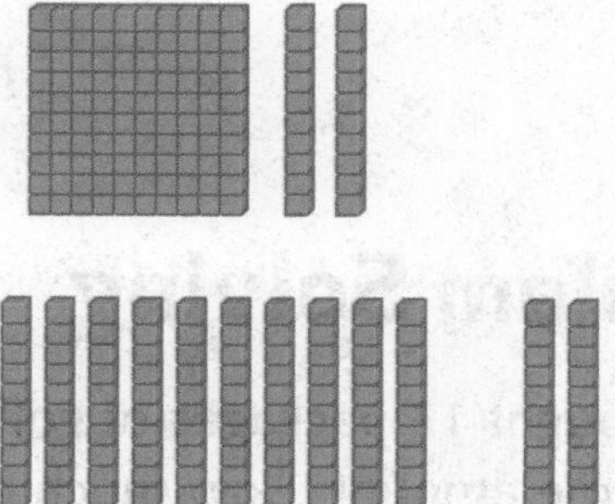

3. The tens were divided equally into 10 groups shown by the circles.
   There is 1 ten in each group.
   There are 2 tens that still need to be divided.

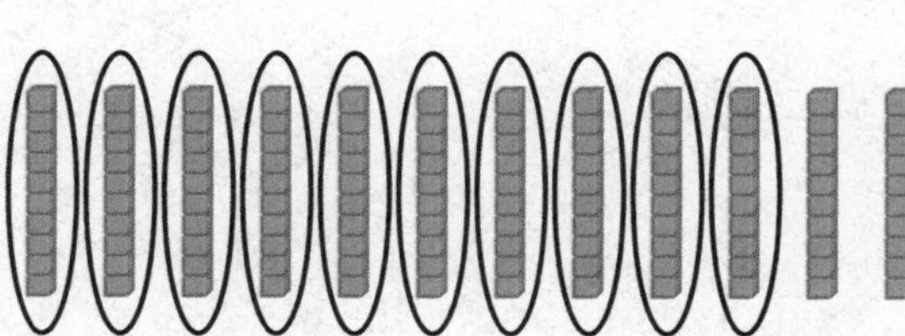

4. The remaining tens were regrouped as ones. They were divided equally into 10 groups.

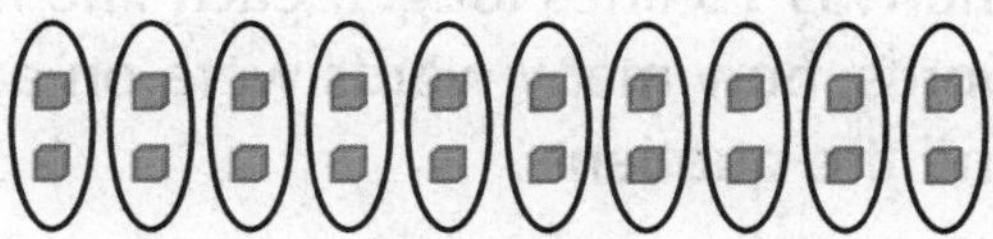

   There are 20 ones.

   There are 2 ones in each group.

   Each group contains 1 ten and 2 ones, or 12.

So, Joey can place 12 stamps on each page.

Check Use multiplication to check your answer. $12 \times 10 = 120$

Practice

Use models to find each quotient. Draw the equal groups.

1. $121 \div 11 =$ ______

2. $153 \div 17 =$ ______

Problem Solving

3. Braydon bought 13 packages of golf balls for $273. Each package cost the same amount. How much does one package cost? Draw models to find the quotient.

______

4. Desirée wrote an essay for school that had a total of 247 words and was 13 lines long. If each line had an equal number of words, how many words were on each line? Draw models to find the quotient.

______

5. Mathematical PRACTICE 5 **Use Math Tools** There were 242 fish in 22 fish tanks at the pet store. If there were an equal number of fish in each fish tank, how many fish were in each tank? Draw models to find the quotient.

______

Name ....

Divide by a Two-Digit Divisor

Lesson 3

ESSENTIAL QUESTION

What strategies can I use to divide by a two-digit divisor?

Math in My World

Example 1

The yearbook committee took 836 photos with a digital camera over 76 days. If they took an equal amount of photos each day, how many photos did they take each day? Check your answer for reasonableness.

SAY CHEESE!

Let p represent the number of photos taken each day. Write an equation to find the value of p.

______ $\div$ ______ $= p$

Estimate $800 \div 80 =$ ______ So, the first digit is in the tens place.

Divide the tens.

$83 \div 76 \approx 1$

Write 1 in the quotient over the tens place.

Multiply. $76 \times 1 = 76$

Subtract. $83 - 76 = 7$

Compare. $7 < 76$

76) 8 3 6

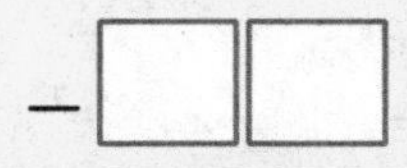

6

3 Bring down 6 ones. There are 76 ones in all.

4 Divide the ones.

$76 \div 76 = 1$

Write 1 in the quotient over the ones place.

$76 \times 1 = 76$

$76 - 76 = 0$

So, $836 \div 76 =$ ______. Since $p =$ ______, the yearbook committee will take ______ photos each day.

Check for Reasonableness ______ $\approx$ ______

Example 2 

Find 751 ÷ 30.

Estimate 750 ÷ 30 = ______

 Divide the tens.

75 ÷ 30 ≈ 2

Write 2 in the quotient over the tens place.

 Multiply. 30 × 2 = 60

Subtract. 75 − 60 = 15

Compare. 15 < 30

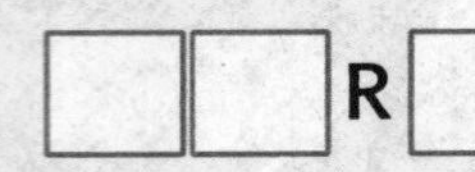

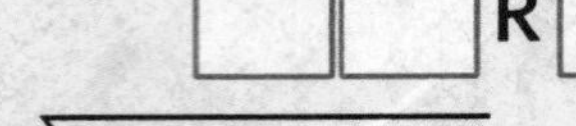

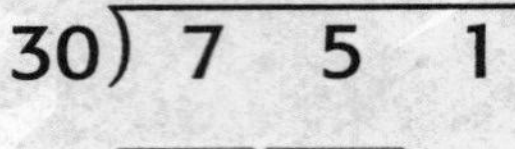

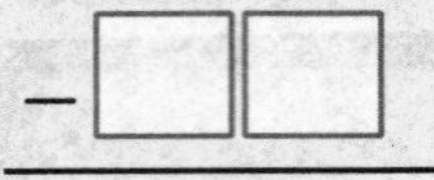

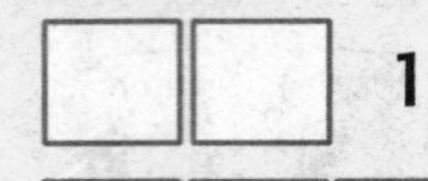

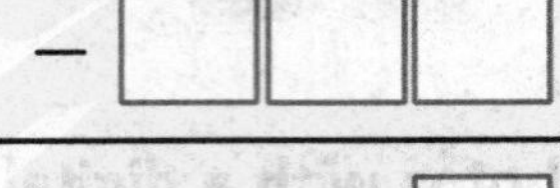

 Bring down 1 one.

There are 151 ones in all.

 Divide the ones.

151 ÷ 30 ≈ 5

Write 5 in the quotient over the ones place.

30 × 5 = 150

151 − 150 = 1

So, 751 ÷ 30 is ______ R ______.

Check for Reasonableness ______ ≈ ______  R

Talk MATH

Explain how estimation is used to help you place the first digit in the quotient.

Guided Practice 

1. Find 176 ÷ 16. **Estimate** 180 ÷ 20 = ______

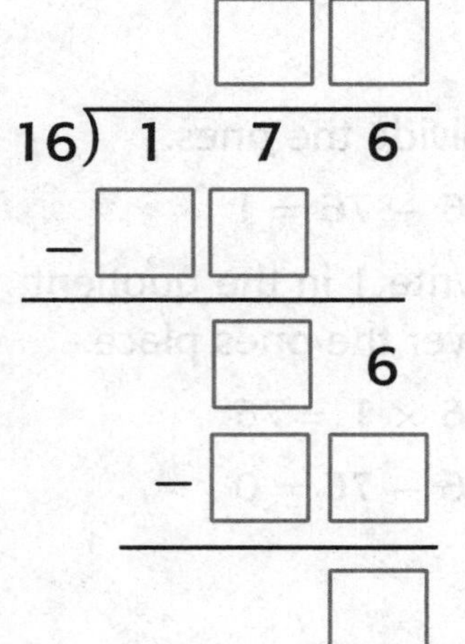

So, 176 ÷ 16 = ______.

Check for Reasonableness ______ ≈ ______

Name

Independent Practice

Divide. Check for reasonableness.

2. $809 \div 62 =$ ______

3. $925 \div 42 =$ ______

4. $210 \div 15 =$ ______

5. $27\overline{)837}$

6. $34\overline{)594}$

7. $12\overline{)155}$

8. $29\overline{)790}$

9. $18\overline{)416}$

10. $42\overline{)624}$

Algebra **Divide to find the variable in each equation.**

11. $840 \div 24 = h$

$h =$ ______

12. $528 \div 12 = b$

$b =$ ______

13. $952 \div 28 = w$

$w =$ ______

Problem Solving

14. Mr. Calzada buys flags for his store. Each flag costs $28. How many flags can he buy for $350? What does the remainder represent?

15. The area of a rectangle is 384 square feet and the width is 24 feet. Find the length.

16. Dreanne uploads 292 pictures to her online album. Her online album shows 12 pictures on each page. How many pages does she scroll through to see all 292 pictures? What does the remainder represent?

17. Mathematical PRACTICE 7 **Identify Structure** In 31 days, Gwen's dog sleeps 496 hours. If she sleeps the same number of hours each night, how many hours does she sleep per night? Find the unknown number in the equation $496 \div 31 = h$.

HOT Problems

18. Mathematical PRACTICE 1 **Make a Plan** Write a division problem that has a two-digit quotient that is greater than 40 but less than 50.

19. **Building on the Essential Question** What is a standard procedure for dividing by a two-digit divisor? Explain.

Name ....

MY Homework

Lesson 3

Divide by a Two-Digit Divisor

Homework Helper

Need help? connectED.mcgraw-hill.com

Find 204 ÷ 12.

Estimate $200 \div 10 = 20$

1 Divide the tens.

$20 \div 12 \approx 1$

Write 1 in the quotient over the tens place.

2 Multiply. $12 \times 1 = 12$

Subtract. $20 - 12 = 8$

Compare. $8 < 12$

$$\begin{array}{r} 17 \\ 12\overline{)204} \\ -\,12 \\ \hline 84 \\ -\,84 \\ \hline 0 \end{array}$$

3 Bring down 4 ones. There are 84 ones in all.

4 Divide the ones.

$84 \div 12 = 7$

Write 7 in the quotient over the ones place.

$12 \times 7 = 84$

$84 - 84 = 0$

So, $204 \div 12$ is 17.

Check for Reasonableness $20 \approx 17$

Practice

Divide. Check for reasonableness.

1. $874 \div 23 =$ ______

2. $988 \div 96 =$ ______

3. $58\overline{)940}$

Problem Solving

My Work!

4. **Algebra** Find the variable in the equation $803 \div 73 = m$.

$m =$ ________

5. **Algebra** Find the variable in the equation $988 \div 76 = d$.

$d =$ ________

6. **Mathematical PRACTICE 2** **Stop and Reflect** Members of the Bladerunners skating club collected $950 from fundraising activities. They want to buy Ultrablade skates, which are $48 a pair. How many pairs of skates can they buy? What does the remainder represent?

____

____

____

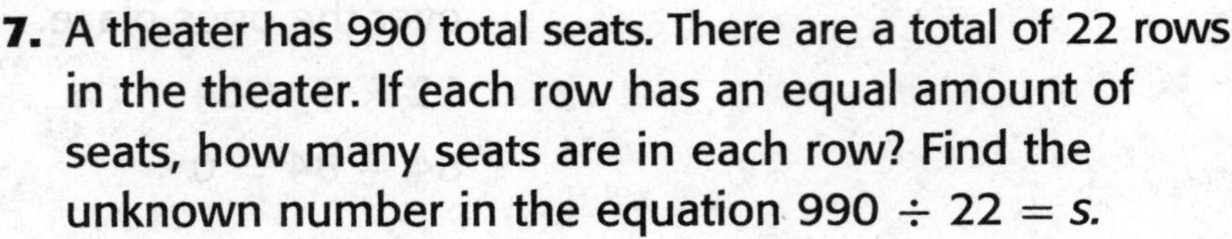

7. A theater has 990 total seats. There are a total of 22 rows in the theater. If each row has an equal amount of seats, how many seats are in each row? Find the unknown number in the equation $990 \div 22 = s$.

____

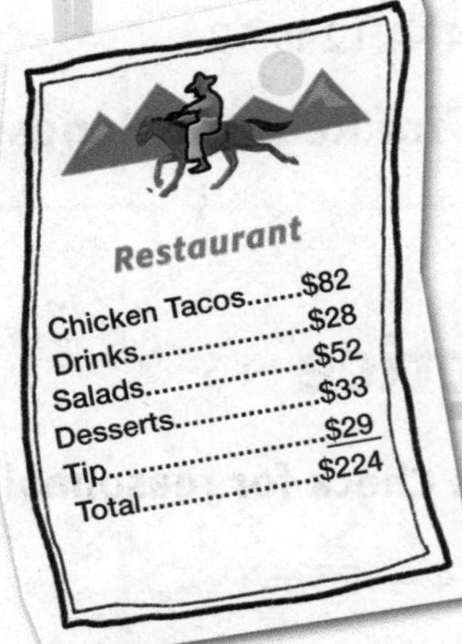

8. A group of friends equally split the dinner bill shown. If each person paid $16, how many friends paid?

____

Test Practice

9. Emily loves to read. She read 527 hours in 31 weeks. If she read an equal number of hours each week, how many hours did she read each week?

Ⓐ 10 hours Ⓒ 15 hours

Ⓑ 14 hours Ⓓ 17 hours

Check My Progress

Vocabulary Check

Draw a line to match each vocabulary term with the correct definition.

1. compatible numbers	• a number close to an exact value which indicates about how much
2. remainder	• the result of a division problem
3. dividend	• the number that is left after one whole number is divided by another
4. estimate	• numbers in a problem that are easy to work with mentally
5. quotient	• a number that is being divided

Concept Check

Estimate. Show how you estimated.

6. $43\overline{)412}$

7. $81\overline{)637}$

8. $595 \div 28$

9. $22\overline{)311}$

Divide. Check for reasonableness.

10. 528 ÷ 16 = ________

11. 821 ÷ 22 = ________

12. $14\overline{)634}$

13. $18\overline{)954}$

Problem Solving

My Work!

14. Algebra Find the variable in the equation 975 ÷ 39 = *k*.

k = ________

15. A movie theater has 576 seats arranged in 36 equal rows. How many seats are in each row?

____

16. The Mitchell family is making payments on the refrigerator to the right. If they pay $41 every month, how many months will it take to pay for the refrigerator?

____

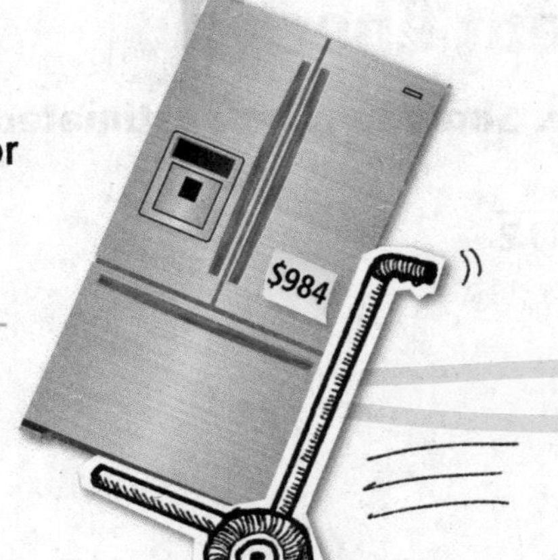

Test Practice

17. Heather bought a package of construction paper that holds 525 pieces with 15 different colors. If there are the same number of pieces per color, how many pieces of each color are there?

Ⓐ 540 Ⓒ 35

Ⓑ 50 Ⓓ 25

Adjust Quotients

Lesson 4

ESSENTIAL QUESTION
What strategies can I use to divide by a two-digit divisor?

When you estimate which digit to place in the quotient, your estimate might be too small or too large. So, you need to adjust the quotient.

Math in My World

Example 1

During lunch, there were 144 students in the cafeteria. The cafeteria has a total of 16 tables. How many students can sit at each table?

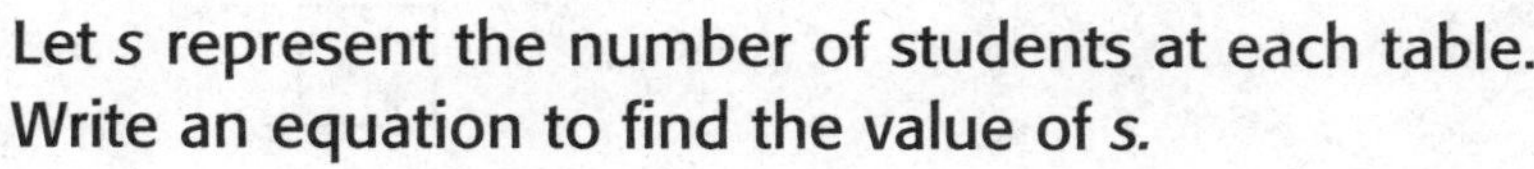

Let *s* represent the number of students at each table. Write an equation to find the value of *s*.

______ ÷ ______ $= s$

Estimate by using compatible numbers. $140 \div 20 =$ ______

Try the estimate.

16) 1 4 4
– 1 1 ☐
☐☐

Since $32 > 16$, the estimated digit is too low.

Adjust. Try 8.

16) 1 4 4
– 1 ☐☐
☐☐

Since $16 = 16$, the estimated digit is too low.

Adjust again. Try 9.

16) 1 4 4
– 1 ☐☐
☐

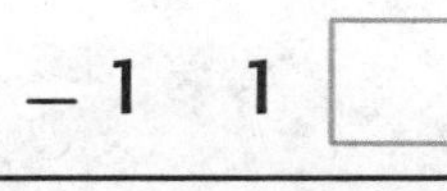

So, $144 \div 16 =$ ______. Since $s =$ ______, ______ students can sit at each table.

Check for Reasonableness ______ ≈ ______

Example 2 

Find 1,252 ÷ 32.

1 Estimate by using compatible numbers.

1,252 ÷ 32

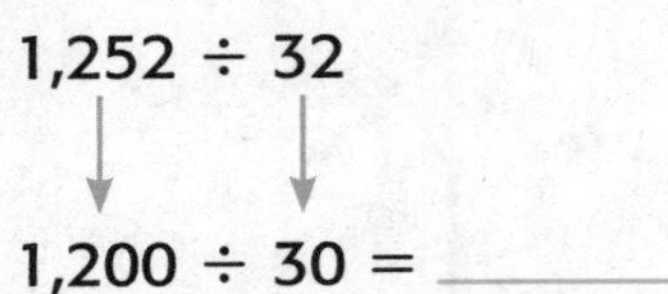

1,200 ÷ 30 = ______

2 Try the estimate.

32) 1, 2 5 2

Since $128 > 125$, the estimated digit is too high.

3 Adjust. Try 3.

32) 1, 2 5 2

$29 < 32$
Continue dividing.

4 Bring down the 2 ones. Try 9.

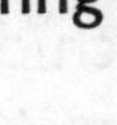 R

32) 1, 2 5 2

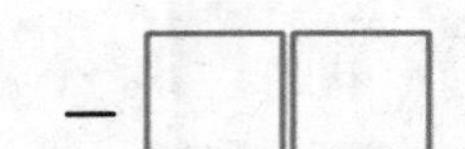

2

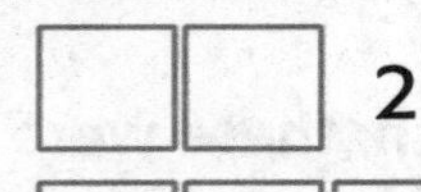

$4 < 32$

So, 1,252 ÷ 32 = ______ R ______.

Check for Reasonableness ______ R ______ ≈ ______

Guided Practice 

1. Sara divided 306 by 34 and got a quotient of 8 R34. Explain and correct her error.

______

______

______

______

Name

Independent Practice

Divide. Check each answer.

2. 1,272 ÷ 53 = ______

3. 548 ÷ 62 = ______

4. 5,243 ÷ 71 = ______

5. 115 ÷ 23 = ______

6. 1,728 ÷ 72 = ______

7. 183 ÷ 19 = ______

8. $57\overline{)413}$

9. $34\overline{)242}$

10. $64\overline{)2{,}712}$

Mathematical PRACTICE 2 **Use Algebra Divide to find the variable in each equation.**

11. 328 ÷ 41 = *m*

m = ______

12. 4,536 ÷ 81 = *w*

w = ______

13. 735 ÷ 15 = *x*

x = ______

Problem Solving

14. Sheila arranged a total of 680 chairs for a school assembly. If she placed an equal amount of chairs in 20 rows, how many chairs are in each row?

____

15. Given the area of a rectangle is 208 square inches, and the length is 26 inches, find the width.

____

16. A crew went net fishing to catch shrimp. They caught 486 shrimp in 54 minutes. How many shrimp did they catch per minute? Find the unknown number in the equation $486 \div 54 = s$.

____

HOT Problems

17. Mathematical PRACTICE 3 **Find the Error** Emma estimated the first digit in the quotient of $2{,}183 \div 42$ as 4. She adjusted the quotient to 3. What did she do wrong? Explain.

$$\begin{array}{r} 4 \\ 42\overline{)2{,}183} \\ -168 \\ \hline 50 \end{array} \quad 50 > 42$$

I'll try 3.

____

____

18. **Building on the Essential Question** How can I adjust a quotient to solve a division problem?

____

____

Name

MY Homework

Lesson 4
Adjust Quotients

Homework Helper

Need help? connectED.mcgraw-hill.com

Find 238 ÷ 62.

1 Estimate by using compatible numbers.

$$238 \div 62$$
$$240 \div 60 = 4$$

2 Try the estimate.

$$\begin{array}{r} 4 \\ 62\overline{)238} \\ -\,248 \end{array}$$

Since 248 > 238, the estimated digit is too high.

3 Adjust. Try 3.

$$\begin{array}{r} 3 \text{ R52} \\ 62\overline{)238} \\ -\,186 \\ \hline 52 \end{array}$$

52 < 62

So, 238 ÷ 62 = 3 R52.

Check for Reasonableness $3 \text{ R52} \approx 4$

Practice

Divide. Check each answer.

1. $48\overline{)1{,}261}$

2. $86\overline{)1{,}204}$

3. 428 ÷ 61 = ______

Algebra **Divide to find the variable in each equation.**

4. $140 \div 28 = t$

$t =$ ______

5. $2,075 \div 83 = c$

$c =$ ______

6. $531 \div 59 = n$

$n =$ ______

Problem Solving

My Work!

7. **Mathematical PRACTICE 2** **Use Algebra** Woodfern School has a raffle each year to raise money for the music program. The raffle needs to sell 1,500 tickets. How many raffle ticket sellers are needed if each seller sells 75 tickets? Find the unknown number in the equation $1,500 \div 75 = t$.

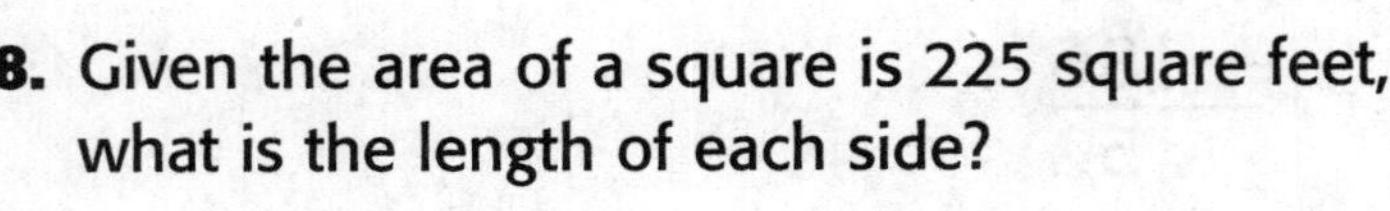

______

8. Given the area of a square is 225 square feet, what is the length of each side?

______

9. The Rodriquez family is taking a 1,430-mile train ride. If the train travels 55 miles per hour, how many hours will the ride last?

______

Are we THERE YET?

Test Practice

10. Alaska has the longest coastline in the United States. Traveling at 60 miles per hour, how many hours would it take to travel along the Pacific Coast?

Alaska Coastline	
Coast	**Miles**
Pacific	5,580
Arctic	1,060

Ⓐ 18 hours

Ⓑ 23 hours

Ⓒ 93 hours

Ⓓ 103 hours

Name ....

Divide Greater Numbers

Lesson 5

ESSENTIAL QUESTION
What strategies can I use to divide by a two-digit divisor?

Math in My World

Example 1

A large city has a total of 22,500 students that ride a bus to school. There are 75 different schools within the city. How many students are dropped off at each school if an equal number of students are dropped off at each school?

Let *s* represent the number of students dropped off at each school. Write an equation to find the value of *s*.

_______ ÷ _______ = *s*

Place the first digit.

225 ÷ 75 = 3

Write 3 in the quotient over the hundreds place.

Multiply. 75 × 3 = 225

Subtract. 225 − 225 = 0

Compare. 0 < 75

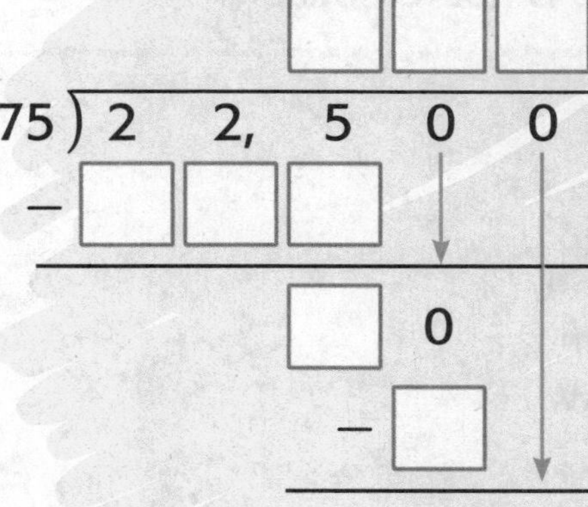

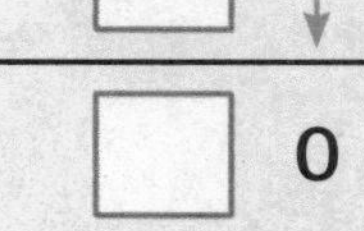

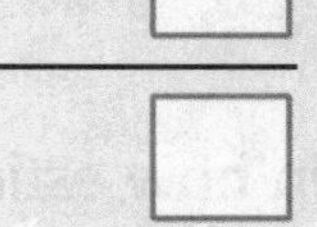

Divide the tens.

0 ÷ 75 = 0

75 × 0 = 0

0 − 0 = 0

0 < 75

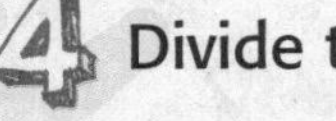

Divide the ones.

0 ÷ 75 = 0

75 × 0 = 0

0 − 0 = 0

0 < 75

So, 22,500 ÷ 75 = _______. Since *s* = _______, _______ students are dropped off at each school.

Example 2

Estimate the quotient of 46,534 and 152. Then divide. Is 36 a reasonable quotient? Explain.

Estimate 45,000 ÷ 150 = ______

1 Place the first digit.

465 ÷ 152 ≈ 3

Write 3 in the quotient over the hundreds place.

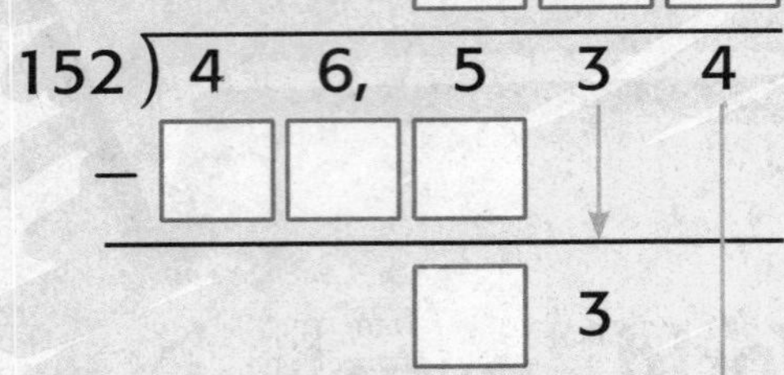

152) 4 6, 5 3 4

2 Multiply.

152 × 3 = 456

Subtract. 465 − 456 = 9

Compare. 9 < 152

3 Divide the tens.

Ninety-three is not divisible by 152, so put a 0 in the quotient over the tens place.

4 Divide the ones.

934 ÷ 152 ≈ 6

152 × 6 = 912

934 − 912 = 22

22 < 152

Check Since the estimate is ______ and the actual quotient is ______, a quotient of 36 is not reasonable.

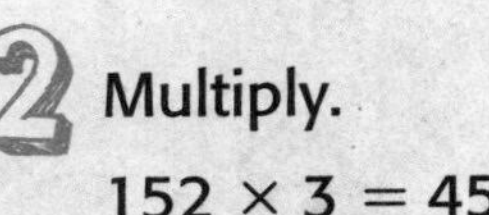

Guided Practice

1. Find the missing number in the division problem below.

```
      1,■12
25)47,800
  − 25
    22 8
  − 22 5
       30
     − 25
       50
     − 50
        0
```

■ = ______

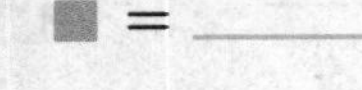

Explain how estimation can be used before, during, and after a division problem.

Name ____________________

Independent Practice

Estimate. Then divide. Check for reasonableness.

2. $51\overline{)91{,}988}$

3. $17\overline{)14{,}637}$

4. $64\overline{)15{,}489}$

5. 36,712 ÷ 52 = ______

6. 43,803 ÷ 93 = ______

7. 26,208 ÷ 28 = ______

8. $42\overline{)25{,}435}$

9. $89\overline{)85{,}978}$

10. $783\overline{)52{,}056}$

Algebra **Divide to find the variable in each equation.**

11. 39,788 ÷ 812 = *y*

y = ______

12. 25,696 ÷ 352 = *g*

g = ______

13. 36,557 ÷ 263 = *d*

d = ______

Real World Problem Solving

14. A farmer plows a corn field in the shape of a rectangle that has an area of 15,840 square yards. If the length of the field is 132 yards, what is the width of the field?

____

15. An average person speaks 35,000 words in one week. Does the average person speak more or less than 2,500 words per day? Find the unknown number in the equation $35{,}000 \div 7 = w$.

____

16. Mathematical PRACTICE 2 **Use Number Sense** The athletic department raised $14,500 to buy new football uniforms. If each uniform costs $258, how many uniforms can they buy? Explain how you interpreted the remainder.

____

____

HOT Problems

17. Mathematical PRACTICE 3 **Draw a Conclusion** Find the unknown number in the equation $30{,}672 \div q = 852$. Explain how you found the unknown.

____

18. **Building on the Essential Question** How can I divide greater numbers using a standard procedure?

____

____

Name ____________________

MY Homework

Lesson 5
Divide Greater Numbers

Homework Helper

eHelp

Need help? connectED.mcgraw-hill.com

At a recent rock concert, $28,440 was made from front-row ticket sales. If the cost of a single ticket was $72, how many people purchased front-row tickets?

Find 28,440 ÷ 72.

Estimate 28,000 ÷ 70 = 400

1 Place the first digit.

284 ÷ 72 ≈ 3

Write 3 in the quotient over the hundreds place.

```
     395
72)28,440
  − 21 6
     6 84
   − 6 48
      360
    − 360
        0
```

2 Multiply. 72 × 3 = 216

Subtract. 284 − 216 = 68

Compare. 68 < 72

3 Divide the tens.

684 ÷ 72 ≈ 9

72 × 9 = 648

684 − 648 = 36

36 < 72

4 Divide the ones.

360 ÷ 72 = 5

72 × 5 = 360

360 − 360 = 0

0 < 72

So, 395 people purchased front-row tickets.

Check Since the estimate is 400, the answer of 395 is reasonable.

Practice

Estimate. Then divide. Check for reasonableness.

1. 21,312 ÷ 36 = ______ **2.** 76,912 ÷ 92 = ______ **3.** 26,878 ÷ 89 = ______

Problem Solving

4. Given the area of a rectangle is 14,628 square millimeters, and the width is 12 millimeters, find the length.

____

5. Turtles and tortoises have long life spans. A tortoise can live for 54,750 days. How many years can a tortoise live? (*hint*: 365 days = 1 year)

____

6. Mathematical PRACTICE 2 **Use Algebra** The new baseball stadium holds 64,506 people. There are 26 gates where people enter the ballpark. The same number of people entered each gate. How many people entered the first gate? Find the unknown number in the equation $64{,}506 \div 26 = p$.

____

Test Practice

7. Joshua works for a computer company at an annual salary of $38,480. He receives 26 equal paychecks during the year. How much does he receive in each paycheck?

Ⓐ $1,370 Ⓒ $1,525

Ⓑ $1,480 Ⓓ $1,560

Name

Real World

Problem-Solving Investigation

STRATEGY: Solve a Simpler Problem

Lesson 6

ESSENTIAL QUESTION

What strategies can I use to divide by a two-digit divisor?

Learn the Strategy

Miguel earns $50 each week. He spends $10 each week and saves the remaining amount. How many weeks will it take until he has saved more than $300?

1 Understand

What facts do you know?

I know that Miguel earns ________ each week, but spends ________ of it.

What do you need to find?

I need to find how long it will take Miguel to save ________.

2 Plan

I can solve the problem by solving a simpler problem.

3 Solve

Miguel saves $50 − $10, or ________ each week.

Divide 300 ÷ 40. Since there is a remainder,

Miguel will need to save for at least ________ weeks.

$$\begin{array}{r} 7\ \text{R}20 \\ 40\overline{)300} \\ -\,280 \\ \hline 20 \end{array}$$

4 Check

Is my answer reasonable? Explain.

Yes. ________ × 8 = ________, and ________ is greater than $300.

Practice the Strategy

Melissa estimates that she watched 104 movies over the past year. On average, about how many movies does she watch per month? (*hint*: 1 year = 52 weeks)

That's awesome!

Understand

What facts do you know?

____

What do you need to find?

____

2 Plan

____

3 Solve

4 Check

Is my answer reasonable? Explain.

____

Name

Apply the Strategy

Solve each problem by solving a simpler problem.

My Work!

1. Mr. Santiago has a flight from New York to Paris that covers a distance of 3,640 miles in 7 hours. If the plane travels at the same speed per hour, how many miles will it have traveled after 4 hours?

2. Josh watches 720 television shows in one year. If he watches the same number of shows each month, how many shows does he watch in 5 months?

3. The football team is raising money to have a new turf field installed. The cost of the turf field is $48,780. The team has 18 months to raise the money. If they raise an equal amount each month, how much will they have raised after one year?

4. Mathematical PRACTICE 1 **Keep Trying** The Clearview Window Washing Company has a contract to wash 3,082 windows on a 23-story building. The company will wash the windows on the first three floors on the first day. If there are the same number of windows on each floor, how many windows will the company wash on the first day?

5. Mr. Thomas is delivering bricks to a construction site. He delivers the same number of bricks with each delivery. The builder has ordered 3,096 bricks and it will take 8 trips to deliver all the bricks. How many bricks will Mr. Thomas have delivered after 7 trips?

Review the Strategies

Use any strategy to solve each problem.

- Solve a simpler problem.
- Determine extra or missing information.
- Make a table.
- Use the four-step plan.

6. An aquarium at a pet store has 18 Black Neon Tetra fish in it. A customer buys 12 Black Neon Tetra fish at the same time the store clerk adds 7 more Black Neon Tetra fish to the tank. How many Black Neon Tetra fish are in the aquarium now?

____

My Work!

7. Kolby read 108 books in one year. If he reads the same number of books each month, how many books does he read in 5 months?

____

8. Mr. Reyes baked 4 batches of muffins for his class. Each batch had 12 muffins. If Mr. Reyes has 24 students, how many muffins will each student receive?

____

9. The quotient of two numbers is 20. Their sum is 84. What are the 2 numbers?

____

10. Mathematical PRACTICE 4 **Model Math** Charity and her friend each want to buy a piece of pizza, a drink, and an ice cream cone. Charity has $10 to pay for her and her friend's meal. Does she have enough money? Explain.

Menu	
Pizza	$3.00
Drink	$1.00
Ice cream cone	$2.00

____

We all scream for ice cream!

Name ....

MY Homework

Lesson 6
Problem Solving: Solve a Simpler Problem

Homework Helper

Need help? connectED.mcgraw-hill.com

Leslie wants to plant flowers in a rectangular garden that has a length of 6 feet and a width of 5 feet. If a tree takes up a square that is 2 feet on each side in the middle of the garden, how much area is left to plant the flowers?

1 Understand

What facts do you know?

I know the length and width of the garden and the space the tree takes up.

What do you need to find?

I need to find the area of the garden remaining for flowers.

2 Plan

You can solve the problem by solving a simpler problem.

3 Solve

Find the area of the entire garden. (*hint*: Area = length × width)

$6 \times 5 = 30$ square feet

Find the area of the square the tree takes up.

$2 \times 2 = 4$ square feet

Subtract the area of the tree from the area of the garden.

$30 - 4 = 26$ square feet

So, Leslie can plant 26 square feet of flowers.

4 Check

Is my answer reasonable? Explain.

area of flower garden + area of tree = total area

26 square feet + 4 square feet = 30 square feet

Problem Solving

Solve each problem by solving a simpler problem.

1. Harold stores 10,656 pounds of grain for his 24 cattle. Each cow eats about 12 pounds of grain each day. How many days would the food supply last?

____

2. Mathematical PRACTICE 1 **Make Sense of Problems** The Pave the Way Company is installing a new walkway. The installer uses 13 paving stones for every foot of walkway. He needs 117 paving stones to complete the project and has used 78 paving stones so far. How many more feet of paving stones will he need to install to finish?

____

3. Harvey collects rare marbles. He has 112 marbles in his collection. Each tray holds an equal number of marbles. If he uses 7 trays to hold all of the marbles, how many marbles will he display on 3 trays?

____

4. Noelle's rectangular deck has a length of 18 feet and a width of 16 feet. If a hot tub takes up a square that is 4 feet on each side in the middle of the deck, how much area is left of the deck?

____

5. Hilary wants to go on the Latin Club trip to Italy. It will cost $2,730 for the trip. The trip is 30 weeks away and she wants to make equal weekly payments. How much money altogether does Hilary need to pay at the end of week 8?

____

My Work!

Name ....

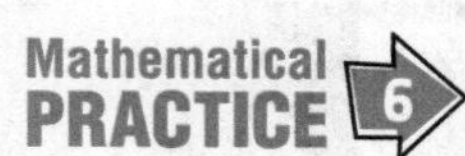

Divide.

1. $12\overline{)432}$ **2.** $40\overline{)850}$ **3.** $29\overline{)493}$ **4.** $18\overline{)535}$

5. $32\overline{)105}$ **6.** $62\overline{)897}$ **7.** $72\overline{)504}$ **8.** $53\overline{)689}$

9. $21\overline{)126}$ **10.** $37\overline{)452}$ **11.** $76\overline{)684}$ **12.** $51\overline{)209}$

13. $67\overline{)938}$ **14.** $41\overline{)868}$ **15.** $34\overline{)646}$ **16.** $15\overline{)879}$

17. $10\overline{)329}$ **18.** $98\overline{)784}$ **19.** $39\overline{)975}$ **20.** $47\overline{)660}$

Name

Fluency Practice

Divide.

1. $13\overline{)32{,}708}$ **2.** $76\overline{)81{,}928}$ **3.** $54\overline{)65{,}775}$ **4.** $83\overline{)72{,}628}$

5. $23\overline{)22{,}890}$ **6.** $61\overline{)61{,}427}$ **7.** $90\overline{)65{,}881}$ **8.** $38\overline{)89{,}148}$

9. $15\overline{)92{,}460}$ **10.** $43\overline{)27{,}998}$ **11.** $27\overline{)61{,}752}$ **12.** $87\overline{)50{,}286}$

13. $216\overline{)16{,}200}$ **14.** $557\overline{)60{,}156}$ **15.** $318\overline{)80{,}779}$ **16.** $861\overline{)89{,}550}$

17. $396\overline{)88{,}704}$ **18.** $221\overline{)82{,}889}$ **19.** $562\overline{)55{,}103}$ **20.** $902\overline{)92{,}004}$

Review

Chapter 4

Divide by a Two-Digit Divisor

Vocabulary Check

Match each word to its definition. Write your answers on the lines provided.

1. compatible numbers ______

2. dividend ______

3. divisor ______

4. quotient ______

5. remainder ______

6. round ______

A. To find the approximate value of a number.

B. A number that is being divided.

C. Numbers in a problem that are easy to compute mentally.

D. The number that is left after one whole number is divided by another.

E. The number that divides the dividend.

F. The result of a division problem.

Estimate. Show how you estimated.

7. $74\overline{)634}$ **8.** $49\overline{)311}$ **9.** $38\overline{)409}$

Divide. Check each answer.

10. $32\overline{)928}$ **11.** $23\overline{)345}$ **12.** $53\overline{)753}$

13. 926 ÷ 71 = ______ **14.** 126 ÷ 17 = ______ **15.** 478 ÷ 93 = ______

Estimate. Then divide. Check for reasonableness.

16. 85,120 ÷ 76 = ______ **17.** 54,184 ÷ 26 = ______ **18.** 25,600 ÷ 25 = ______

19. $61\overline{)37{,}520}$ **20.** $41\overline{)16{,}859}$ **21.** $53\overline{)75{,}578}$

Name

Problem Solving

22. Tito purchased 11 tickets to a baseball game for $374. About how much did one ticket cost?

____

23. Tonya saved $564 in one year. If she saved the same amount each month, how much did Tonya save in two months?

____

24. Bennett is placing 104 photographs in the school yearbook. He will put the same number of photos on each of the 13 pages. If he can put 4 pictures in a row, how many rows will be on each page?

____

25. Trevon is making payments on a computer that costs $1,548. If he makes 12 equal payments, how much is three of his payments?

____

Test Practice

26. A jet ski costs $3,204. If you want to make 36 equal payments, how much is each payment?

Ⓐ $79 Ⓒ $89

Ⓑ $88 Ⓓ $99

Reflect

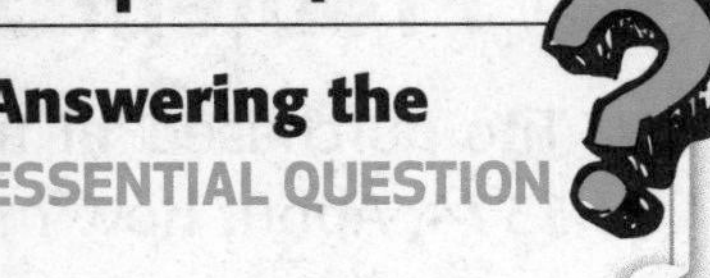

Use what you learned about dividing whole numbers to complete the graphic organizer.

Write the Example

Real-World Example

Estimate

ESSENTIAL QUESTION
What strategies can I use to solve division problems with whole numbers?

Vocabulary

Now reflect on the ESSENTIAL QUESTION  **Write your answer below.**

Chapter

5 Add and Subtract Decimals

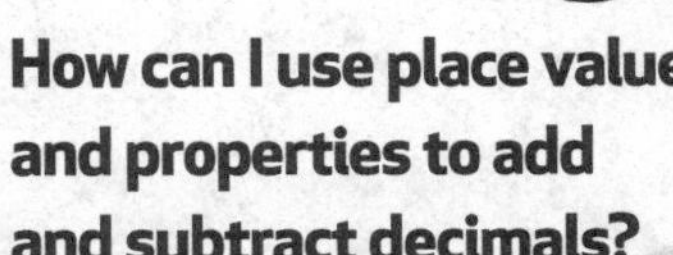

ESSENTIAL QUESTION

How can I use place value and properties to add and subtract decimals?

Let's Explore Technology!

Watch

Watch a video!

MY Common Core State Standards

Number and Operations in Base Ten

5.NBT.4 Use place value understanding to round decimals to any place.

5.NBT.7 Add, subtract, multiply, and divide decimals to hundredths, using concrete models or drawings and strategies based on place value, properties of operations, and/or the relationship between addition and subtraction; relate the strategy to a written method and explain the reasoning used.

I'll be able to get this - no problem!

Standards for Mathematical PRACTICE

1. Make sense of problems and persevere in solving them.
2. Reason abstractly and quantitatively.
3. Construct viable arguments and critique the reasoning of others.
4. Model with mathematics.
5. Use appropriate tools strategically.
6. Attend to precision.
7. Look for and make use of structure.
8. Look for and express regularity in repeated reasoning.

= focused on in this chapter

Name ....

Am I Ready?

← Go online to take the Readiness Quiz

Name the place-value position of each highlighted digit.

1. 52 ____________ 2. 138 ____________ 3. 4.3 ____________

4. 901 ____________ 5. 1.216 ____________ 6. 2,785 ____________

Round each number to the highlighted place.

7. 19 = ________ 8. 681 = ________ 9. 735 = ________

10. 3,705 = ________ 11. 106,950 = ________ 12. 5,750 = ________

Add.

13. 38 + 16 = ________ 14. 151 + 218 = ________

15. 260 + 398 = ________ 16. 235 + 68 = ________

17. The Pham family and the Weber family have many pets. How many more pets does the Pham family have than the Weber family?

____

Pets	
Pham	**Weber**
3 dogs	2 dogs
1 cat	3 gerbils
6 fish	1 turtle

How Did I Do?

Shade the boxes to show the problems you answered correctly.

1	2	3	4	5	6	7	8	9	10	11	12	13	14	15	16	17

Name

MY Math Words

Vocab

Review Vocabulary

greater than (>) less than (<) equal to (=)

Making Connections

Compare the numbers in each row. Use the review vocabulary to compare the two numbers in each row using >, <, or =.

Greater Than, Less Than, or Equal To

57.2		57.02
12.01		12.1
12.6		12.60
24.56		24.5
6.99		6.89

Describe how you used place value to complete the chart.

MY Vocabulary Cards

Mathematical PRACTICE

Lesson 5-7

Associative Property of Addition

$(15.19 + 25.05) + 88 =$
$15.19 + (25.05 + 88)$

Lesson 5-7

Commutative Property of Addition

$15.9 + 8.42 = 8.42 + 15.9$

Lesson 5-7

Identity Property of Addition

$12.7 + 0 = 12.7$

Lesson 5-10

inverse operations

$1.73 - 0.87 = 0.86$

$0.87 + 0.86 = 1.73$

Ideas for Use

- Write a tally mark on each card every time you read the word in this chapter or use it in your writing. Challenge yourself to use at least 10 tally marks for each card.
- Use the blank cards to write your own vocabulary cards.

The order in which numbers are added does not change the sum.

Commutative means "something that involves substitution." How does this definition relate to the Commutative Property?

The way in which numbers are grouped does not change the sum.

Compare the example on this card with the example on the Commutative Property card. What differences do you notice?

Operations that undo each other.

What are the inverse operations for addition and division?

The sum of any number and 0 equals the number.

Explain how one of the three properties in this chapter can help you add mentally.

MY Foldable

FOLDABLES® Follow the steps on the back to make your Foldable.

Estimate

1.29 + 1.55 = ______

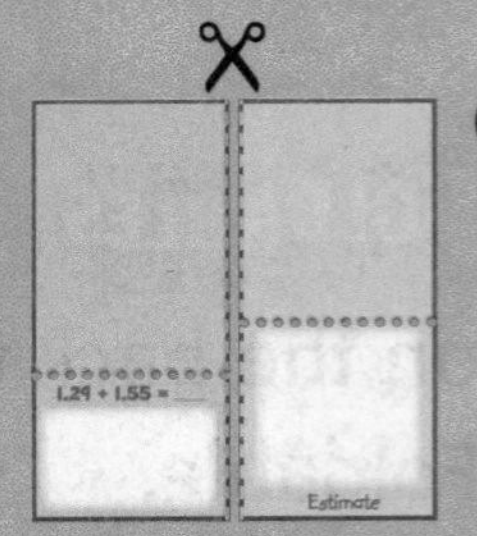

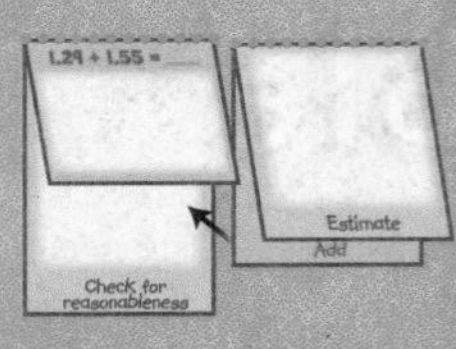

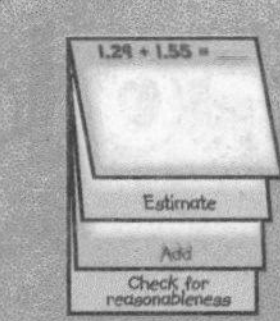

Check for Reasonableness

Add

Round Decimals

Lesson 1

ESSENTIAL QUESTION
How can I use place value and properties to add and subtract decimals?

Recall that numbers with digits in the tenths place, hundredths place, or beyond are called decimals. When you round a decimal, you find its approximate value.

Math in My World

Example 1

Andrew's laptop has a processor with a speed of 2.8 gigahertz. Round the processing speed of the laptop to the nearest whole number.

Use a number line to round 2.8 to the nearest whole number.

1. Draw 10 equal increments between 2 and 3 on the number line.

(number line from 2 to 3)

2. Place a dot at 2.8 and label it.

Determine if 2.8 is closer to 2 or 3.

There are 2 equal increments between 2.8 and 3. There are 8 equal increments between 2 and 2.8.

2.8 is closer to ______.

So, round 2.8 to ______.

Key Concept Rounding Decimals

- Underline the digit in the place to which you want to round.
- Look at the digit to its right. If the digit is 4 or less, keep the underlined digit. If the digit is 5 or greater, round the underlined digit up to the next greater digit.
- Drop the digit to the right of the underlined digit.

Example 2

Round 46.73 to the nearest tenth.

1. Underline the digit in the tenths place, ______.

46.73

2. Look at the digit to the right of 7.

3 is ______ than 5

3. So, keep the underlined digit. Drop the digit to its right.

46.73 → 46.___

So, 46.73 rounds to ______.

Guided Practice

1. Round 8.74 to the nearest one. Underline the digit in the ones place.

8.74

Look at the ______, the digit to the right of 8.

So, 8.74 rounds to ______.

Name

Independent Practice

Round each decimal to the place indicated.

2. 5.476; hundredths

3. 983.625; hundredths

4. 28.6; ones

5. 4.35; tenths

6. 110.079; hundredths

7. 67.142; ones

8. 1.8; ones

9. 7.358; hundredths

10. 48.32; ones

11. 9.045; tenths

12. 19.25; ones

13. 8.17; tenths

Problem Solving

14. What is the length of the 10-dollar bill to the nearest whole number?

15. The weight of a new touch screen device is 1.6 pounds. What is the weight to the nearest whole number?

My Work!

HOT Problems

16. Mathematical PRACTICE 1 **Keep Trying** Write two different numbers that when rounded to the nearest tenth will give you 18.3.

17. Mathematical PRACTICE 2 **Use Number Sense** Explain what happens when you round 9,999.999 to any place.

18. **Building on the Essential Question** In what real-world situations would you want to round a number?

Name ....

MY Homework

Lesson 1
Round Decimals

Homework Helper

Need help? connectED.mcgraw-hill.com

An ice sheet that covers most of Antarctica is about 1.34 miles thick. To the nearest tenth of a mile, how thick is the ice? Round 1.34 to the nearest tenth.

1. The digit in the tenths place is 3.

1.34

2. Look at the digit to the right of 3. 4 is less than 5

3. So, keep the 3 and drop the digit to its right, 4.

1.34 → 1.3

So, the thickness of the ice rounded to the nearest tenth is 1.3 miles.

Practice

Round each decimal to the place indicated.

1. 5.476; hundredths ______

2. 4.35; tenths ______

3. 1.8; ones ______

4. 0.79; ones ______

5. 1.049; hundredths ______

6. 17.92; tenths ______

Real World

Problem Solving

Use the information in the table to solve Exercises 7–9. Round each number to the place indicated.

Place	Area (square miles)
Florida	65,754.59
Georgia	59,424.77
Alabama	52,419.02
South Carolina	32,020.20

7. What is the area of Florida rounded to the nearest tenth?

8. What is the area of South Carolina rounded to the nearest tenth?

9. What is the area of Georgia rounded to the nearest square mile?

10. The African bush elephant weighs between 4.4 tons and 7.7 tons. What are its least and greatest weights, rounded to the nearest ton?

11. Mathematical PRACTICE 6 **Be Precise** The price of a gallon of milk is $3.75. Circle the price of a gallon of milk rounded to the nearest dollar.

$3.70 $3.75 $4.00 $3.80

Test Practice

12. Holly bought blank CDs and spent $34.57. What is that amount rounded to the nearest dollar?

Ⓐ $35
Ⓑ $34.60
Ⓒ $34.50
Ⓓ $34

Name ....

Estimate Sums and Differences

Lesson 2

ESSENTIAL QUESTION
How can I use place value and properties to add and subtract decimals?

One way to estimate is to use rounding. If you round numbers to a lesser place value, you are likely to get an estimate that is closer to the exact answer.

Math in My World

Example 1

Hadassah used a digital thermometer to find the morning temperature and afternoon temperature. She found the morning temperature was 31.3°F and the afternoon temperature was 37.6°F. Estimate the difference in average temperatures.

One Way **Round to the nearest ten.**

Round 37.6 to the nearest ten.

37.6 → ________

Round 31.3 to the nearest ten.

31.3 → ________

Subtract.

________ − ________ = ________

Another Way **Round to the nearest one.**

Round 37.6 to the nearest one.

37.6 → ________

Round 31.3 to the nearest one.

31.3 → ________

Subtract.

________ − ________ = ________

The difference is about ________°F or about ________°F.

The actual difference is 6.3°F. So, rounding to the nearest ________ gave the more accurate estimate.

Tutor

Example 2

Estimate 5.26 + 1.93 by rounding to the nearest one.

5.26 → ________

1.93 → ________

Add.

________ + ________ = ________

The sum is about ________.

Helpful Hint

You will learn how to find the actual sum in a later lesson. The actual sum is 7.19, so 7 is a fairly accurate estimate.

Guided Practice

Round each decimal to the nearest one. Then add or subtract.

1. 2.8 + 1.3

Round to the nearest one.

2.8 → ________

1.3 → ________

Add.

________ + ________ = ________

So, 2.8 + 1.3 is about ________.

2. 5.98 − 1.03

Round to the nearest one.

5.98 → ________

1.03 → ________

Subtract.

________ − ________ = ________

So, 5.98 − 1.03 is about ________.

Describe a real-world example when it might be appropriate to estimate rather than get the exact answer.

Name ....

Independent Practice

Round each decimal to the nearest one. Then add or subtract.

3. 10.08 + 5.6 = ________

4. 10.4 + 32.8 = ________

5. \$42.01 − \$5.92 = ________

6. 75.2 + 82.3 = ________

7. $\begin{array}{r} 1.509 \\ +\ 3.106 \\ \hline \end{array}$

8. $\begin{array}{r} 8.058 \\ -\ 3.181 \\ \hline \end{array}$

9. $\begin{array}{r} 3.872 \\ +\ 1.249 \\ \hline \end{array}$

Round each decimal to the nearest ten. Then add or subtract.

10. 23.78 + 10.45 = ________

11. 83.69 − 55.41 = ________

12. $\begin{array}{r} 37.58 \\ -\ 21.25 \\ \hline \end{array}$

13. $\begin{array}{r} 32.56 \\ +\ 6.7 \\ \hline \end{array}$

14. $\begin{array}{r} 25.21 \\ -\ 12.47 \\ \hline \end{array}$

Problem Solving

Solve Exercises 15–17 by rounding to the nearest one.

15. The weights of Marisa and Toni's televisions are shown in the table. About how much more does Marisa's television weigh than Toni's?

Student	Television Weight (lb)
Marisa	52.7
Toni	43.9

____

16. Mathematical PRACTICE 4 **Model Math** Sophia has $20. She buys a hair band for $3.99, gum for $1.29, and a brush for $6.75. Not including tax, estimate how much change she should receive. Show your work.

____

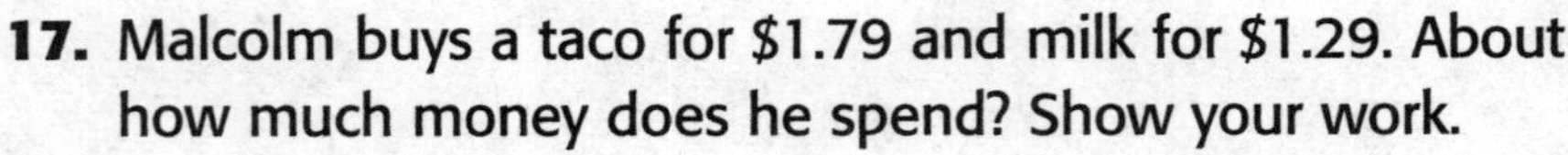

17. Malcolm buys a taco for $1.79 and milk for $1.29. About how much money does he spend? Show your work.

____

HOT Problems

18. Mathematical PRACTICE 3 **Find the Error** Kim wants to estimate 5.494 + 1.108 by first rounding to the nearest hundredth. Find her mistake and correct it.

5.494	→	5.50
+ 1.108	→	+ 1.11

____

19. **Building on the Essential Question** When is estimating an effective tool?

____

____

Name ....

MY Homework

Lesson 2
Estimate Sums and Differences

Homework Helper

eHelp

Need help? connectED.mcgraw-hill.com

The results of a recent skateboard competition are shown. About how many more points did Steamer have than Dal Santo?

Athlete	Points
Elissa Steamer	87.83
Marisa Dal Santo	81.50
Amy Caron	80.00

Round each decimal to the nearest ten. Then subtract.

Steamer → 87.83 → 90
Dal Santo → 81.50 → − 80
10

So, Steamer scored about 10 more points than Dal Santo.

Practice

Round each decimal to the nearest one. Then add or subtract.

1. $3.85 − $2.17

2. 52.85 − 9.09

3. 19.83 + 9.93

4. 3.872 + 2.409 = ________

5. 9.086 − 2.419 = ________

Problem Solving

Solve Exercises 6–9 by rounding to the nearest one.

6. The table shows the average speeds of two airplanes in miles per hour. About how much faster is the Foxbat than the Hawkeye? Show your work.

Plane	Speed (mph)
Hawkeye	375.52
Foxbat	1,864.29

____

7. Aluminum and tin are both metals. The standard atomic weight for aluminum is 26.98. The standard atomic weight for tin is 118.71. Estimate the difference between the standard atomic weights of these two metals. Show your work.

____

8. Lorena and her cousin are fishing at the lake. They caught two largemouth bass. One fish weighs 71.27 ounces and the other fish weighs 38.86 ounces. Estimate the total weight of the two fish. Show your work.

____

9. Mathematical PRACTICE 4 **Model Math** The table shows the lengths of four trails at a horseback riding camp. Estimate the total length of all the trails. Show your work.

Trail	A	B	C	D
Length (mi)	2.6	1.8	4.2	3.3

____

Test Practice

10. Mr. Dixon bought a whiteboard that was on sale for $1,989.99. The regular price was $2,499.89. Which is the best estimate of the amount of money Mr. Dixon saved by buying the whiteboard on sale?

Ⓐ $500 Ⓒ $3,000

Ⓑ $1,000 Ⓓ $4,000

Name

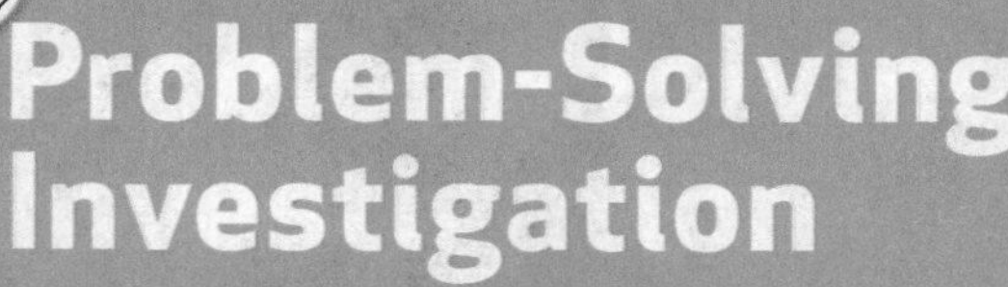

Problem-Solving Investigation

STRATEGY: Estimate or Exact Answer

Lesson 3

ESSENTIAL QUESTION
How can I use place value and properties to add and subtract decimals?

Learn the Strategy

Mrs. Trainer needs to purchase a classroom response clicker and a new calculator. A clicker costs $31.99 and a calculator costs $11.75. About how much money will it cost to buy both?

1 Understand

What facts do you know?

Mrs. Trainer needs to buy a clicker for $31.99 and a calculator for $11.75.

What do you need to find?

about how much it will _______ to buy the clicker and calculator

2 Plan

You can _______ the cost of the clicker and calculator.

3 Solve

Round the costs of the clicker and calculator to the nearest whole dollar.

$31.99 ⟶ _______ $11.75 ⟶ _______

Next, add to find the estimated total cost.

_______ + _______ = _______

So, Mrs. Trainer will need to spend about _______ for the two items.

4 Check

Is my answer reasonable? Explain.

Estimate another way. $30 + $10 = _______

_______ ≈ _______

Practice the Strategy

Madison drove to her grandparents' house. They drove 58.6 miles in the first hour, 67.2 miles in the second hour, and 60.5 miles in the third hour. They followed the same route to return home. About how far did Madison's family travel?

1 Understand

What facts do you know?

What do you need to find?

---

2 Plan

---

3 Solve

4 Check

Is my answer reasonable? Explain.

Name

Apply the Strategy

Determine whether you need an estimate or an exact answer to solve each problem. Then solve.

My Work!

1. A restaurant can make 95 dinners each night. The restaurant has been sold out for 7 nights in a row. How many dinners were sold during this week?

2. Mathematical PRACTICE 1 **Plan Your Solution** A gardener has 35 feet of fencing to enclose the garden shown. About how much fencing will be left over after the garden is enclosed?

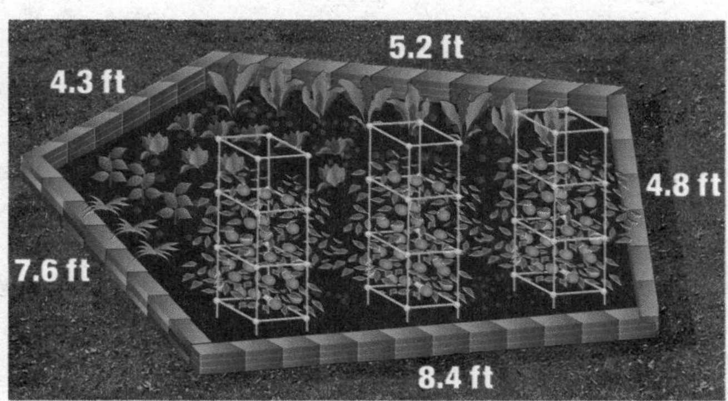

3. A family is renting a cabin for $59.95 a day for 3 days. About how much will they pay for the cabin?

4. Four friends ordered two pizzas. The cost of each pizza is $13.80. About how much will it cost for both pizzas?

5. Ann bought two shirts for $28.95 each and a skirt for $33.95. The sales tax was $3.71. About how much did she pay altogether?

Review the Strategies

Use any strategy to solve each problem.

- Determine an estimate or exact answer.
- Make a table.
- Solve a simpler problem.
- Determine extra or missing information.

6. **Mathematical PRACTICE 1 Make Sense of Problems** Students at a high school filled out a survey. The results showed that out of 640 students, 331 speak more than one language. How many students speak only one language?

7. Pablo is dividing a dish of brownies for a party. The brownie mix costs $2.89. He cuts the brownies into squares that measure 3 inches by 3 inches. If the pan is 18 inches long and 12 inches wide, how many brownies did he cut?

My Work!

8. Evita has 9 quarters, 7 dimes, and 5 nickels. Does she have enough money to buy a box of crayons for $3.25?

9. The fisherman with the longest fish wins a fishing competition. About how much longer is the first-place fish than the third-place fish?

Place	Fish Length (cm)
1st	68.7
2nd	59.8
3rd	58.2

10. Tammi is planning to buy a video game system for $310. Each month she doubles the amount she saved the previous month. If she saves $10 the first month, in how many months will Tammi have enough money to buy the video game system?

Name

MY Homework

Lesson 3
Problem Solving: Estimate or Exact Answer

Homework Helper

Need help? connectED.mcgraw-hill.com

Each school raffle ticket costs \$8.50. The Jordan family purchased 3 tickets. About how much did it cost the Jordan family to buy the raffle tickets?

1 Understand

What facts do you know?

Each ticket costs \$8.50. The Jordan family purchased 3 tickets.

What do you need to find?

about how much it will cost for all 3 tickets

2 Plan

You can estimate to find the total cost.

3 Solve

Round \$8.50 to the nearest whole dollar.

$\$8.50 \longrightarrow \$9$

Next, add to find the estimated total cost.

$\$9 + \$9 + \$9 = \$27$

So, the Jordan family spent about \$27 to purchase the raffle tickets.

4 Check

Is my answer reasonable? Explain.

Estimate another way.

$\$10 + \$10 + \$10 = \$30$

$\$30 \approx \$27$

Problem Solving

Determine whether you need an estimate or an exact answer to solve each problem. Then solve.

My Work!

1. A library wants to buy a new computer that costs $989.99. So far, the library has collected $311.25 in donations. About how much more money does the library need to buy the computer?

____

____

2. On Friday, a museum had 185 visitors. On Saturday, there were twice as many visitors as Friday. On Sunday, 50 fewer people visited than Saturday. How many people visited the museum during these three days?

____

3. Tomás orders a meal that costs $7.89. Lisa's meal costs $9.05. About how much is the combined cost of their meals?

____

4. Mathematical PRACTICE 1 **Make Sense of Problems** The ski team has a race in 3 hours and it is 120 miles away. The team drives 50 miles each hour. Will they arrive at the race on time?

____

5. At the beginning of last year, there were 368 students at the elementary school. By the beginning of this year, 72 of those students had moved. About how many students started the school year this year?

____

Check My Progress

Vocabulary Check

Draw a line to match each word to its correct place in the number below.

1. ones **2. hundreds** **3. hundredths** **4. tens** **5. tenths**

1 2 3 . 4 5

Concept Check

Round each decimal to the place indicated.

6. 8.067; hundredths ________

7. 178.03; tens ________

8. 7.48; ones ________

9. 26.138; tenths ________

Round each decimal to the nearest one. Then add or subtract.

10. 32.9 + 17.2 = ________

11. 7.91 + 21.9 = ________

12.
$$\begin{array}{r} 48.21 \\ -\ 12.64 \\ \hline \end{array}$$

13.
$$\begin{array}{r} 107.14 \\ -\ 76.87 \\ \hline \end{array}$$

14.
$$\begin{array}{r} 6.239 \\ -\ 1.750 \\ \hline \end{array}$$

15.
$$\begin{array}{r} 3.902 \\ +\ 2.017 \\ \hline \end{array}$$

Problem Solving

For Exercises 16–19, determine whether you need an estimate or an exact answer. Then solve.

16. Book World receives 12 boxes of books. Each box contains 16 copies of the new best-seller, *Norton's Last Laugh.* How many copies of *Norton's Last Laugh* does the store receive?

____

17. Students at Tuscan School filled out a survey. The survey showed that of 374 students, 195 speak a second language. How many students speak only one language?

____

18. Pensacola, Florida, gets an average of 64.28 inches of precipitation per year. Key West, Florida, gets an average of 38.94 inches per year. About how many more inches of precipitation does Pensacola get than Key West?

____

19. It costs Marc $29.75 a week to feed his dog. About how much does it cost him to feed his dog for 3 weeks?

____

____

Test Practice

20. The average price of gasoline in a recent year was $2.38. Which is the price per gallon of gasoline to the nearest tenth of a dollar?

Ⓐ $2.00　　Ⓒ $2.38

Ⓑ $2.30　　Ⓓ $2.40

Name ....

Hands On
Add Decimals Using Base-Ten Blocks

Lesson 4

ESSENTIAL QUESTION
How can I use place value and properties to add and subtract decimals?

Build It

Ruben downloaded two ringtones to his cell phone. One cost him $1.30 and the other one was discounted to $0.50. How much did it cost for both ringtones?

Find $1.30 + $0.50. Use base-ten blocks.

1 **Model 1.3 and 0.5.**

	Ones	Tenths	Hundredths
1.3 →	[1 hundred flat]	[3 tenth rods]	
0.5 →		[5 tenth rods]	

2 **Combine the base-ten blocks. Draw the result.**

How many tenths are there? ________

$1.30 + $0.50 = ________

So, Ruben spent ________ for both ringtones.

Helpful Hint
3 tenths + 5 tenths = 8 tenths

My Drawing!

Check Use estimation to check for reasonableness.

$1 + $1 = $2 and ________ ≈ $2

Try It

Find 1.42 + 0.87. Use base-ten blocks.

Model 1.42 and 0.87.

Ones	Tenths	Hundredths

My Drawing!

Combine the base-ten blocks.
Since there are 12 tenths, you need to regroup. You can regroup 12 tenths as 1 whole and 2 tenths. Draw the result.

How many ones are there? ______

How many tenths are there? ______

How many hundredths are there? ______

So, 1.42 + 0.87 = ______.

Check Use estimation to check for reasonableness.

1 + 1 = 2 and ______ ≈ 2

Talk About It

1. Mathematical PRACTICE 2 **Reason** How is adding decimals similar to adding whole numbers?

______

______

______

Name ....

Practice It

Add. Use base-ten blocks. Draw each result in the table.

2. 0.3 + 0.4 = ______

Ones	Tenths	Hundredths

3. 2.4 + 0.5 = ______

Ones	Tenths	Hundredths

4. 1.52 + 0.37 = ______

Ones	Tenths	Hundredths

5. 3.71 + 1.53 = ______

Ones	Tenths	Hundredths

6. 1.2 + 0.9 = ______

Ones	Tenths	Hundredths

7. 2.71 + 0.45 = ______

Ones	Tenths	Hundredths

Apply It

8. Mathematical PRACTICE 5 **Use Math Tools** Nathan ran 2.35 miles on Tuesday and 3.15 miles on Thursday. What total distance did he run? Use base-ten blocks to help you solve.

9. Valerie had $1.87 in her bank at home. She adds $2.67 in change. How much does she have now?

10. Mathematical PRACTICE 3 **Find the Error** Tom used base-ten blocks to find 1.72 + 1.77. Find his mistake and circle the correct sum.

Ones	Tenths	Hundredths

Which is the correct sum?

2.29 **3.14**

3.29 **3.49**

Write About It

11. How can you use base-ten blocks to add decimals?

Name ....

MY Homework

Lesson 4
Hands On: Add Decimals Using Base-Ten Blocks

Homework Helper

Need help? connectED.mcgraw-hill.com

Jessica purchased a cup of hot chocolate for $1.29 and a granola bar for $1.55. How much did it cost for both items?

Find $1.29 + $1.55. Use base-ten blocks.

1 Model 1.29 and 1.55.

Ones	Tenths	Hundredths

2 Combine the base-ten blocks.

Since there are 14 hundredths, you need to regroup. You can regroup 14 hundredths as 1 tenth and 4 hundredths. The model shows the result.

Ones	Tenths	Hundredths

There are 2 ones, 8 tenths, and 4 hundreths.

So, the total for the hot chocolate and granola bar was $2.84.

Check Use estimation to check for reasonableness.

$1 + $2 = $3 and $2.84 ≈ $3

Practice

Add. Use base-ten blocks. Draw each result in the table.

1. 1.83 + 0.36 = ______

Ones	Tenths	Hundredths

2. 3.1 + 1.34 = ______

Ones	Tenths	Hundredths

Problem Solving

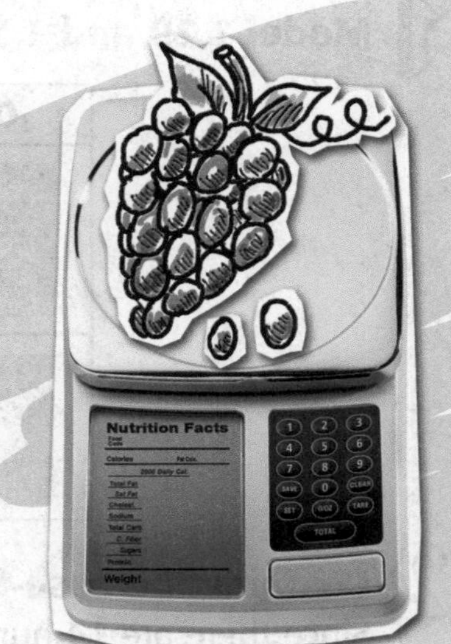

3. Lucas had 2.3 pounds of grapes left over from his class party. The class ate 1.9 pounds of the grapes. How many pounds of grapes did Lucas buy? Use models to find the sum.

______

4. Mathematical **PRACTICE** 3 **Justify Conclusions** Janice wanted to download the following items for her cell phone: A ringtone for $3.10, a picture for $1.95 and a game for $2.05. She has $8. Not including tax, does Janice have enough money to buy all three items? Explain.

______

5. Kelli had $0.55 in her pocket to buy a snack after school. Renee had $1.64 in her pocket. They decided to share a snack. How much can they spend on a snack altogether? Use models.

______

Name ____________________

Hands On
Add Decimals Using Models

Lesson 5
ESSENTIAL QUESTION
How can I use place value and properties to add and subtract decimals?

Build It

Find 1.2 + 0.7. Use models to find the sum.

Use 10-by-10 grids to model 1.2.

To show 1.2, shade one whole 10-by-10 grid and two-tenths of a second grid.

Two-tenths is equal to ________ small squares.

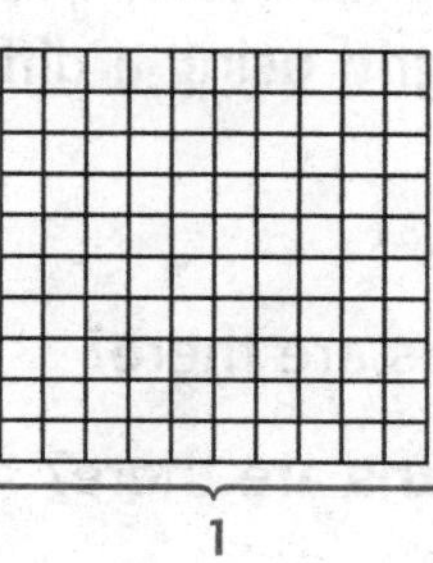

1

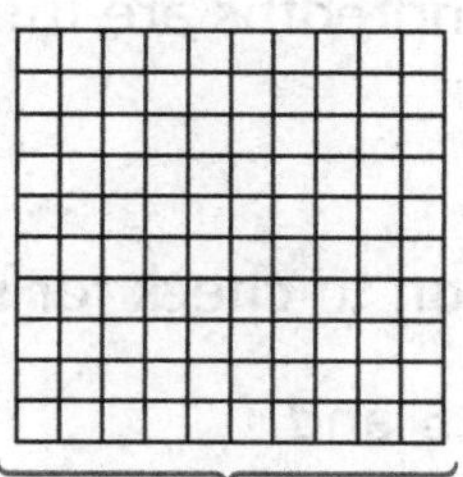

0.2 + 0.7

Model 0.7.

To show 0.7, shade seven-tenths of the second grid using a different color.

3 Add.

How many ones are there? ________

How many tenths are there? ________

So, 1.2 + 0.7 = ________.

Check Use estimation to check for reasonableness.

1 + 1 = 2 and ________ ≈ 2

Try It

Find 1.08 + 0.45. Use models.

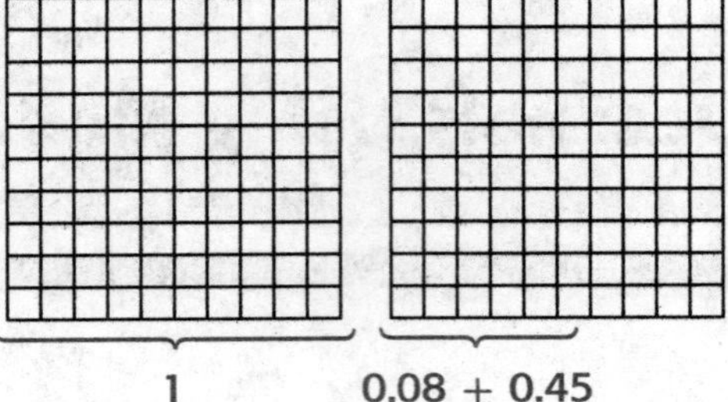

1 Model 1.08.

To show 1.08, shade one whole 10-by-10 grid and eight-hundredths of a second grid.

Eight-hundredths is equal to

________ small squares.

Model 0.45.

To show 0.45, shade forty-five hundredths of the second grid using a different color.

Helpful Hint

1 one, 5 tenths, and 3 hundredths is equal to 1.53.

3 Add.

How many ones are there? ________

How many tenths are there? ________

How many hundredths are there? ________

So, 1.08 + 0.45 = ________.

Check Use estimation to check for reasonableness.

1 + 0.5 = 1.5 and ________ ≈ 1.5

Talk About It

1. Explain how to use grid paper to model 0.2 and 0.02. Describe any differences.

____

____

____

2. Mathematical PRACTICE 4 **Model Math** What decimal is modeled by a 10-by-10 grid with 82 small squares shaded? Explain.

____

Name ....

CCSS

Practice It

Add. Shade the decimal models.

3. 2.46 + 1.13 = ________

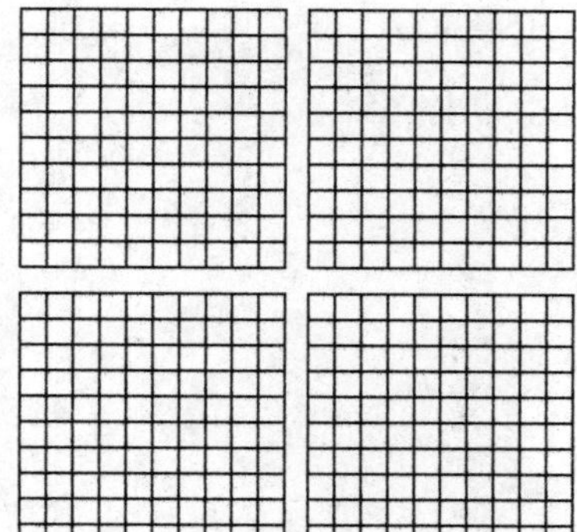

4. 2.05 + 1.87 = ________

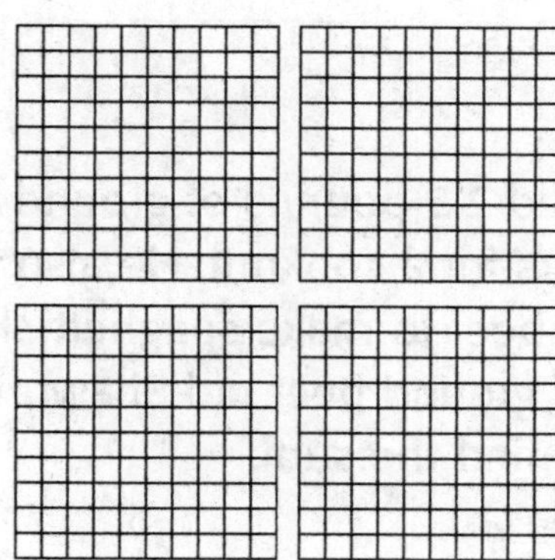

5. 2.91 + 1.8 = ________

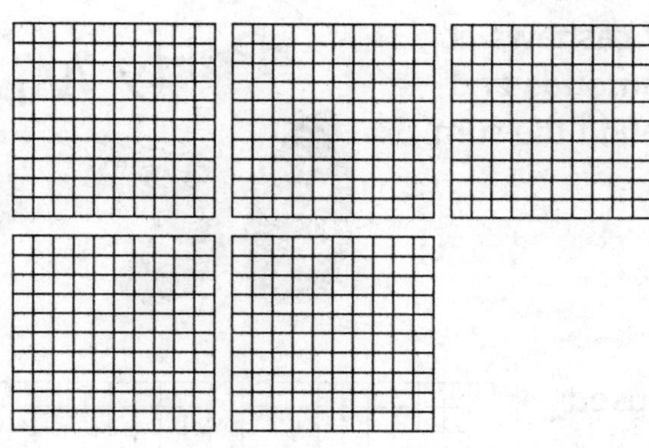

6. 1.34 + 1.15 = ________

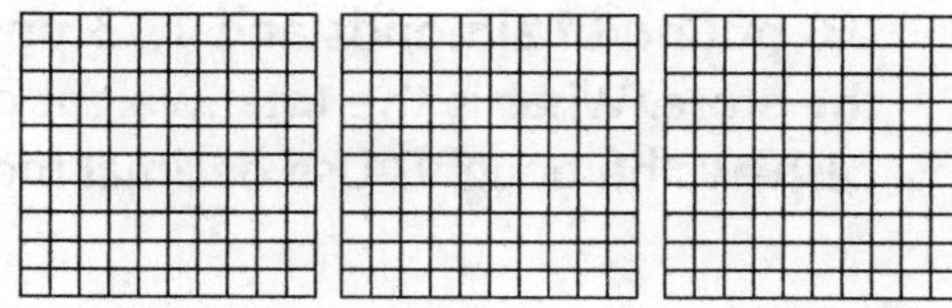

7. 1.74 + 0.36 = ________

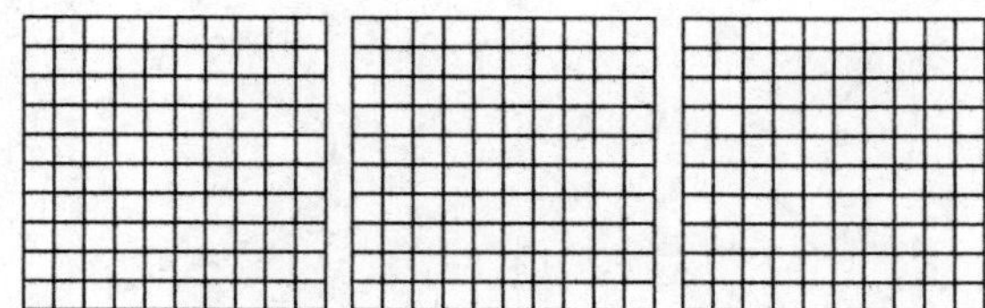

8. 2.05 + 1.12 = ________

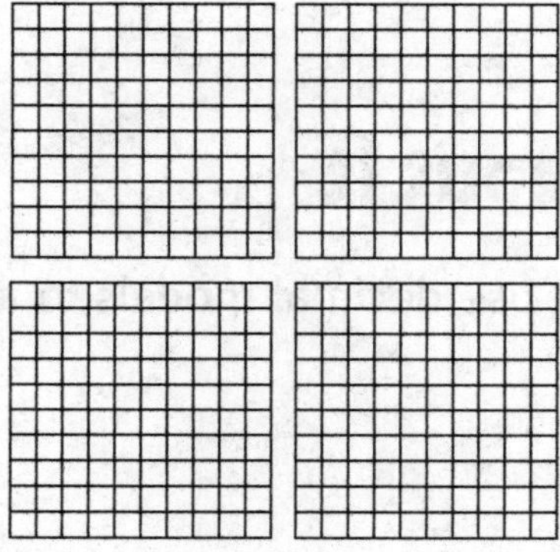

Apply It

9. The length of a nickel is 2.1 centimeters. What is the combined length of two nickels lying side by side? Use decimal models to find the sum.

____

10. Shawn used 2.5 pounds of ground beef to make hamburgers for a cookout. He also used 1.32 pounds of ground beef to make spaghetti. How many total pounds of ground beef did Shawn use? Use decimal models to find the sum.

____

11. Mathematical PRACTICE 5 **Use Math Tools** Marvin bought 0.6 pound of almonds and 1.73 pounds of cashews at the store. What is the total weight of the almonds and cashews he bought? Use decimal models to find the sum.

____

12. Mathematical PRACTICE 3 **Find the Error** Melanie used decimal models to add 1.03 and 0.4. Her result was 1.07. Shade the models to find the correct sum.

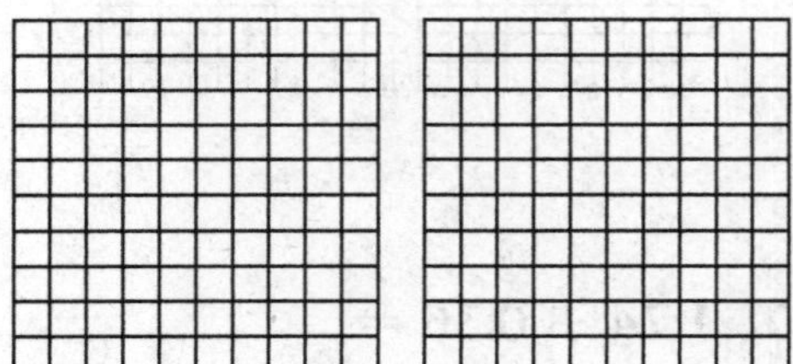

The correct sum is ______.

Write About It

13. How can I use decimal models to add decimals?

____

____

____

MY Homework

Lesson 5
Hands On: Add Decimals Using Models

Homework Helper

Need help? connectED.mcgraw-hill.com

Find 0.43 + 0.81. Use models.

1 Use 10-by-10 grids to model 0.43.

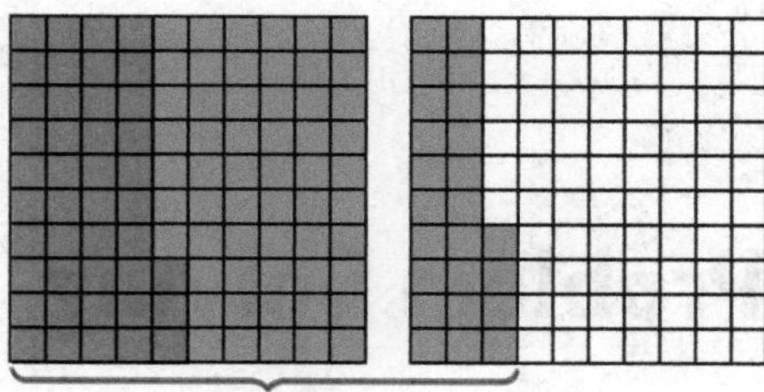

To show 0.43, shade forty-three hundredths of a 10-by-10 grid.

Forty-three hundredths is equal to 43 small squares.

2 Model 0.81.

To show 0.81, shade an additional eighty-one hundredths using a different color.

3 Add.

There is 1 one.
There are 2 tenths.
There are 4 hundredths.

Helpful Hint

1 one, 2 tenths, and 4 hundredths is equal to 1.24.

So, 0.43 + 0.81 = 1.24.

Check Use estimation to check for reasonableness.
$0 + 1 = 1$ and $1.24 \approx 1$

Practice

Add. Shade the decimal models.

1. 0.51 + 0.63 = ______

2. 1.93 + 1.73 = ______

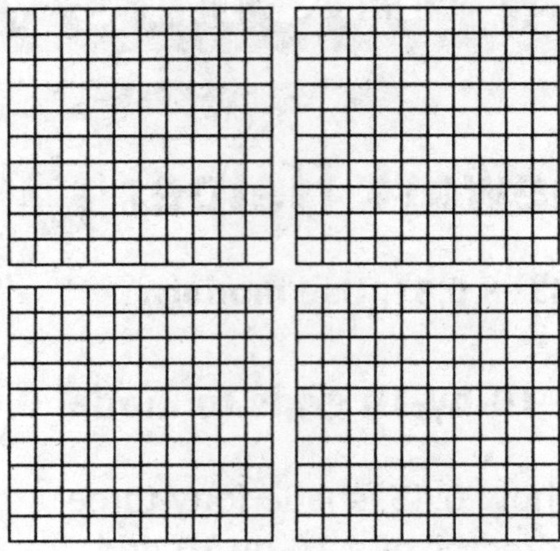

Problem Solving

3. Norah wants to download two different applications on her cell phone. One used 2.42 megabytes and the other one used 1.76 megabytes. How many total megabytes of memory will Norah use to download the two applications? Draw decimals models to find the sum.

______

4. Mathematical PRACTICE 4 **Model Math** Write a real-world problem that can be solved by adding 1.39 and 2.45 using decimal models. Then solve.

______

______

______

5. Dorian feeds his dog 7.5 pounds of food in a week. He feeds his cat 3.75 pounds of food in a week. How many total pounds of food do his pets eat in a week? Draw decimal models to find the sum.

______

Add Decimals

Lesson 6

ESSENTIAL QUESTION
How can I use place value and properties to add and subtract decimals?

To add decimals, line up the decimal points and add the digits in the same place-value position.

Math in My World

Example 1

Mr. Jacobson uses a digital scale to measure 44.2 milligrams of sodium for a chemistry experiment. During the second experiment, he used 33.1 milligrams of sodium. What is the total amount of sodium used?

Find 44.2 + 33.1.

Estimate 44 + 33 = ______

1. Line up the decimal points.
2. Add the digits in the same place-value positions.
3. Bring the decimal point straight down in the sum.

$$\begin{array}{r} 44.2 \\ +\ 33.1 \\ \hline \square\square.\square \end{array}$$

So, the total amount of sodium used is ______ milligrams.

	Tens	Ones	Tenths
	4	4	2
+	3	3	1
	7	7	3

Check for Reasonableness

Compare to the estimate, ______.

______ ≈ ______

You can annex a zero to one of the numbers in an addition problem so that both numbers end in the same place value.

Example 2

Find 19.6 + 4.31.

Estimate 20 + 4 = ______

Line up the decimal points. Annex a 0 so that both numbers end in the same place value.

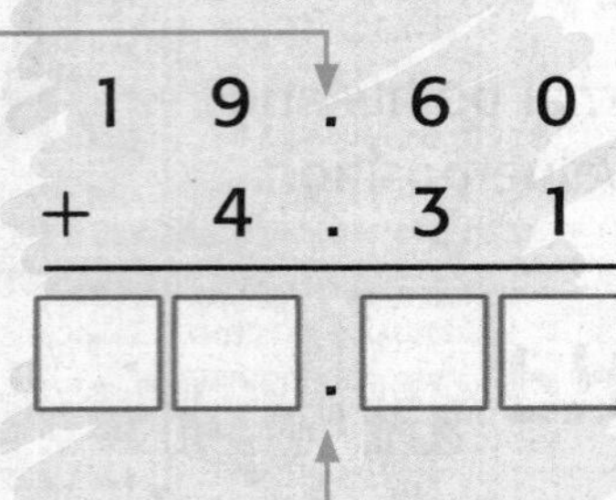

```
   1 9 . 6 0
+    4 . 3 1
-----------
  [ ][ ].[ ][ ]
```

Add the digits in the same place-value positions. Rename as necessary.

Bring the decimal point straight down in the sum.

So, 19.6 + 4.31 = ______.

Check for Reasonableness Compare to the estimate, ______.

______ ≈ ______

Talk MATH

Explain how annexing zeros might be helpful when adding decimals.

Guided Practice

Add. Check for reasonableness.

1.

```
   6 . 3 2
+  1 . 4 6
----------
  [ ].[ ][ ]
```

Estimate 6 + 1 = ______

So, 6.32 + 1.46 = ______.

Check for Reasonableness

______ ≈ ______

2.

```
   0 . 8 9
+  0 . 0 3
----------
  [ ].[ ][ ]
```

Estimate 1 + 0 = ______

So, 0.89 + 0.03 = ______.

Check for Reasonableness

______ ≈ ______

Name

Independent Practice

Add. Check for reasonableness.

3. $\begin{array}{r} 0.54 \\ +\ 7.8 \\ \hline \end{array}$

4. $\begin{array}{r} 14.8 \\ +\ 10.26 \\ \hline \end{array}$

5. $\begin{array}{r} 25 \\ +\ 8.46 \\ \hline \end{array}$

6. $\begin{array}{r} 35.08 \\ +\ 11.9 \\ \hline \end{array}$

7. $\begin{array}{r} 0.8 \\ +\ 0.22 \\ \hline \end{array}$

8. $\begin{array}{r} 9.14 \\ +\ 2.05 \\ \hline \end{array}$

9. 6.57 + 1.2 = ______

10. 19.21 + 11.03 = ______

11. 3.08 + 1.64 = ______

Algebra **Find each unknown.**

12. $8.9 + 0.15 = x$

$x =$ ______

13. $42.2 + 7.69 = d$

$d =$ ______

14. $5.63 + 1.22 = w$

$w =$ ______

Real World Problem Solving

15. Horacio bought a logic puzzle and batteries from a toy store. Use the table at the right to find the total cost of the two items, not including tax.

Item	Cost ($)
logic puzzle	14.95
batteries	10.39
carrying case	12.73

My Work!

16. An athlete training for the Olympics swims each lap of a four-lap race in the following times: 54.73, 54.56, 54.32, and 54.54 seconds. What is the total time it takes her to swim the four laps?

17. Terrance is bicycling on a trail. He bicycles for 12.6 miles and takes a break. Then he bicycles for 10.7 miles. How many miles has Terrance bicycled in all?

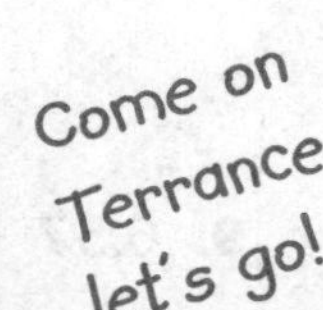

HOT Problems

18. Mathematical PRACTICE 1 **Make a Plan** Write a real-world word problem that can be solved by adding 34.99 and 5.79. Then solve.

19. Mathematical PRACTICE 2 **Use Number Sense** Write two different pairs of decimals whose sums are 8.69. One pair should involve regrouping.

20. **Building on the Essential Question** How does place value help you add decimals?

Name ....

MY Homework

Lesson 6
Add Decimals

Homework Helper

Need help? connectED.mcgraw-hill.com

Brett's social studies book weighs 5.34 pounds. His science book weighs 4.78 pounds. Suppose Brett only has these books in his bookbag. How much weight is he carrying, not including the weight of his bookbag?

Find 5.34 + 4.78.

Estimate 5 + 5 = 10

1. Line up the decimal points.
2. Add the digits in the same place-value positions. Rename as necessary.
3. Bring the decimal point straight down in the sum.

$$\begin{array}{r} \overset{1}{5}.\overset{1}{3}4 \\ +\,4.78 \\ \hline 10.12 \end{array}$$

So, Brett is carrying 10.12 pounds in books.

Check for Reasonableness 10.12 ≈ 10

Practice

Add. Check for reasonableness.

1. $2.72 + $3.83

2. 12.03 + 0.14

3. 26.76 + 2.99

Problem Solving

4. Devin has a new cell phone that holds 1.5 gigabytes of memory for music. He has already used 1.35 gigabytes of the memory. Will he have enough space to download a song that will use 0.12 gigabyte of memory? Explain.

____

5. A large bag of sand weighs 48.5 pounds. A small bag of sand weighs 24.6 pounds. If Mrs. Waggoner buys a large bag and a small bag, how many pounds of sand did she purchase altogether?

____

6. Mathematical PRACTICE 6 **Explain to a Friend** Grady wants to buy a basketball video game that costs $59.95, including tax. He has $45.50 in cash and a gift certificate for $15.25. Is that enough to buy the video game? Explain to a friend.

____

____

7. Mathematical PRACTICE 7 **Identify Structure** Lakshmi wants to start saving coins in a piggy bank. Her mother gave her three quarters and two pennies on Monday and two dimes and one nickel on Tuesday. Fill in the amounts of money that Lakshmi's mother gave her in the space provided on the bar diagram below.

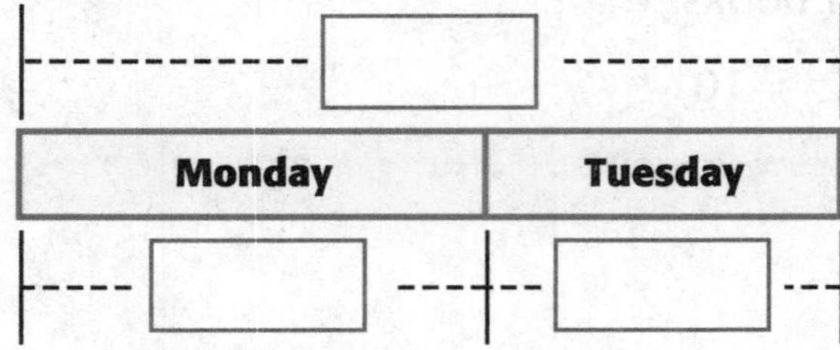

Test Practice

8. Marcus entered a race that involves swimming and running. He will need to swim 0.72 mile and run 1.65 miles. How far will Marcus travel in all during the race?

Ⓐ 2.37 miles　　Ⓒ 2.07 miles

Ⓑ 2.17 miles　　Ⓓ 1.37 miles

Name ....

Addition Properties

Lesson 7

ESSENTIAL QUESTION
How can I use place value and properties to add and subtract decimals?

Movie Night!

You can use properties of addition to find sums of whole numbers and decimals mentally. When there are no parentheses, add in order from left to right.

Math in My World

Example 1

Elijah recorded the number of different movies he watched last month. Use properties of addition to mentally find the total number of movies.

Movie Genre	Number of Movies Watched
Comedy	5
Action	27
Drama	15

Find 5 + 27 + 15.

You can easily add 5 and 15. So, change the order and group those numbers together.

The **Commutative Property of Addition** states that the order in which numbers are added does not change the sum.

The **Associative Property of Addition** states that the way in which numbers are grouped does not change the sum.

5 + 27 + 15 = ______ + 5 + 15 — Commutative Property

= 27 + (______ + ______) — Associative Property

= 27 + ______ — Add mentally. 5 + 15 = ______

= ______ — Add mentally. 20 + 27 = ______

So, Elijah watched a total of ______ movies.

The **Identity Property of Addition** states that the sum of any number and 0 equals the number.

Example 2

Use properties to find 1.8 + 2.6 + 0 mentally.

1.8 + 2.6 + 0 = (______ + 0.8) + (______ + 0.6) + 0

= 1 + ______ + 0.8 + ______ + 0 — Commutative Property

= (1 + ______) + (0.8 + ______) + 0 — Associative Property

= ______ + ______ + 0 — Add.

= ______ + 0 — Identity Property

= ______ — Add.

Guided Practice

1. Use properties of addition to find each sum mentally. Show your steps and identify the properties that you used.

9 + 27 + 1 = ______ + 9 + 1 — Commutative Property

= 27 + (______ + ______) — Associative Property

= 27 + ______ — Add.

= ______ — Add.

Name ______________________________

Independent Practice

Identify the properties that are used to find each sum.

2. $69 + 22 = (60 + 9) + (20 + 2)$

$= 60 + 20 + 9 + 2$ ______________________

$= (60 + 20) + (9 + 2)$ ______________________

$= 80 + 11$ Add.

$= 91$ Add.

3. $11 + 7.7 + 4.3 + 0 = 11 + (7.7 + 4.3) + 0$ ______________________

$= 11 + 12 + 0$ Add.

$= 23 + 0$ ______________________

$= 23$ Add.

4. $37 + 26 + 53 = 26 + 37 + 53$ ______________________

$= 26 + (37 + 53)$ ______________________

$= 26 + 90$ Add.

$= 116$ Add.

5. Use properties of addition to find the sum mentally. Show your steps and identify the properties that you used.

$10.9 + 3 + 0.1 =$ ________

Problem Solving

6. **Mathematical PRACTICE 2 Use Number Sense** Casey spent $2.50 on a snack, $1.24 on gum, $3.76 on a comic book, and $5.50 on lunch. Use mental math to find the total amount that he spent.

7. In one week, a classroom collected 43, 58, 62, 57, and 42 cans. Find the total number of cans the classroom collected using mental math. Explain how you found the sum.

8. The table shows the cost of a cheerleading uniform. Use properties of addition to find the total cost of the uniform mentally.

Cost of a Cheerleading Uniform	
Shoes	$65
Pom-poms	$18
T-shirt and shorts	$35

HOT Problems

9. **Mathematical PRACTICE 4 Model Math** Write a real-world problem that can be solved using the Associative Property of Addition. Solve the problem. Explain how you found the sum.

10. **Building on the Essential Question** How can properties help me add whole numbers and decimals?

MY Homework

Lesson 7
Addition Properties

Homework Helper

Need help? connectED.mcgraw-hill.com

Mackenzie practices violin four days a week. One week, she practiced for 21, 39, 45, and 25 minutes. Use mental math to find the total amount of time she practiced.

You can easily add 21 and 39. You can easily add 45 and 25. So, group those numbers together.

$21 + 39 + 45 + 25 = (21 + 39) + (45 + 25)$	Associative Property
$= 60 + 70$	Add. $21 + 39 = 60$ and $45 + 25 = 70$
$= 130$	Add. $60 + 70 = 130$

So, Mackenzie practiced violin for 130 minutes.

Practice

Use properties of addition to find each sum mentally. Show your steps and identify the properties that you used.

1. $9 + 6 + 31 =$ ______

2. $12.5 + 0 + 1 + 43.5 =$ ______

Real World Problem Solving

3. Sasha spent $1.05 on a soda, $5.25 on a sandwich, $0.75 on a piece of fruit, and $4.95 on a magazine. Use mental math to find the total amount she spent.

____

4. Jessie went to the mall and bought a CD for $12.98, a skirt for $17.50, a T-shirt for $8.50, and a bottle of water for $1.02. Use mental math to find the total amount she spent.

____

5. Gary played soccer for 1 hour and tennis for 2 hours. Tanya played tennis for 2 hours and soccer for 1 hour. Who played sports longer? Explain.

____

____

____

6. Mathematical PRACTICE 5 **Use Math Tools** Without calculating, would 0.4 + (2 + 0.6) be less than, greater than, or equal to 3? Explain.

____

____

Test Practice

7. Paula was reading a novel. She read 13 pages on Sunday, 12 pages on Tuesday, 17 pages on Friday, and 8 pages on Saturday. Use mental math to find the total number of pages she read.

Ⓐ 40 pages
Ⓑ 42 pages
Ⓒ 50 pages
Ⓓ 60 pages

Check My Progress

Vocabulary Check

Match each word to its correct definition.

1. **Commutative Property of Addition**
2. **Associative Property of Addition**
3. **Identity Property of Addition**

- The sum of any number and 0 equals the number.
- The way in which numbers are grouped does not change the sum.
- The order in which numbers are added does not change the sum.

Concept Check

Add.

4. $3.15 + 1.2 =$ ______

5. $68.9 + 7.1 =$ ______

6. $4.67 + 1.70$

7. $25.39 + 18.68$

8. Use properties of addition to find $37 + 58$ mentally. Show your steps and identify the properties that you used.

$37 + 58 =$ ______

Real World Problem Solving

9. What is the combined cost of the sweatshirt and hat shown?

____

10. Francis earned the following scores on his math quizzes: 86, 84, 90. Jillian earned the following scores on her math quizzes: 84, 90, 86. Who earned a greater total score? Which addition property can you use to find the answer?

____

____

11. Dean received the following amounts for mowing his neighbors' lawns: $12.50, $16.75, $20.50, $33.25. Use mental math to find the total amount he earned.

____

12. Explain how you would find the sum of 4.2 and 2.14.

____

____

Test Practice

13. Tammy made a bracelet using red, white, and blue string. The red string is 2.4 centimeters long, the white string is 2.1 centimeters long, and the blue string is 2.6 centimeters long. What is the total length of the three strings?

Ⓐ 7.1 centimeters
Ⓑ 6.1 centimeters
Ⓒ 5.5 centimeters
Ⓓ 4.27 centimeters

Name ....

Hands On
Subtract Decimals Using Base-Ten Blocks

Lesson 8

ESSENTIAL QUESTION
How can I use place value and properties to add and subtract decimals?

Build It
Tools

Find 1.8 − 0.4. Use base-ten blocks.

1 **Model 1.8.**

Ones	Tenths	Hundredths

2 **Take 0.4, or four tenths, away.**

Count the remaining base-ten blocks. Draw the result.

My Drawing!

How many ones are left? ______

How many tenths are left? ______

So, 1.8 − 0.4 = ______.

Check Use addition to check your answer.

______ + 0.4 = 1.8

Try It

Ones	Tenths	Hundredths

Find 2.25 − 0.75. Use base-ten blocks.

Model 2.25.

2 **Subtract 0.75. Count the remaining base-ten blocks. Draw the result.**

To take away 0.75, take away 7 tenths and 5 hundredths. But you cannot subtract 7 tenths from 2 tenths. So, regroup the ones block as 10 tenths. Then subtract.

How many ones are left? ______

How many tenths are left? ______

How many hundredths are left? ______

Helpful Hint

2 ones − 1 one = 1 one

12 tenths − 7 tenths = 5 tenths

5 hundredths − 5 hundredths = 0 hundredths

So, 2.25 − 0.75 = ______.

Check Use addition to check your answer.

______ + 0.75 = 2.25

Talk About It

1. Mathematical PRACTICE 6 **Explain to a Friend** Explain to a friend or classmate when you should regroup when subtracting decimals with base-ten blocks.

______

______

______

2. How many tenths blocks will you need to subtract 0.35 from 1.2? Explain.

______

Name ....

Practice It

Subtract. Use base-ten blocks. Draw each result in the table.

3. 0.8 − 0.3 = ________

Ones	Tenths	Hundredths

4. 2.8 − 0.7 = ________

Ones	Tenths	Hundredths

5. 1.43 − 0.31 = ________

Ones	Tenths	Hundredths

6. 2.17 − 1.9 = ________

Ones	Tenths	Hundredths

7. 2.52 − 1.48 = ________

Ones	Tenths	Hundredths

8. 3.85 − 2.19 = ________

Ones	Tenths	Hundredths

Apply It

9. Vaughn's MP3 player had a memory of 2.5 gigabytes. He has used 1.76 gigabytes of memory. How much memory space does Vaughn's MP3 player have left? Use base-ten blocks to help you solve.

____

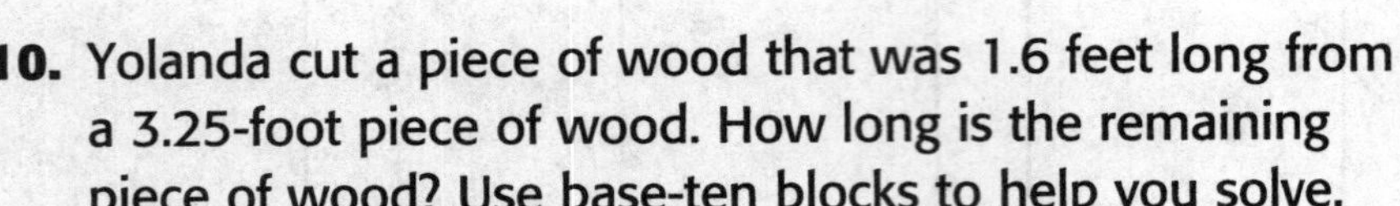

10. Yolanda cut a piece of wood that was 1.6 feet long from a 3.25-foot piece of wood. How long is the remaining piece of wood? Use base-ten blocks to help you solve.

____

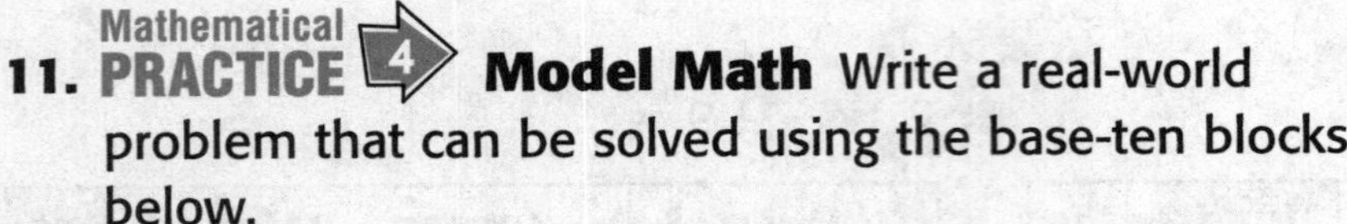

11. Mathematical PRACTICE 4 **Model Math** Write a real-world problem that can be solved using the base-ten blocks below.

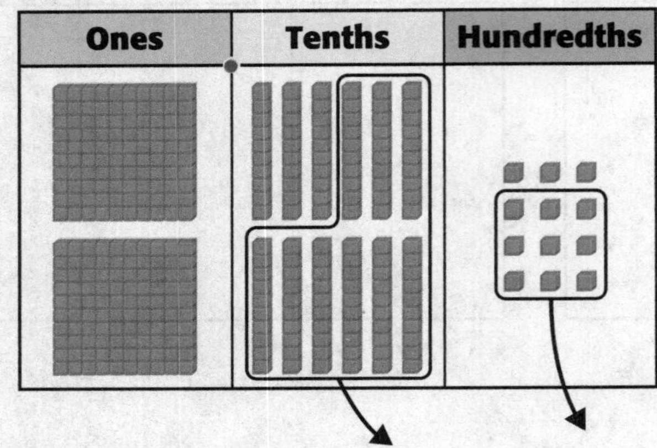

____

____

Write About It

12. How can I use base-ten blocks to subtract decimals?

____

____

____

Name ....

MY Homework

Lesson 8
Hands On: Subtract Decimals Using Base-Ten Blocks

Homework Helper

Need help? connectED.mcgraw-hill.com

Charlotte bicycled 1.6 miles on Tuesday and 0.9 mile on Wednesday. How much farther did she bike on Tuesday than on Wednesday?

Find 1.6 − 0.9. Use base-ten blocks.

1 Model 1.6.

Ones	Tenths	Hundredths

2 Subtract 0.9. Count the remaining base-ten blocks.

Ones	Tenths	Hundredths

You need to take away 9 tenths, but you cannot subtract 9 tenths from 6 tenths. So, regroup the ones block as 10 tenths. Then subtract.
There are no ones left.
There are seven tenths left.

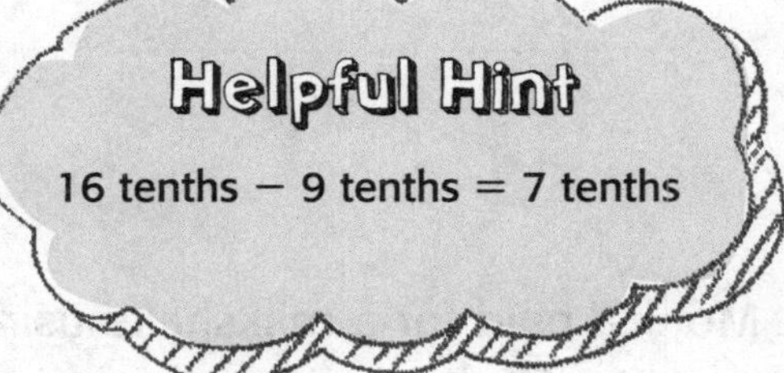

So, 1.6 − 0.9 = 0.7.

Charlotte biked 0.7 mile farther on Tuesday.

Check Use addition to check your answer.
0.7 + 0.9 = 1.6

Practice

Subtract. Use base-ten blocks. Draw each result in the table.

1. 1.3 − 0.28 = ______

Ones	Tenths	Hundredths

2. 3.52 − 1.39 = ______

Ones	Tenths	Hundredths

Problem Solving

3. Mathematical PRACTICE 4 **Model Math** The temperature at 6:00 A.M. was 0.8°F. By 3:00 P.M., the temperature had increased to 2.4°F. Find the difference between the two temperatures. Draw models to find the difference.

______

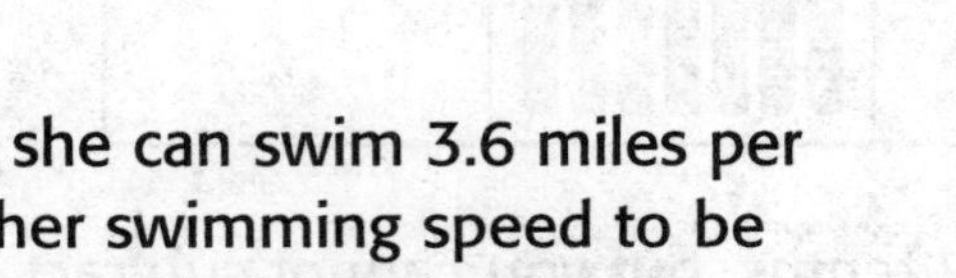

4. Melissa calculated that she can swim 3.6 miles per hour. Avery calculated her swimming speed to be 2.8 miles per hour. How much faster was Melissa's swimming speed? Draw models to find the difference.

______

5. Monica paid for a milkshake using $3. If the milkshake costs $2.67, how much change will she receive? Draw models to find the difference.

______

Name ______________________

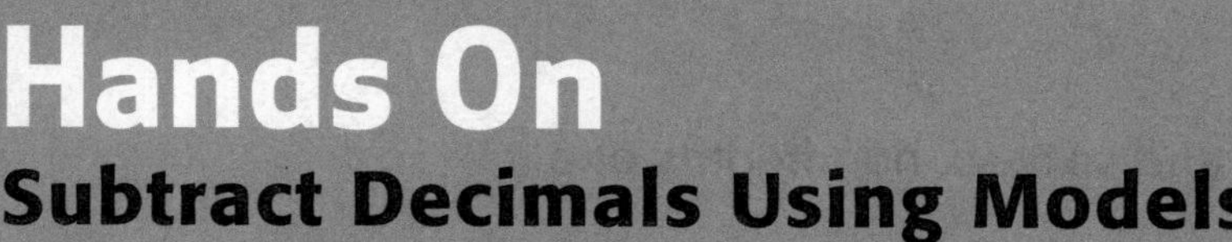

Hands On
Subtract Decimals Using Models

Lesson 9

ESSENTIAL QUESTION
How can I use place value and properties to add and subtract decimals?

Build It

Find 2.4 − 1.07. Use models.

1 Use 10-by-10 grids to model 2.4.
To show 2.4, shade two whole grids and four tenths of a third grid.

Four tenths is equal to ________ small squares.

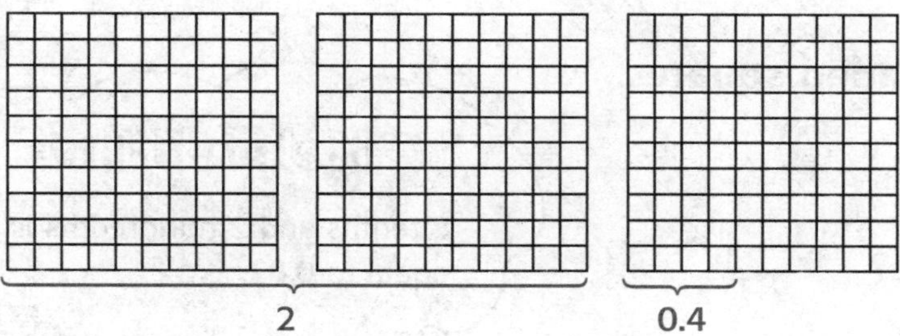

2 Subtract 1.07.
To subtract 1.07, cross out 1 whole grid and 7 hundredths of the third grid.

3 Count the remaining shaded squares.

How many ones are there? ________

How many tenths are there? ________

How many hundredths are there? ________

So, 2.4 − 1.07 = ________.

Check Use addition to check your answer.

1.07 + ________ = 2.4

Try It

Find 1.66 − 0.84. Use models.

Model 1.66.
To show 1.66, shade one grid and sixty-six hundredths of a second grid.

Sixty-six hundredths is equal to ______ small squares.

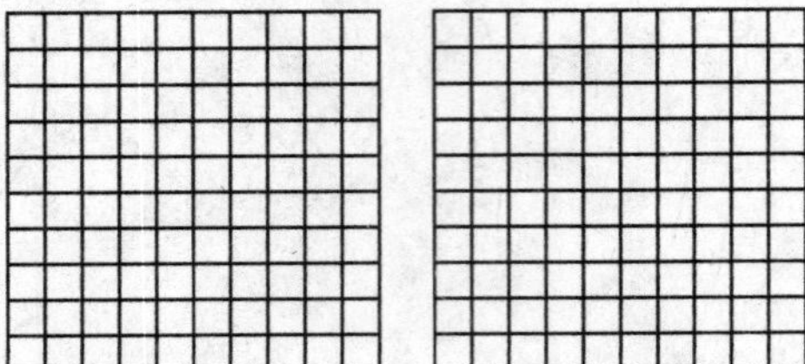

2 Subtract 0.84.
To subtract 0.84, cross out 4 hundredths and 8 tenths.

3 Count the remaining shaded squares.

How many ones are there? ______

How many tenths are there? ______

How many hundredths are there? ______

So, 1.66 − 0.84 = ______.

Check Use addition to check your answer.

0.84 + ______ = 1.66

Talk About It

1. Mathematical PRACTICE 5 **Use Math Tools** Explain how using models to find 2.4 − 1.07 is similar to using models to find 240 − 107.

____

____

____

Name ....

CCSS

Practice It

Subtract. Use decimal models.

2. 0.93 − 0.7 = ________

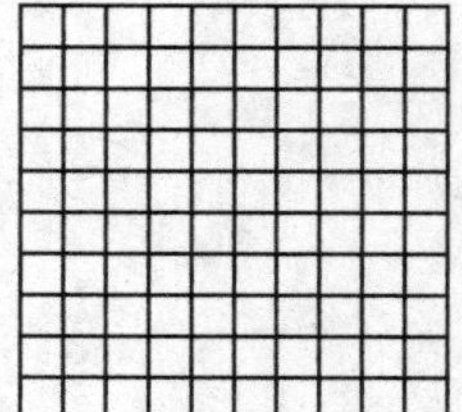

3. 1.53 − 1.41 = ________

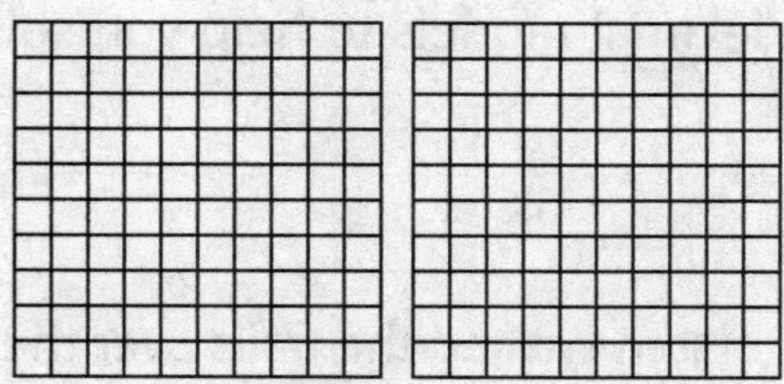

4. 0.9 − 0.3 = ________

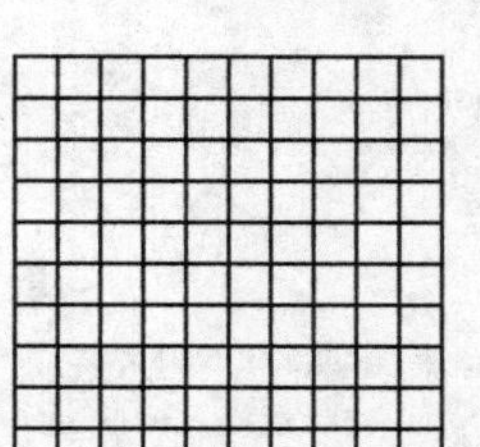

5. 3.94 − 0.4 = ________

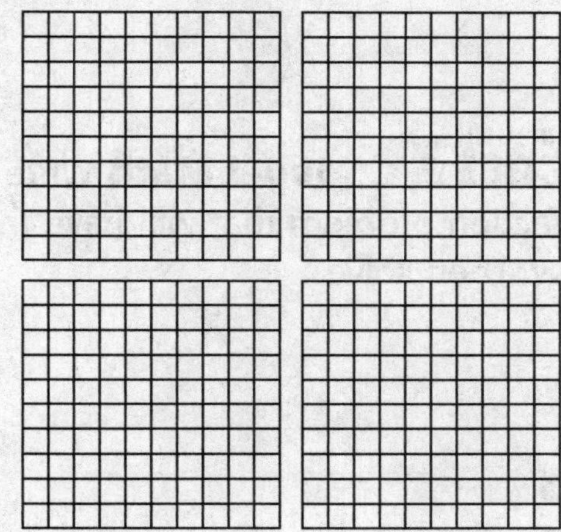

6. 3.55 − 0.1 = ________

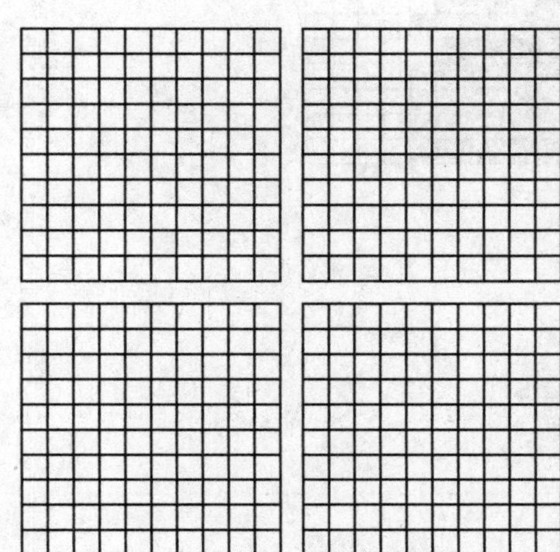

7. 2.4 − 0.9 = ________

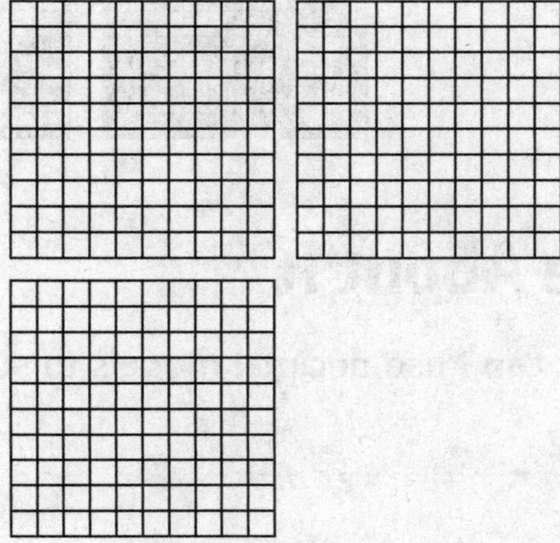

Apply It

8. Mathematical PRACTICE 5 **Use Math Tools** Nora wrote two papers for English class. One uses 0.65 megabyte of memory and the other uses 0.92 megabyte of memory. How much more memory does the one paper use? Use decimal models to help you solve.

____

9. Mr. Dorsten walked the trails over the weekend. He hiked 2.3 miles. His friend went hiking and walked 1.6 miles. How many more miles did Mr. Dorsten hike compared to his friend? Use decimal models to help you solve.

____

10. Mathematical PRACTICE 4 **Model Math** Write a real-world subtraction problem that will have the result modeled below. Then solve.

____

____

____

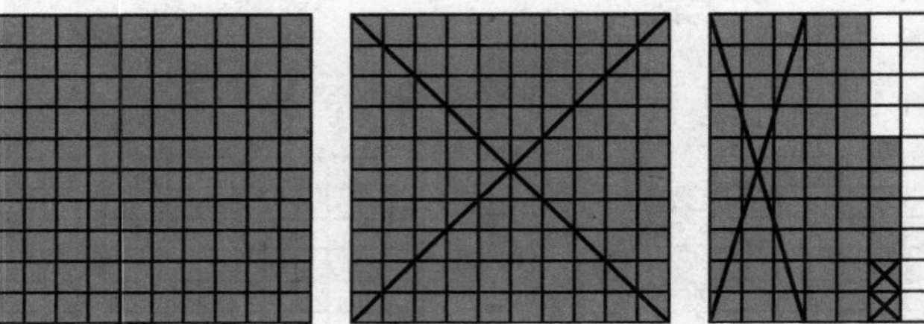

Write About It

11. How can I use decimal models to subtract decimals?

____

____

____

MY Homework

Lesson 9
Hands On: Subtract Decimals Using Models

Homework Helper

Need help? connectED.mcgraw-hill.com

Marcy measured the lengths of two insects. The first insect was 2.43 centimeters long. The second insect was 1.05 centimeters long. What is the difference between the lengths of the two insects?

Find 2.43 - 1.05.

Use 10-by-10 grids to model 2.43.

To show 2.43, shade two grids and forty-three hundredths of a grid.

Forty-three hundredths is equal to 43 small squares.

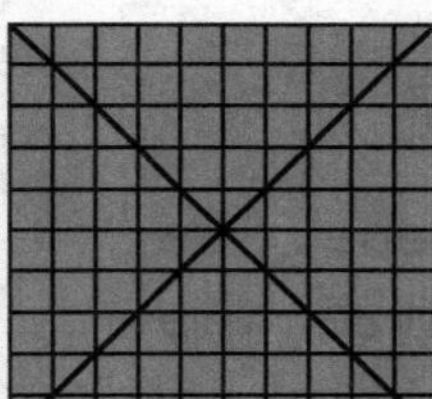
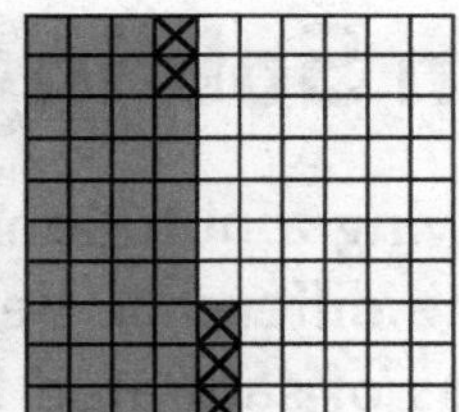

2 Subtract 1.05.

To subtract 1.05, cross out 1 whole grid and 5 hundredths of the third grid.

3 Count the remaining shaded squares.

There is 1 one.

There are 3 tenths.

There are 8 hundredths.

Helpful Hint

1 one, 3 tenths, and 8 hundredths is equal to 1.38.

So, 2.43 − 1.05 = 1.38.

The difference in the lengths of the two insects is 1.38 centimeters.

Check Use addition to check your answer.

1.05 + 1.38 = 2.43

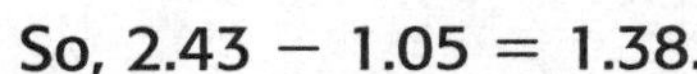

Practice

Subtract. Use decimal models.

1. $3.8 - 2.3 =$ ______

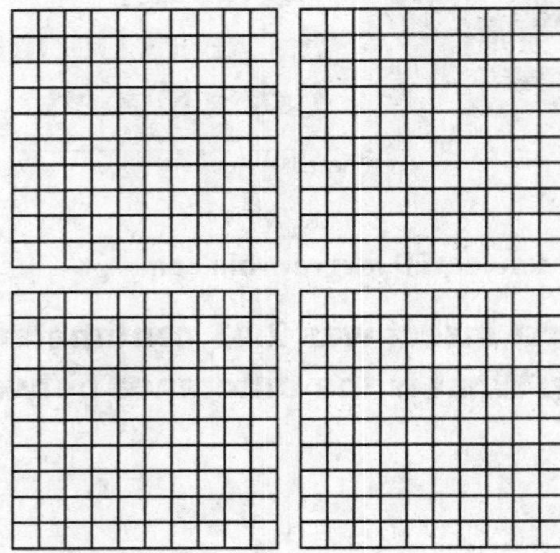

2. $2.13 - 1.7 =$ ______

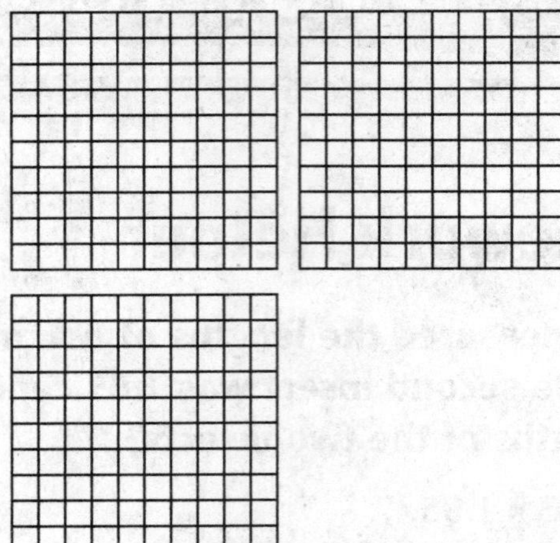

Problem Solving

3. Bridget is reviewing a nutrition label. She wants to find the difference between the amount of grams of saturated fat and protein. How many more grams are there of saturated fat than protein? Draw decimal models to find the difference.

______

Nutrition Facts
Serving Size: 1 tbsp (15 g)

Amount Per Serving	
Calories 190	**Calories from Fat** 1
	% Daily Value*
Total Fat 1.48 g	0%
Saturated Fat 1.08 g	0%
Trans Fat	
Cholesterol 0 mg	0%
Sodium 164.4 mg	7%
Potassium 1.4 mg	0%
Total Carbohydrate 5.17 g	2%
Dietary Fiber 0.48 g	2%
Sugars	
Sugar Alcohols	
Protein 0.49 g	

4. Mathematical PRACTICE 5 **Use Math Tools** Leanne is building a model tower that needs to be 1.42 meters tall. The model currently is only 0.68 meter tall. How many more meters does she need to build in order to finish the model tower? Draw decimal models to find the difference.

______

Name

Subtract Decimals

Lesson 10

ESSENTIAL QUESTION
How can I use place value and properties to add and subtract decimals?

To subtract decimals, line up the decimal points. Then subtract digits in the same place-value position.

Math in My World

Example 1

The table shows the average lengths of the three longest bones in the human body. How much longer is the average femur than the average tibia?

Longest Bones in the Human Body	
Bone	**Length (in.)**
Femur (upper leg)	19.8
Tibia (inner lower leg)	16.9
Fibula (outer lower leg)	15.9

Find 19.8 − 16.9.

Estimate 20 − 17 = ______

Line up the decimal points.

Subtract the digits in the same place-value positions. Rename as necessary.

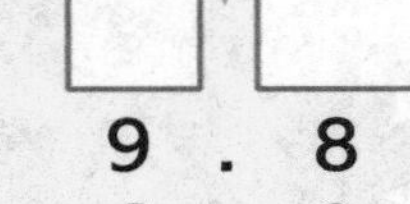

$$\begin{array}{r} 19.8 \\ -\ 16.9 \\ \hline \square.\square \end{array}$$

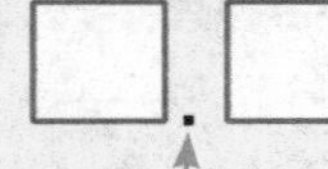

3 Bring the decimal point straight down in the difference.

So, the average femur is ______ inches longer than the average tibia.

Subtraction and addition are inverse operations. **Inverse operations** are operations that undo each other.

Check Use addition to check your answer.

______ + 16.9 = 19.8.

Sometimes both numbers in a subtraction problem do not end in the same place value. Annex zeros if necessary.

Example 2

Find 6.3 − 4.78.

Estimate 6 − 5 = ______

1. Line up the decimal points. Annex a 0 so that both numbers end in the same place value.
2. Subtract the digits in the same place-value positions. Rename as necessary.
3. Bring the decimal point straight down in the difference.

$$\begin{array}{r} 6.30 \\ -\ 4.78 \\ \hline \square.\square\square \end{array}$$

So, 6.3 − 4.78 = ______.

Check Use addition to check your answer.

______ + 4.78 = 6.3

Guided Practice

Talk MATH

Explain the strategy used in Exercise 1 to subtract the decimals.

Subtract. Use addition to check your answer.

1.

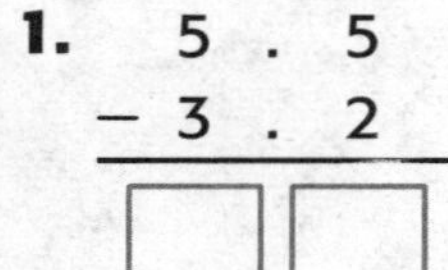

$$\begin{array}{r} 5.5 \\ -\ 3.2 \\ \hline \square.\square \end{array}$$

Estimate 6 − 3 = ______

So, 5.5 − 3.2 = ______.

Check

______ + 3.2 = 5.5

2.

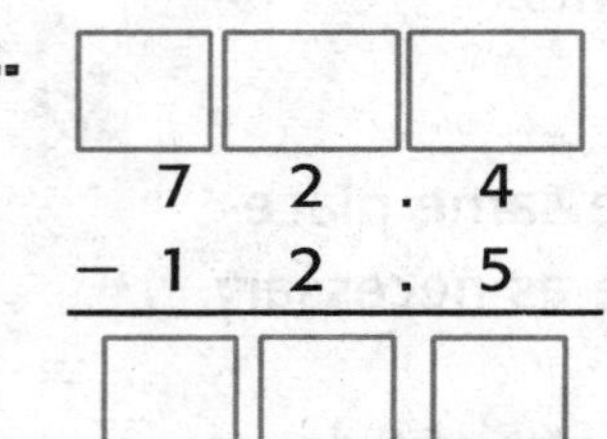

$$\begin{array}{r} 72.4 \\ -\ 12.5 \\ \hline \square\square.\square \end{array}$$

Estimate 70 − 10 = ______

So, 72.4 − 12.5 = ______.

Check

______ + 12.5 = 72.4

Name

Independent Practice

Subtract. Use addition to check your answer.

3. $\begin{array}{r} 29.34 \\ -\ 9 \\ \hline \end{array}$

4. $\begin{array}{r} 0.4 \\ -\ 0.2 \\ \hline \end{array}$

5. $\begin{array}{r} \$9.67 \\ -\ \$2.35 \\ \hline \end{array}$

6. $\begin{array}{r} 97 \\ -\ 16.98 \\ \hline \end{array}$

7. $\begin{array}{r} 42.28 \\ -\ 1.52 \\ \hline \end{array}$

8. $\begin{array}{r} 8 \\ -\ 5.78 \\ \hline \end{array}$

9. 36 − 7.3 = ______

10. 5.6 − 3.5 = ______

11. 19.86 − 9.94 = ______

Algebra **Find each unknown.**

12. \$15.00 − \$6.24 = *b*

b = ______

13. 8.2 − 6.72 = *k*

k = ______

14. 58.67 − 28.72 = *t*

t = ______

Problem Solving

15. Use the table to find out how many more people there are per square mile in Iowa than in Colorado.

Population Density	
State	People Per Square Mile
Colorado	41.5
Iowa	52.4

16. Mathematical PRACTICE 6 **Explain to a Friend** You decide to buy a hat for $10.95 and a T-shirt for $14.20. How much change will you receive if you pay with a $50 bill? Explain to a friend.

17. Coach Trainer timed his athletes running the 40-yard dash. He used an electronic timer for accuracy. The fastest player on his team ran the 40-yard dash in 4.58 seconds and the slowest time was 5.75 seconds. Find the difference between the fastest and slowest times.

HOT Problems

18. Mathematical PRACTICE 2 **Reason** Explain how place value could be used to find the difference of 4.23 and 2.75.

19. **Building on the Essential Question** How does place value help me subtract decimals?

Name ....

MY Homework

Lesson 10
Subtract Decimals

Homework Helper

Need help? connectED.mcgraw-hill.com

Stephen's father gave him \$10 to buy lunch at the concession stand. If his lunch cost \$7.74, how much change should Stephen give his father?

Find \$10 − \$7.74.

Estimate 10 − 8 = 2

1. Line up the decimal points. Annex zeros so that both numbers end in the same place value.
2. Subtract the digits in the same place-value positions. Rename as necessary.
3. Bring the decimal point straight down in the difference.

```
    0 9  9 10
  $ 1 0 . 0 0
− $   7 . 7 4
  $   2 . 2 6
```

So, Stephen should give his father \$2.26.

Check Use addition to check your answer.
\$2.26 + \$7.74 = \$10.00

Practice

Subtract. Use addition to check your answer.

1.
```
  9.03
− 4.09
```

2. 71.65 − 55.12 = ______

3.
```
  $18.94
− $3.25
```

Problem Solving

4. Roberto uses his handheld GPS device to determine that he hikes 21.48 miles in one weekend. The next weekend he hikes 30 miles. How much less did he hike the first weekend than the second weekend?

____

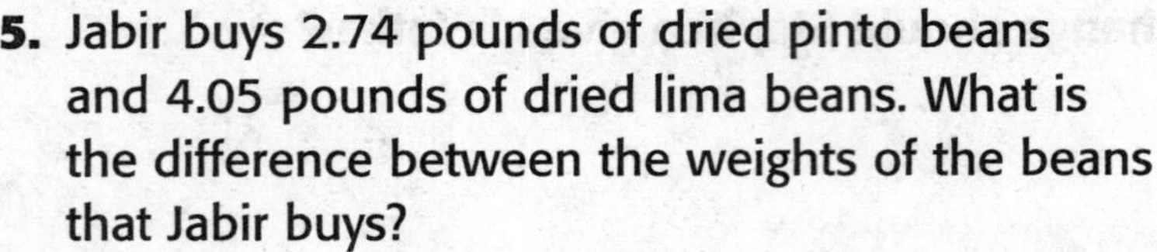

5. Jabir buys 2.74 pounds of dried pinto beans and 4.05 pounds of dried lima beans. What is the difference between the weights of the beans that Jabir buys?

____

6. Rebekah's baby brother weighs 7.71 pounds. Her newborn kitten weighs 0.24 pound. How much more does Rebekah's baby brother weigh than her kitten?

____

7. Mathematical PRACTICE 2 **Use Number Sense** Marshall compares water tanks at the home improvement store. One tank holds 20.2 gallons of water. A second tank holds 28.1 gallons of water. How much more water does the second tank hold than the first tank?

____

Test Practice

8. Kamal buys a pen that costs $1.09 and a tablet of paper that costs $2.50. How much more does the paper cost than the pen?

Ⓐ $1.31   Ⓒ $1.59

Ⓑ $1.41   Ⓓ $3.59

Review

Chapter 5

Add and Subtract Decimals

Vocabulary Check

Use the words in the words bank to complete each sentence.

Associative Property	**Commutative Property**	**Identity Property**
inverse operations	**estimate**	**difference**
exact answer	**place value**	**regroup**

1. The ______________________ states that you can add numbers in any order.

2. When you round a number, you find its ______________________.

3. The value given to a digit by its position in a number is its ______________________.

4. To ______________________ is to form into a new or restructured group or grouping.

5. Subtracting results in finding the ______________________ between two numbers.

6. The ______________________ states that the sum of any number and 0 equals the number.

7. The ______________________ states that the way in which numbers are grouped does not change the sum or product.

8. When computing using actual numbers, you find a(n) ______________________.

Concept Check

Check

Round each decimal to the place indicated.

9. 0.781; hundredths ________

10. 17.93; ones ________

Round each decimal to the nearest tenth. Then add or subtract.

11. 17.16 − 9.01 = ________

12. 62.84 + 74.2 = ________

Add.

13. 6.07 + 1.34

14. $1.06 + $4.89

15. 4.15 + 20.68 = ________

16. Use properties of addition to find 37 + 48 mentally. Show your steps and identify the properties that you used.

37 + 48 = ________

Subtract. Use addition to check your answer.

17. 4.15 − 1.29

18. $13.78 − $5.41

19. 8.3 − 4.7 = ________

Name

Real World Problem Solving

For Exercises 20 and 21, determine whether you need an estimate or an exact answer. Then solve.

20. A family is staying in a hotel for 2 nights. The hotel costs $91.35 a night. About how much will they pay for the hotel?

21. Jesse had three long jumps during a track meet of 17 feet, 18 feet, and 15 feet. Andrew had three long jumps during a track meet of 17 feet, 15 feet, and 18 feet. Overall, who had a greater total jump distance? Which addition property can you use to find the answer?

22. You have to be 48 inches tall to ride the roller coaster. Eva is 45 inches tall. If she grows 2 inches each year, will she be able to ride the roller coaster in 2 years?

My Work!

Test Practice

23. Micah had $10.52 after he left the book store. If he bought a book for $6.39, how much money did he have before he went to the book store?

Ⓐ $4.13　　Ⓒ $14.91

Ⓑ $8.23　　Ⓓ $16.91

Reflect

Chapter 5

Answering the ESSENTIAL QUESTION

Use what you learned about adding and subtracting decimals to complete the graphic organizer.

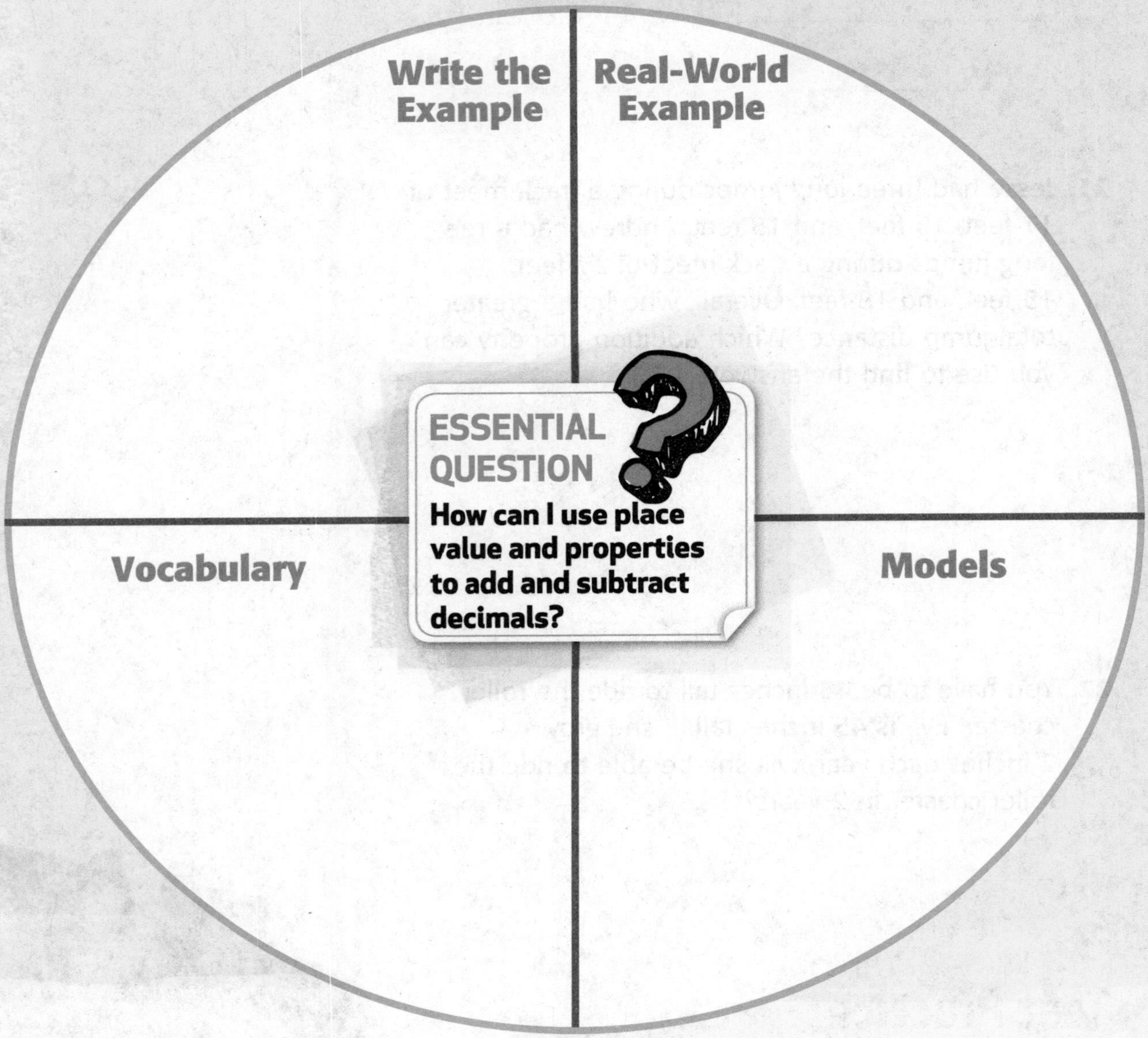

Write the Example

Real-World Example

ESSENTIAL QUESTION

How can I use place value and properties to add and subtract decimals?

Vocabulary

Models

Now reflect on the ESSENTIAL QUESTION  **Write your answer below.**

Chapter

6 Multiply and Divide Decimals

ESSENTIAL QUESTION

How is multiplying and dividing decimals similar to multiplying and dividing whole numbers?

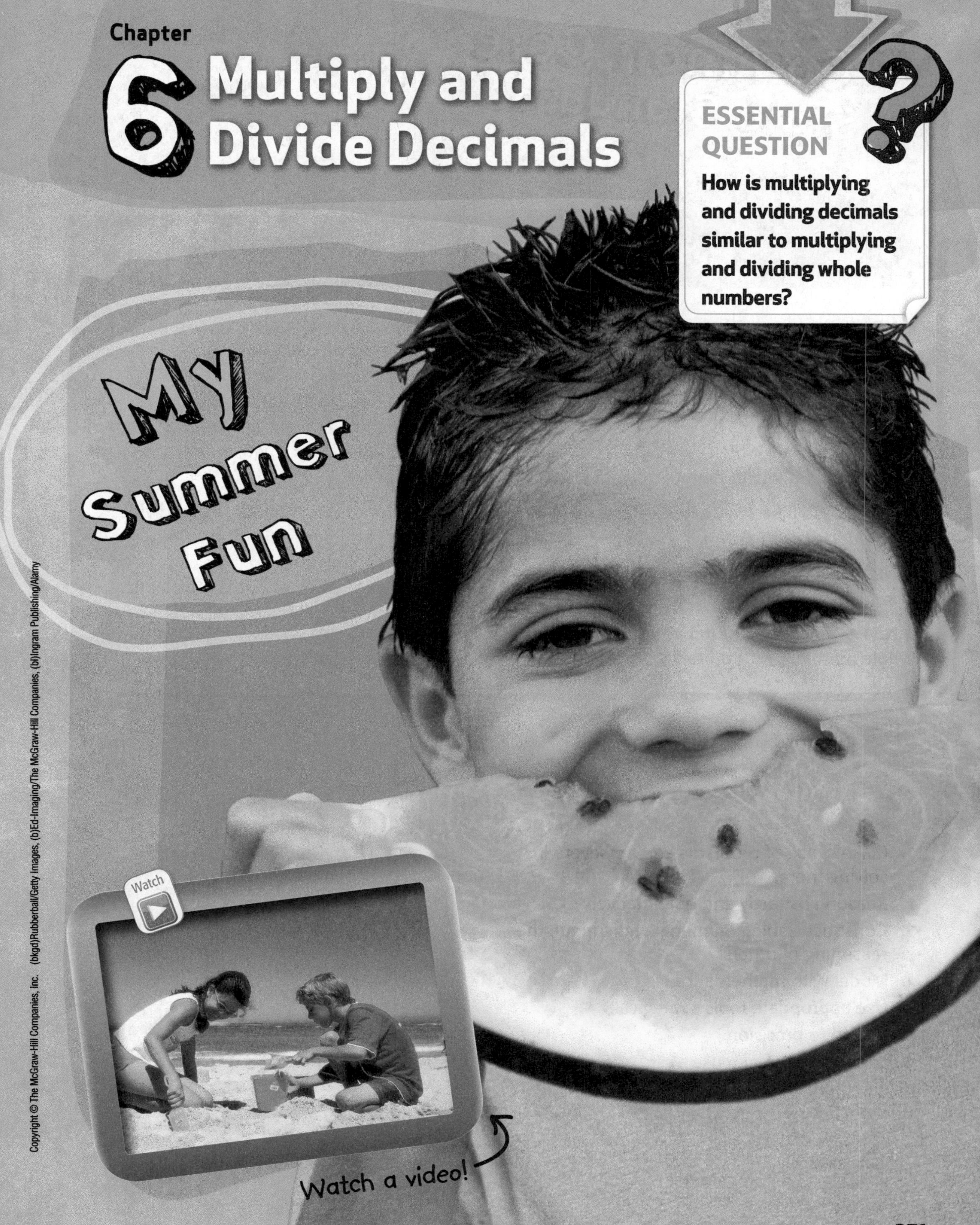

MY Common Core State Standards

Number and Operations in Base Ten

5.NBT.2 Explain patterns in the number of zeros of the product when multiplying a number by powers of 10, and explain patterns in the placement of the decimal point when a decimal is multiplied or divided by a power of 10. Use whole-number exponents to denote powers of 10.

5.NBT.4 Use place value understanding to round decimals to any place.

5.NBT.5 Fluently multiply multi-digit whole numbers using the standard algorithm.

5.NBT.6 Find whole-number quotients of whole numbers with up to four-digit dividends and two-digit divisors, using strategies based on place value, the properties of operations, and/or the relationship between multiplication and division. Illustrate and explain the calculation by using equations, rectangular arrays, and/or area models.

5.NBT.7 Add, subtract, multiply, and divide decimals to hundredths, using concrete models or drawings and strategies based on place value, properties of operations, and/or the relationship between addition and subtraction; relate the strategy to a written method and explain the reasoning used.

Wow! Lots of these make sense to me!

Standards for Mathematical PRACTICE

1. Make sense of problems and persevere in solving them.
2. Reason abstractly and quantitatively.
3. Construct viable arguments and critique the reasoning of others.
4. Model with mathematics.
5. Use appropriate tools strategically.
6. Attend to precision.
7. Look for and make use of structure.
8. Look for and express regularity in repeated reasoning.

= focused on in this chapter

Name ....

Am I Ready?

← Go online to take the Readiness Quiz

Multiply.

1. 126 × 30 = ________ **2.** 12 × 28 = ________ **3.** 320 × 10 = ________

4. How much will it cost for Gary to reserve 13 lanes?

Bowling Reservations	
1 lane	$12

____

Divide.

5. 114 ÷ 6 = ________ **6.** 84 ÷ 4 = ________ **7.** 1,500 ÷ 10 = ________

8. Hugo spent $216 on 4 sweaters. If each sweater cost the same amount, find the cost of one sweater.

____

Round each decimal to the nearest whole number.

9. 2.7 ________ **10.** 0.7 ________ **11.** 18.2 ________

12. 6.34 ________ **13.** 9.8 ________ **14.** 9.4 ________

Shade the boxes to show the problems you answered correctly.

How Did I Do?

1	2	3	4	5	6	7	8	9	10	11	12	13	14

Name

MY Math Words

Vocab

Review Vocabulary

composite number	decimal point	divide	estimate
hundredths	multiply	ones	place value
power	tenths	thousands	

Making Connections

Use the Venn diagram to categorize the review vocabulary.

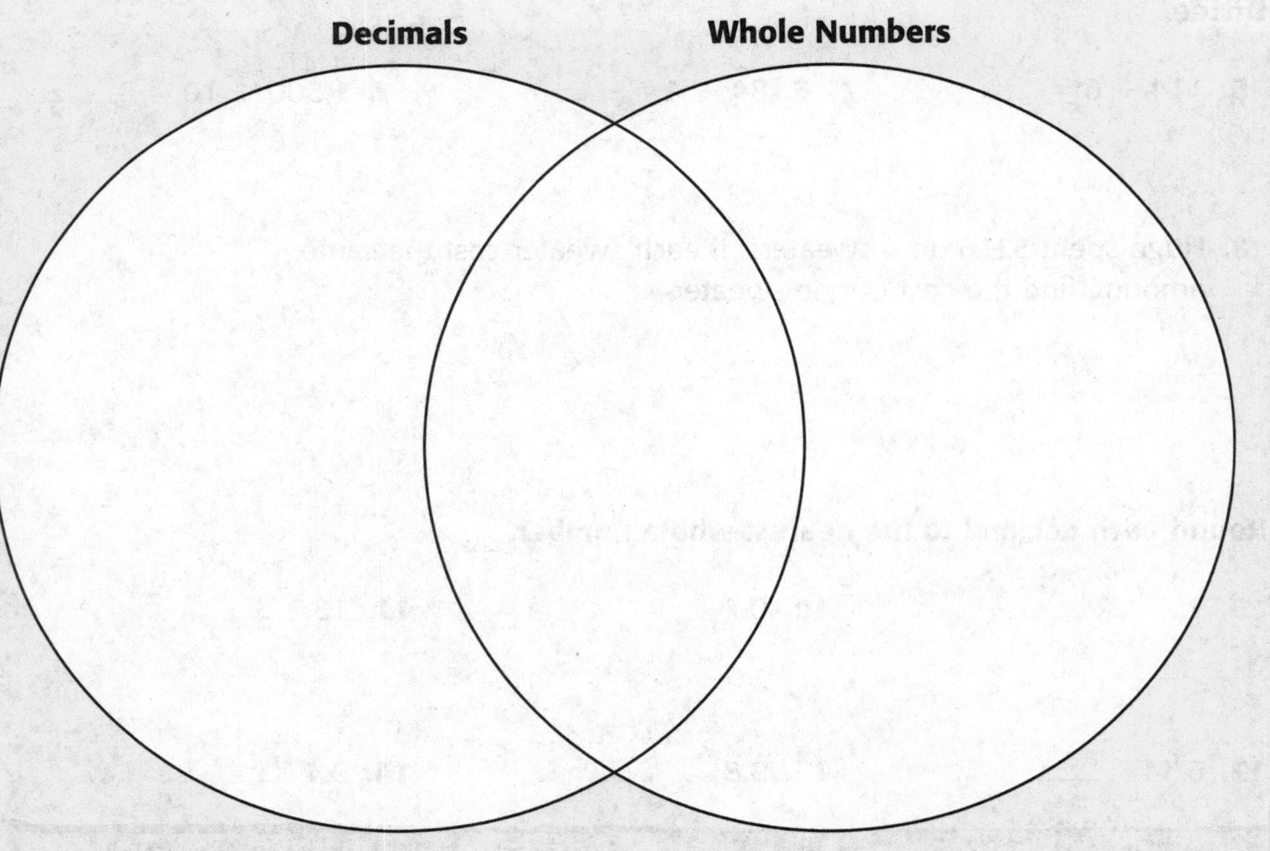

MY Vocabulary Cards

Lesson 6-8

Associative Property of Multiplication

$(8 \times 7) \times 10 = 8 \times (7 \times 10)$

Lesson 6-8

Commutative Property of Multiplication

$8.92 \times 455 = 455 \times 8.92$

Lesson 6-8

Identity Property of Multiplication

$42.08 \times 1 = 42.08$

Ideas for Use

- Use blank cards to group like ideas you find throughout the chapter, such as multiplying using decimals and dividing using decimals. Discuss with a friend strategies to understand these concepts.
- Use a blank card to write this chapter's essential question. Use the back of the card to write examples that help you answer the question.

The order in which factors are multiplied does not change the product.

How is the Commutative Property of Multiplication different from the Identity Property of Multiplication?

The way in which factors are grouped does not change the product.

How can the word *associate* help you remember this property?

The product of any factor and 1 equals the factor.

Write a number sentence that is an example of the Identity Property of Multiplication.

MY Foldable

FOLDABLES® Follow the steps on the back to make your Foldable.

Power of 10

exponent

power of 10

5.1

5.1×10^1 ▶ 51 ◀ 5.1×10

5.1×10^2 ▶ 510 ◀ 5.1×100

5.1×10^3 ▶ 5,100 ◀ $5.1 \times 1{,}000$

There is 1 zero in 10.
So, the exponent is 1.
$10 = 10^1$

There are 2 zeros in 100.
So, the exponent is 2.
$100 = 10^2$

There are 3 zeros in 1,000.
So, the exponent is 3.
$1{,}000 = 10^3$

FOLDABLES®
Study Organizer

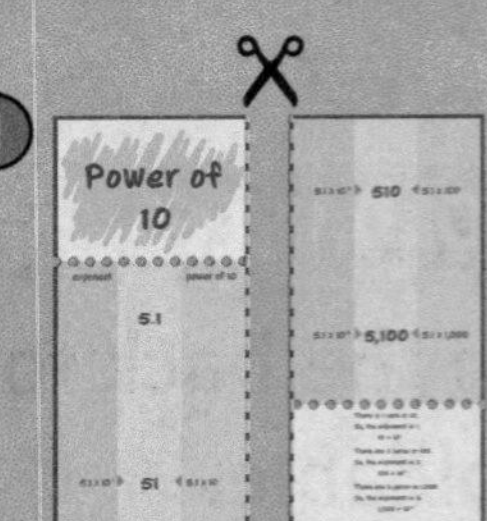
1
Power of 10

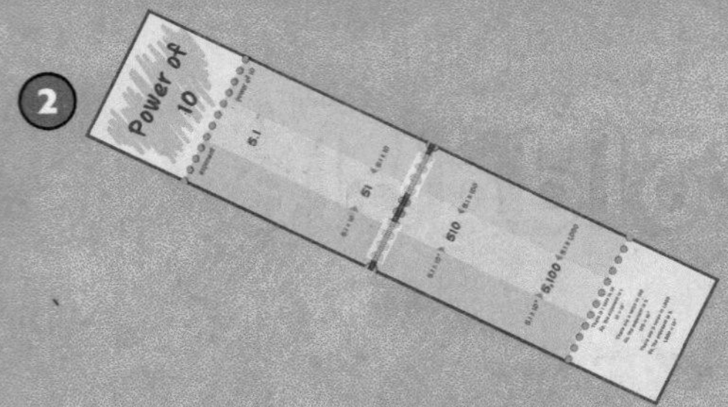
2
Power of 10

3
Power of 10

Name ____________________

Estimate Products of Whole Numbers and Decimals

Lesson 1

ESSENTIAL QUESTION
How is multiplying and dividing decimals similar to multiplying and dividing whole numbers?

Math in My World

Example 1

Mrs. James buys about 15.8 gallons of gas each week for her car. About how many gallons of gas will she buy in 5 weeks?

Estimate the product of 15.8 and 5.

Round.

$$\begin{array}{r} 15.8 \\ \times \quad 5 \\ \hline \end{array} \longrightarrow \begin{array}{r} \square \\ \times \quad 5 \\ \hline \end{array}$$

Then multiply to estimate the product. ⟶ $\square$

So, 15.8 × 5 is about ____.

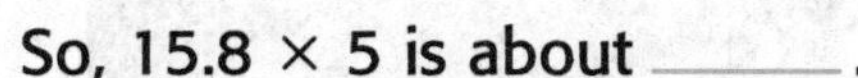

Mrs. James buys about ____ gallons of gas in 5 weeks.

Helpful Hint

Since 15.8 rounded *up* to 16, the estimated product is greater than the actual product.

Check Use repeated addition.

16 + 16 + 16 + 16 + 16 = ____ or

15.8 + 15.8 + 15.8 + 15.8 + 15.8 = ____

____ is close to the estimate, ____, so the answer is reasonable.

Example 2 

A worker loads 16 packages onto a truck. Each package weighs 58.5 pounds. About how many pounds are loaded onto the truck?

Estimate the product of 58.5 and 16.

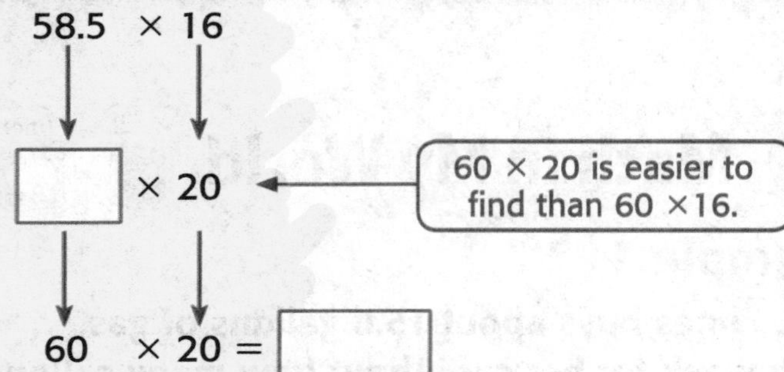

 Use compatible numbers.

58.5 × 16

☐ × 20 ← 60 × 20 is easier to find than 60 × 16.

 Multiply.

60 × 20 = ☐

So, 58.5 × 16 is about ______.

Talk MATH

Explain how to round 18.9 to the nearest whole number.

Guided Practice 

Estimate each product.

1. 2 × \$4.10

______

2. 18.4 × 10

______

3. 19.8 × 3

______

4. 5 × \$2.14

______

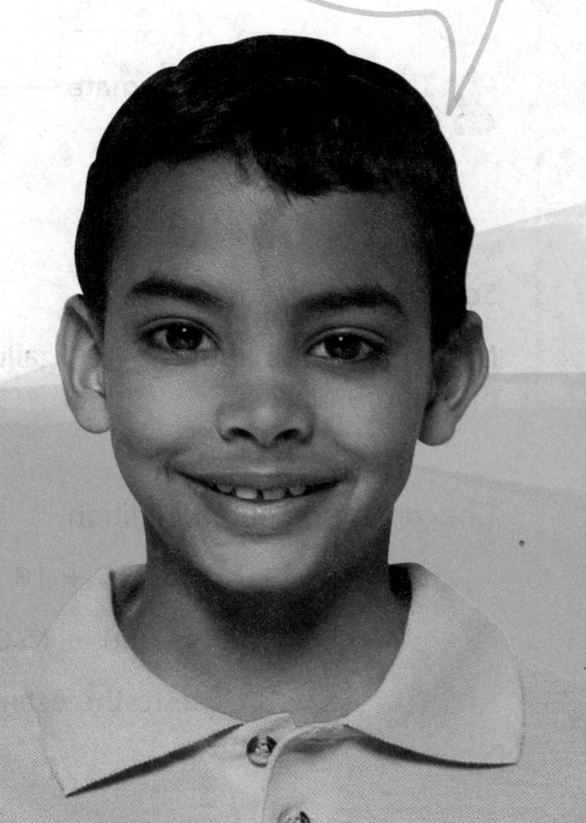

Name ____________________

Independent Practice

Estimate each product.

5. $4 \times \$4.62$ ____________

6. $3 \times \$23.07$ ____________

7. $\$15.50 \times 6$ ____________

8. $\$16.85 \times 9$ ____________

9. 7.2×5 ____________

10. 14.5×3 ____________

11. 9×19.7 ____________

12. 11×26.2 ____________

13. $\$0.89 \times 14$ ____________

14. 18.8×13 ____________

Problem Solving

Use the items below to estimate the cost of each order in Exercises 15 and 16.

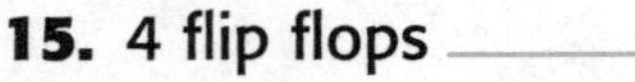

15. 4 flip flops ____________

16. 3 sunglasses and 2 flip flops

____

17. Mathematical PRACTICE 2 **Use Number Sense** About 57 people rent roller blades at the skating rink each day. It costs $4.25 to rent roller blades. About how much does the rink make in skate rentals in one day?

____

My Work!

HOT Problems

18. Mathematical PRACTICE 4 **Model Math** Write a real-world multiplication problem involving whole numbers and decimals in which you would use estimation to solve. Then estimate to solve the problem.

____

____

____

19. **Building on the Essential Question** Why is it important to use estimation when solving problems?

____

____

Name ....

MY Homework

Lesson 1

Estimate Products of Whole Numbers and Decimals

Homework Helper

Need help? connectED.mcgraw-hill.com

Estimate the product of 44.5 and 11.

Use compatible numbers.

44.5 × 11

↓ ↓

45 10

2 Multiply.

45 × 10 = 450

So, 44.5 × 11 is about 450.

Practice

Estimate each product.

1. \$27.64 × 3 ____________

2. 11.9 × 21 ____________

3. 87.2 × 41 ____________

4. 7.7 × 4 ____________

5. \$63.44 × 6 ____________

6. 51.7 × 9 ____________

7. 33.3 × 33 ____________

8. \$17.55 × 9 ____________

9. 17.6 × 51 ____________

Problem Solving

10. Mathematical PRACTICE 1 **Make Sense of Problems** Andrea earns $32.25 a day. After 9 days, about how much will she have earned?

11. Janie skates at the local skating rink for about 2.4 hours every day. If the skating rink is open 54 days a year, about how many hours does Janie skate each year?

12. Each Sunday during his nine-week summer vacation, Ray buys a newspaper. The Sunday paper costs $1.85. About how much does Ray spend on the Sunday newspaper during his vacation?

13. A group of 5 people goes to the theater. If each ticket costs $6.50, is $50 a reasonable estimate for the cost of 5 tickets? Explain.

14. An amusement park charges $35.50 for admission. On Saturday, 6,789 people visited the park. About how much money did the park earn from admission that day?

My Work!

Test Practice

15. In Spanish class, Kevin learns an average of 34.5 new words per month. If he takes Spanish for 9 months, which is the best estimate for about how many words he will learn?

Ⓐ 350 words
Ⓑ 355 words
Ⓒ 360 words
Ⓓ 365 words

Name

Hands On
Use Models to Multiply

Lesson 2

ESSENTIAL QUESTION
How is multiplying and dividing decimals similar to multiplying and dividing whole numbers?

Draw It

Find 0.4 × 2 using decimal models.

Shade 4 rows of each model to represent 0.4.

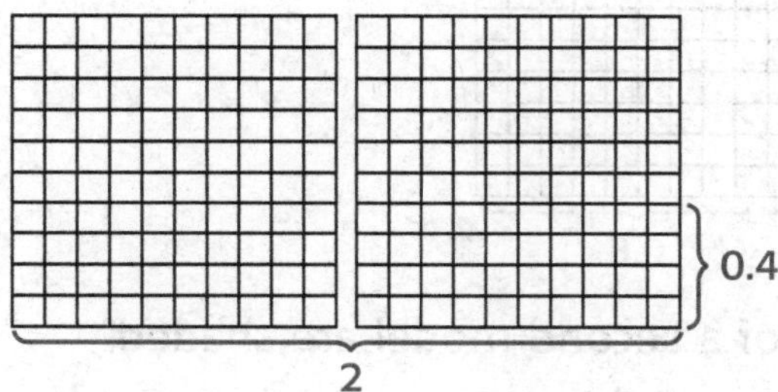

2 Combine both shaded parts onto one model.

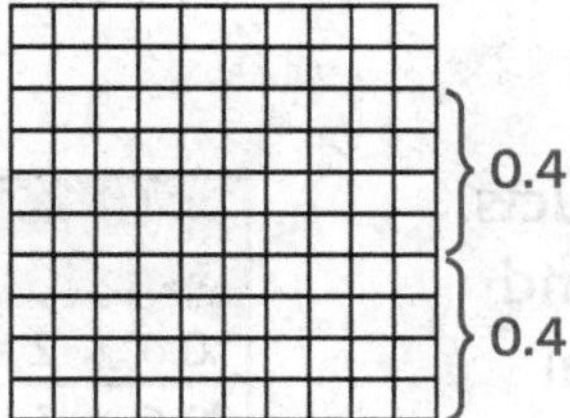

The amount shaded is ______. *Eight tenths* of the model is shaded.

So, 0.4 × 2 = ______.

Check Use repeated addition.

$$\begin{array}{r} 0.4 \\ +\,0.4 \\ \hline 0.8 \end{array}$$

Try It

Find 0.6 × 3 using decimal models.

Shade 6 rows of each decimal model to represent 0.6.

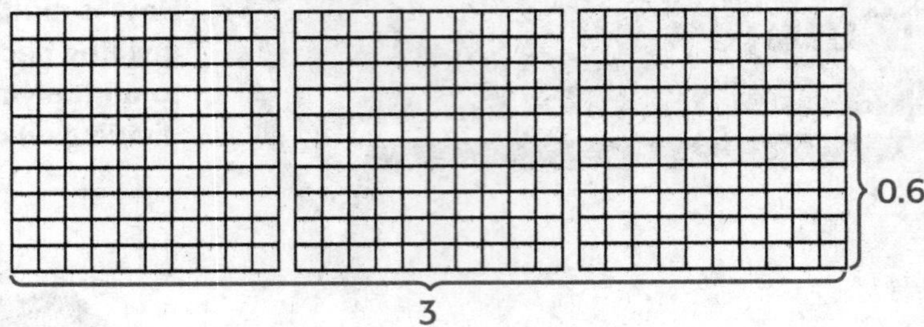

2 Combine all three shaded parts.

Since the combined shaded parts will not fit into one model, use two models.

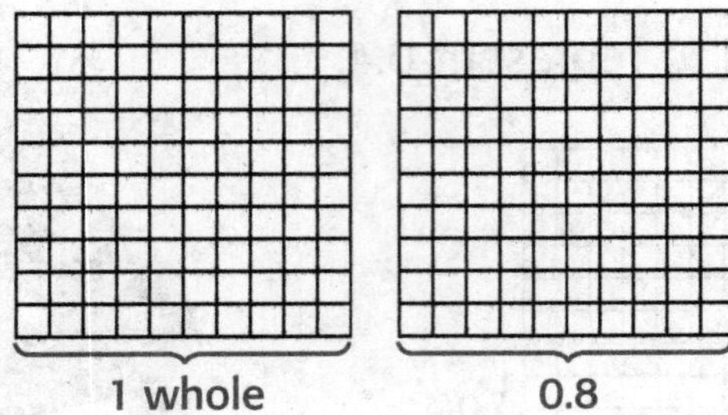

One whole model and *eight tenths* of a second model are shaded.

The total amount shaded is ______.

So, 0.6 × 3 = ______.

Talk About It

1. The table shows some factors and their products. Study the table. Write a rule you can use to find the product of a whole number and a decimal without using models.

Decimals	Whole Numbers
0.4 × 2 = 0.8	4 × 2 = 8
0.6 × 3 = 1.8	6 × 3 = 18
0.7 × 5 = 3.5	7 × 5 = 35

____

____

2. Mathematical PRACTICE 3 **Justify Conclusions** Use your rule from Exercise 1 to find 0.4 × 4 without using models. Explain the process you used.

____

____

Name ....

Practice It

Shade the models to find each product.

3. 0.3 × 2 = ______

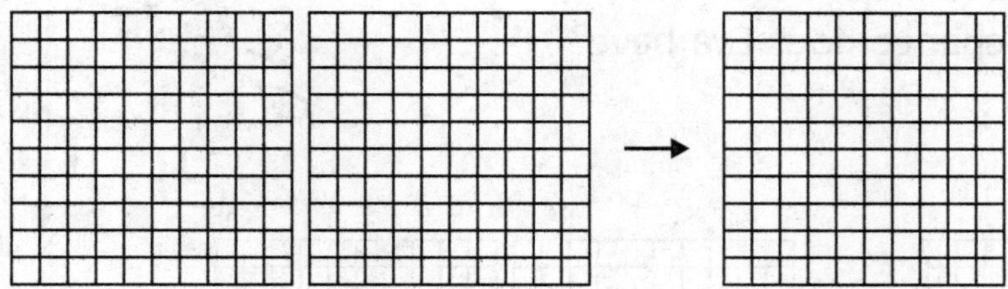

4. 2 × 0.7 = ______

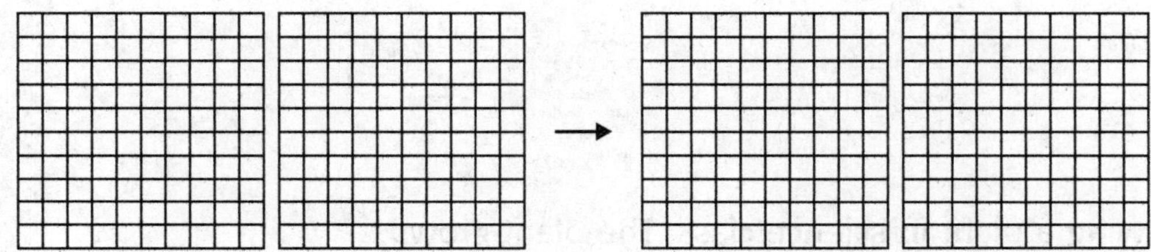

5. 0.3 × 3 = ______

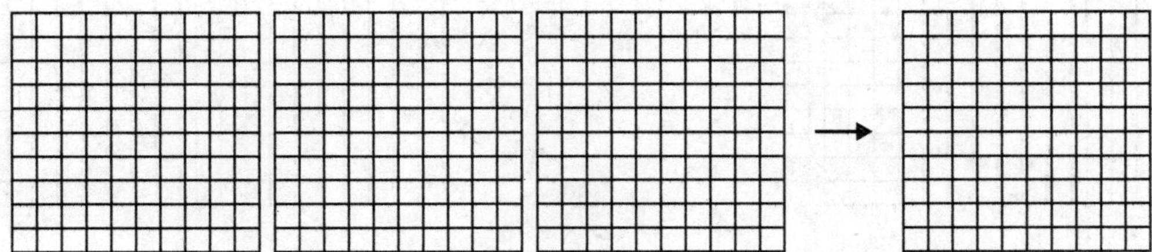

6. 3 × 0.1 = ______

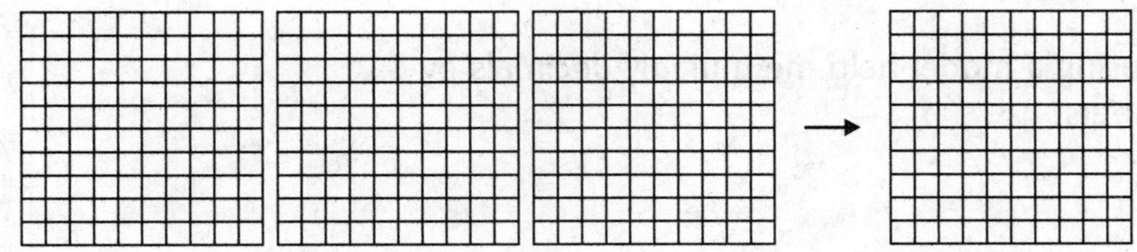

Apply It

Shade decimal models to solve each problem.

7. Mathematical PRACTICE 5 **Use Math Tools** Eva has change to buy bottled water after gymnastics class for herself and her friend. Each bottle of water is $0.70. How much change does Eva have? Use repeated addition.

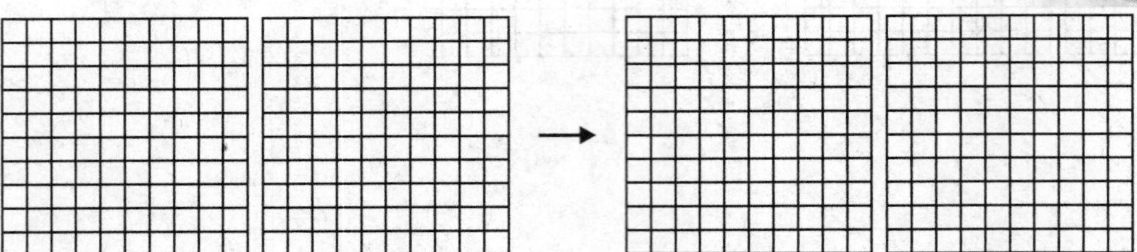

8. Caleb is growing a plant in science class. The plant grows 0.4 centimeters each week. How much will the plant grow in 3 weeks?

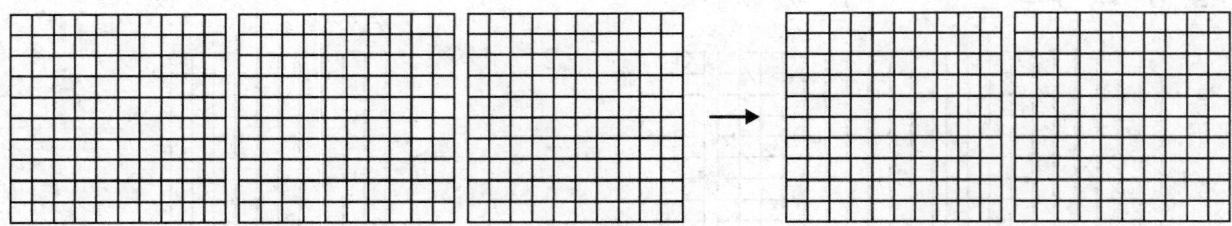

Write About It

9. How does using a model help me multiply decimals by whole numbers?

Name ....

MY Homework

Lesson 2
Hands On: Use Models to Multiply

Homework Helper

Need help? connectED.mcgraw-hill.com

Find 0.2 × 3 using decimal models.

1 The two rows of each model that are shaded represent 0.2.

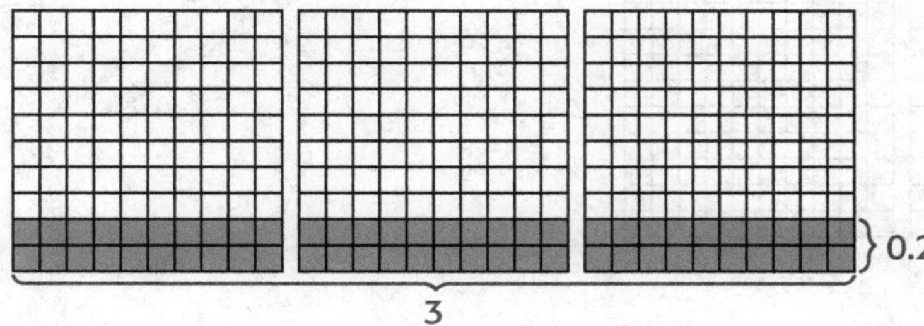

2 The shaded parts are combined onto one model.

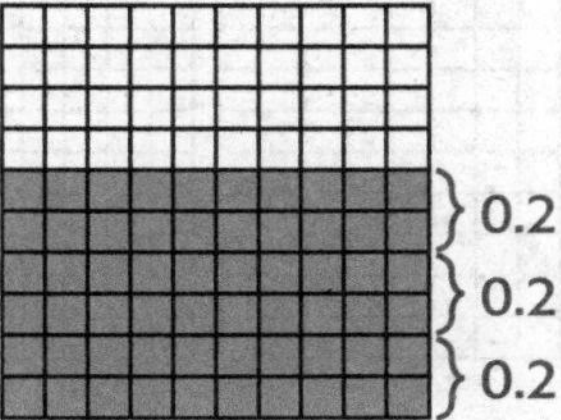

Six tenths of the model is shaded.

So, 0.2 × 3 = 0.6.

Practice

1. Shade the models to find 3 × 0.7.

3 × 0.7 = ______

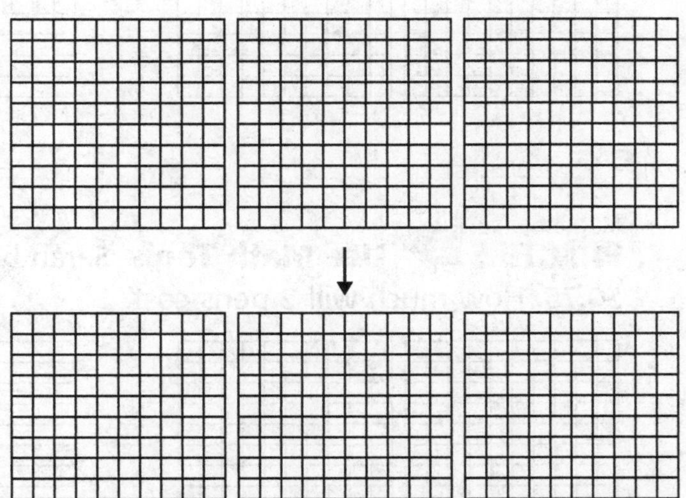

Real World Problem Solving

Shade decimal models to solve each problem.

2. Tyrell lives in a town that is 0.8 kilometer above sea level. He works at a ski resort that is 3 times higher above sea level. How many kilometers above sea level is the ski resort?

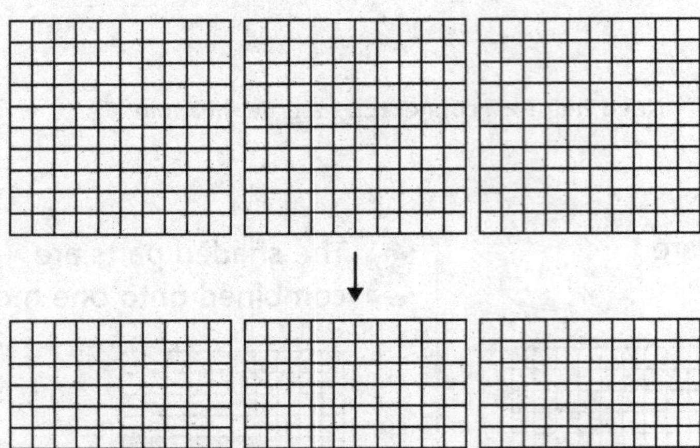

3. Poncio walks 0.5 mile to school each day. Multiply the distance by 2 to find the total distance he walks each day.

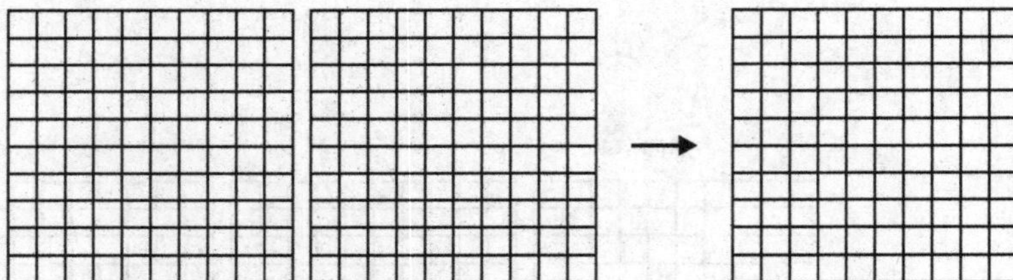

4. Mathematical PRACTICE 5 **Use Math Tools** Sarah buys a pen that costs \$0.75. How much will 2 pens cost?

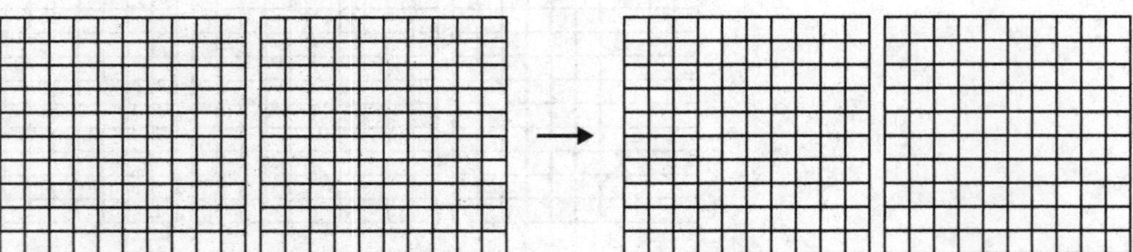

Name ....

Multiply Decimals by Whole Numbers

Lesson 3

ESSENTIAL QUESTION
How is multiplying and dividing decimals similar to multiplying and dividing whole numbers?

Math in My World

Example 1

Find the area of a rectangular board that is 4 feet by 3.62 feet.

One Way **Use repeated addition.**

4×3.62 means ______ + ______ + ______ + ______.

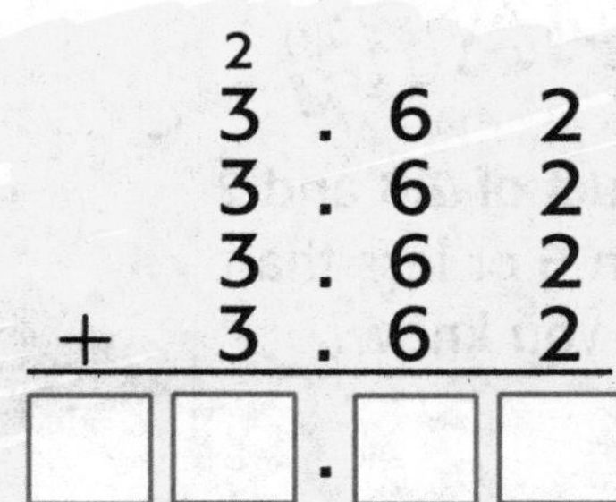

$$\begin{array}{r} \overset{2}{3}.62 \\ 3.62 \\ 3.62 \\ +\ 3.62 \\ \hline \square\square.\square\square \end{array}$$

The area of a rectangle is found by multiplying the length by the width.

Another Way **Multiply as with whole numbers. Then count decimal places.**

$$\begin{array}{r} \overset{2}{3}.62 \\ \times\ 4 \\ \hline 14.48 \end{array}$$

There are ______ places to the right of the decimal point.

Count ______ decimal places from right to left in the product.

Check The answer is the same whether you use repeated addition or multiplication. So, the answer is reasonable.

Example 2 

Find 3 × 0.96.

Estimate 3 × 1 = ______

1. Multiply as with whole numbers.

Write the number with the most digits on top. → 0.96 × 3

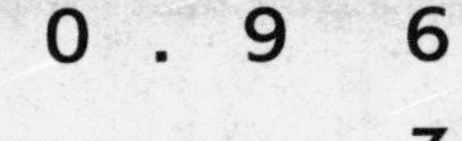

2. Count the decimal places. There are 2 places to the right of the decimal point in 0.96. So, count 2 places from right to left in the product.

So, 3 × 0.96 = ______.

Check The product, ______, is close to the estimate, ______.

The answer is reasonable.

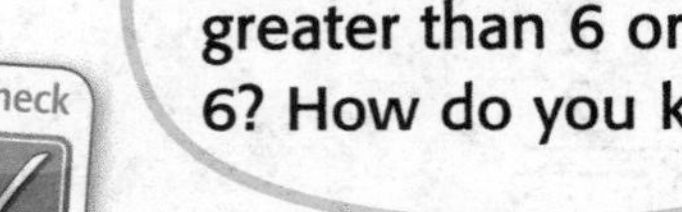

Is the product of 2.8 and 2 greater than 6 or less than 6? How do you know?

Guided Practice 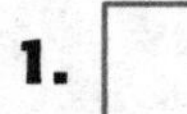

Multiply. Check for reasonableness.

1. 0.5 × 6 = ☐.☐

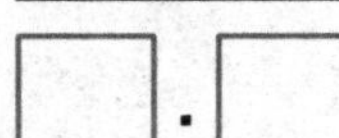

2. 2.6 × 4 = ☐☐.☐

Name

Independent Practice

Multiply. Check for reasonableness.

3. 2.49×3

4. 1.59×7

5. 3.4×7

6. $2 \times 1.3 =$ ______

7. $3 \times 0.5 =$ ______

8. $1.8 \times 9 =$ ______

9. 0.48×3

10. 2.4×8

11. 0.02×4

12. $0.66 \times 5 =$ ______

13. $67 \times 4.3 =$ ______

14. $52 \times 2.1 =$ ______

Algebra **Find each unknown.**

15. $0.8 \times 9 = h$

$h =$ ______

16. $6 \times 0.05 = x$

$x =$ ______

17. $1.48 \times 7 = m$

$m =$ ______

Problem Solving

18. Mr. Thomas' car can travel 17.6 miles per gallon of gas. If his gas tank holds 11 gallons, how far can he travel on a full tank of gas?

19. Tom and Sonia both mowed lawns last week. Tom mowed five lawns and charged $10.75 per lawn. Sonia mowed four lawns but charged $2.25 more per lawn than Tom. Who made the most money?

HOT Problems

20. Mathematical PRACTICE 1 **Plan Your Solution** Write a multiplication problem in which the product has two decimal places.

21. Mathematical PRACTICE 2 **Use Number Sense** Place the decimal point in the answer below to make it correct. Explain your reasoning.

498 × 8.32 = 4 1 4 3 3 6

22. **Building on the Essential Question** How is multiplying a decimal by a whole number similar to and different from multiplying two whole numbers?

Name ....

MY Homework

Lesson 3

Multiply Decimals by Whole Numbers

Homework Helper

Need help? connectED.mcgraw-hill.com

Find 3 × 0.85.

One Way **Use repeated addition.**

3 × 0.85 means 0.85 + 0.85 + 0.85.

$$\begin{array}{r} {}^{2\,1} \\ 0.85 \\ 0.85 \\ +\ 0.85 \\ \hline 2.55 \end{array}$$

Another Way **Multiply as with whole numbers. Then count decimal places.**

$$\begin{array}{r} {}^{2\,1} \\ 0.85 \\ \times\ \ 3 \\ \hline 2.55 \end{array}$$

There are two places to the right of the decimal point.

Count two decimal places from right to left in the product.

Practice

Multiply. Check for reasonableness.

1. $\begin{array}{r} 1.7 \\ \times 4 \\ \hline \end{array}$

2. $\begin{array}{r} 0.62 \\ \times\ 2 \\ \hline \end{array}$

3. $\begin{array}{r} 0.5 \\ \times 9 \\ \hline \end{array}$

4. 3.6 × 8 = ________

5. 5.1 × 7 = ________

6. 4 × 2.3 = ________

Algebra **Find each unknown.**

7. 2 × 0.33 = p

p = ________

8. 5 × 2.4 = n

n = ________

9. 7 × 8.1 = s

s = ________

Problem Solving

10. Justin bought 9 packs of trading cards. Each pack of cards costs \$3.27. What was the total cost for all 9 packs?

11. Richard paints a picture on a rectangular canvas that is 3 feet by 2.64 feet. What is the area of Richard's painting?

12. Mathematical PRACTICE 6 **Explain to a Friend** Taylor wants to buy a bicycle that costs \$48.75. She works for her mom after school to earn the money. If Taylor is paid \$4.50 per hour, will she be able to buy the bicycle after working 9 hours? Explain to a friend.

13. Janita practices on the track field 1.5 hours each day before a track meet. If there are 6 days until the track meet, how many hours will Janita practice?

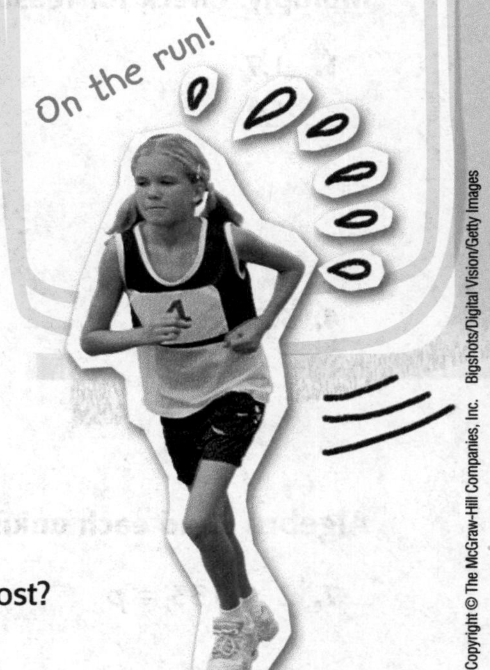

Test Practice

14. Kent buys a pen that costs \$1.21. How much will 4 pens cost?

Ⓐ \$3.63
Ⓑ \$3.84
Ⓒ \$4.21
Ⓓ \$4.84

Name

Hands On
Use Models to Multiply Decimals

Lesson 4

ESSENTIAL QUESTION
How is multiplying and dividing decimals similar to multiplying and dividing whole numbers?

Draw It

Tools

Find 0.3 × 0.7 using a decimal model.

1 Use one 10-by-10 decimal model.

2 Shade a rectangle that is 0.3 unit wide and 0.7 unit long.

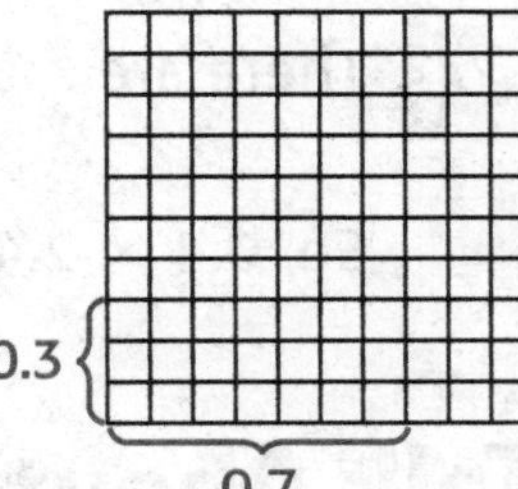

3 There are ______ hundredths in the shaded region. The shaded region represents the area of a rectangle with a length of 0.7 unit and a width of 0.3 unit. So, 0.3 × 0.7 = ______.

Talk About It

1. How are the equations 3 × 7 = 21 and 0.3 × 0.7 = 0.21 similar? How are they different?

______

______

2. Mathematical PRACTICE 5 **Use Math Tools** How is using a model to multiply 0.3 by 0.7 similar to finding the area of a rectangle?

______

______

Try It

Find 0.4 × 2.4 using decimal models.

1 Use three 10-by-10 decimal models side by side.

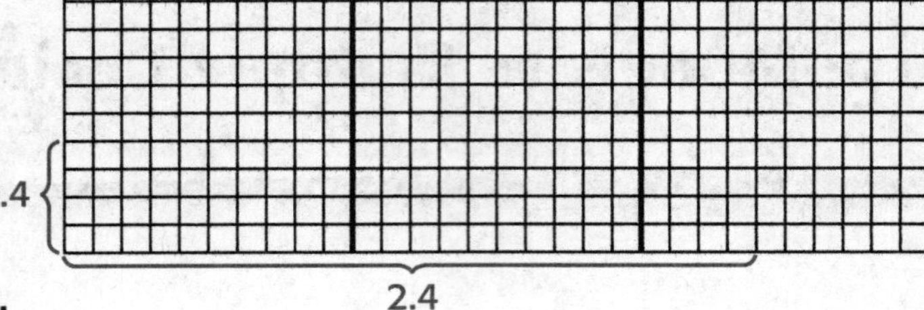

2 Shade a rectangle that is 0.4 unit wide and ______ units long.

3 Shade the same amount of squares on a 10-by-10 grid.

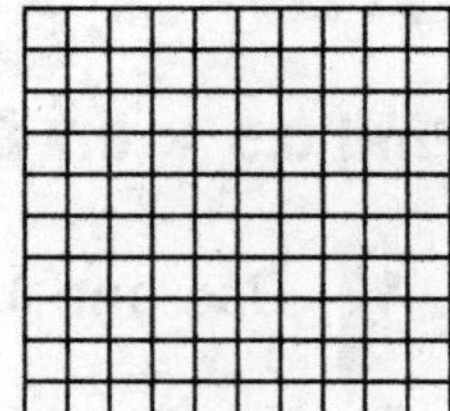

4 There are ______ hundredths in the shaded region.

So, 0.4 × 2.4 = ______.

Talk About It

3. Without using models, explain why 2 tenths times 3 tenths is equal to 6 *hundredths.* Use place value in your explanation.

____

4. The table shows some factors and their products. Study the table. Write a rule you can use to find the product of two decimals, both to the tenths place, without using models.

Decimals	Whole Numbers
0.3 × 0.7 = 0.21	3 × 7 = 21
0.4 × 2.4 = 0.96	4 × 24 = 96
0.5 × 1.1 = 0.55	5 × 11 = 55

____

____

5. Mathematical PRACTICE 2 **Stop and Reflect** Without using models, find 0.4 × 0.8. Explain how you found the product.

____

____

Name ....

Practice It

Shade the decimal models to find each product.

6. 0.4 × 0.8 = ________

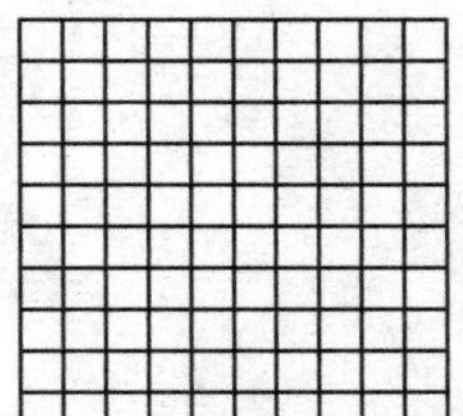

7. 0.5 × 0.6 = ________

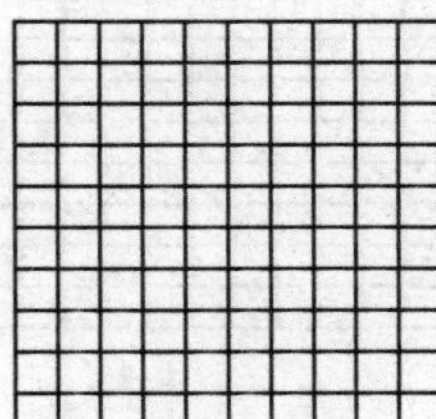

8. 0.2 × 0.9 = ________

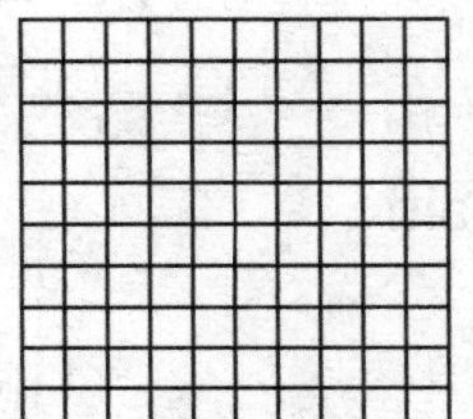

9. 0.4 × 0.6 = ________

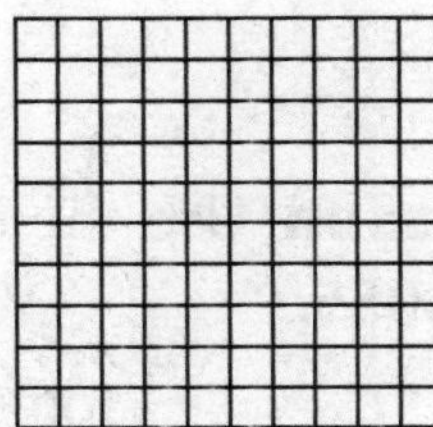

10. 0.3 × 1.8 = ________

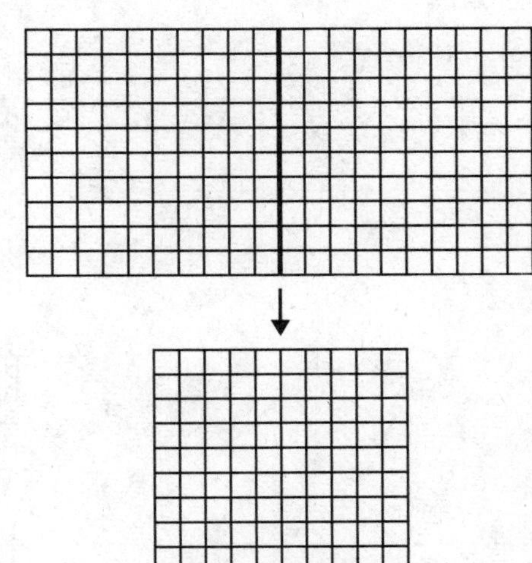

11. 0.2 × 1.4 = ________

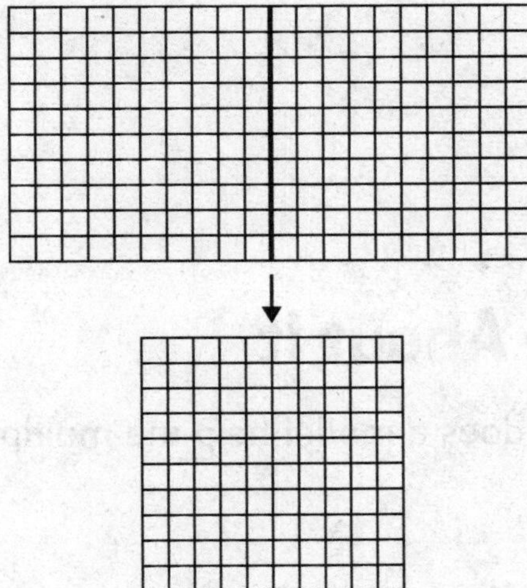

Apply It

12. Diana's house has a rectangular window on the door with a height of 1.5 feet and a width of 0.8 feet. What is the area of the window? Shade the decimal models to solve the problem.

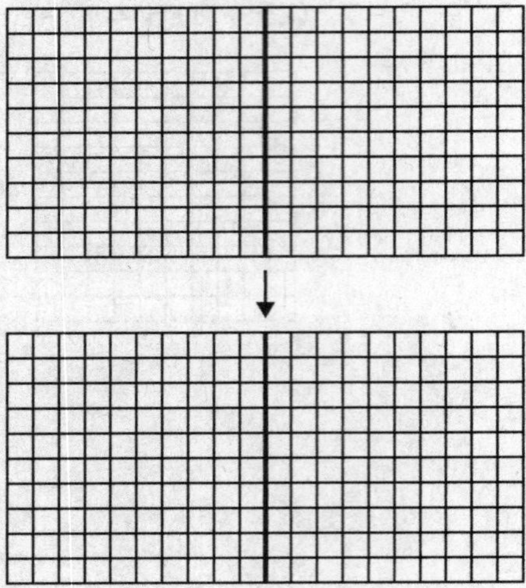

____

13. Mathematical PRACTICE 2 **Reason** Use the decimal model below to explain why 0.5 × 0.11 is 0.055.

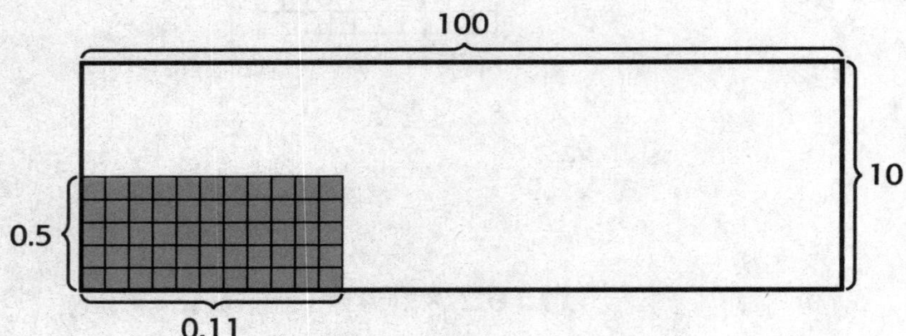

____

____

Write About It

14. How does a model help me multiply decimals?

____

____

Name

MY Homework

Lesson 4

Hands On: Use Models to Multiply Decimals

Homework Helper

Need help? connectED.mcgraw-hill.com

Find 0.4 × 0.6 using a decimal model.

1 Use one 10-by-10 decimal model.

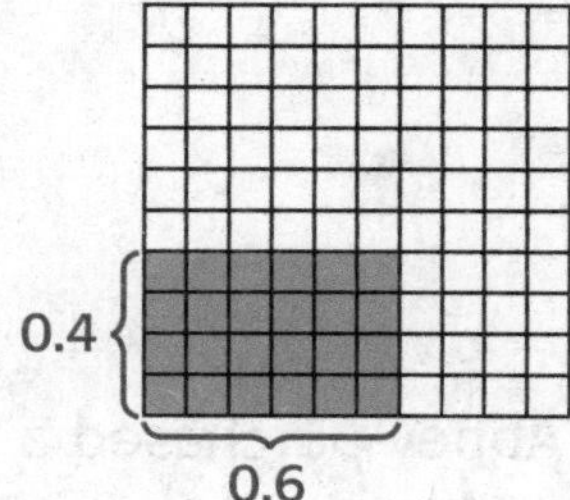

2 The shaded rectangle represents 0.4 unit wide and 0.6 unit long.

3 There are twenty-four hundredths in the shaded region.

So, 0.4 × 0.6 = 0.24.

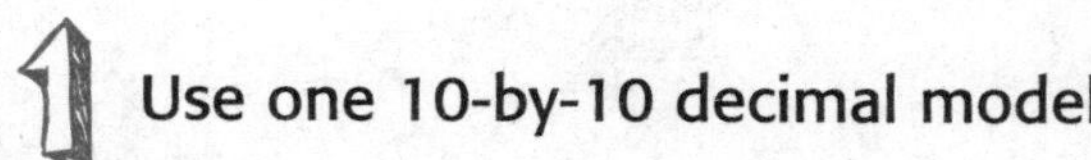

Practice

Shade the decimal models to find each product.

1. 0.7 × 0.5 = ______

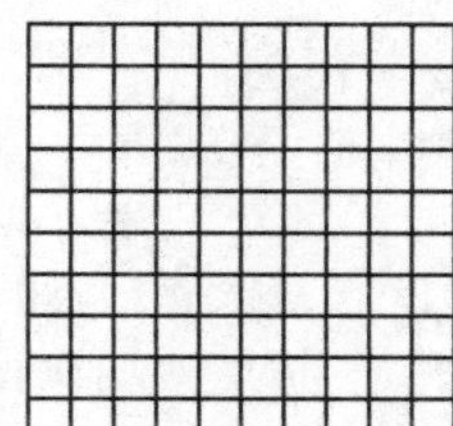

2. 0.8 × 0.2 = ______

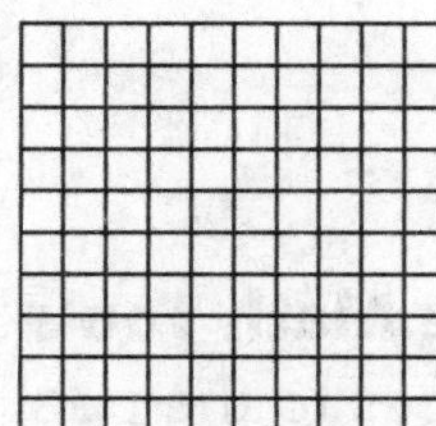

3. 0.6 × 2.2 = ______

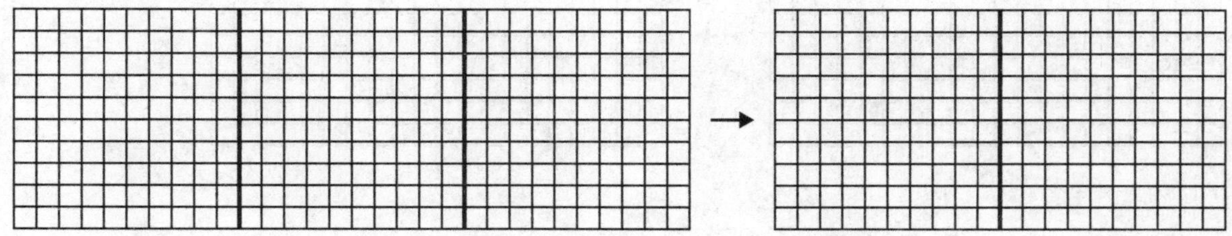

Real World Problem Solving

Shade the decimal models to solve each problem.

4. Lanny bought 1.4 pounds of rice that cost $0.40 per pound. How much does Lanny pay for the rice?

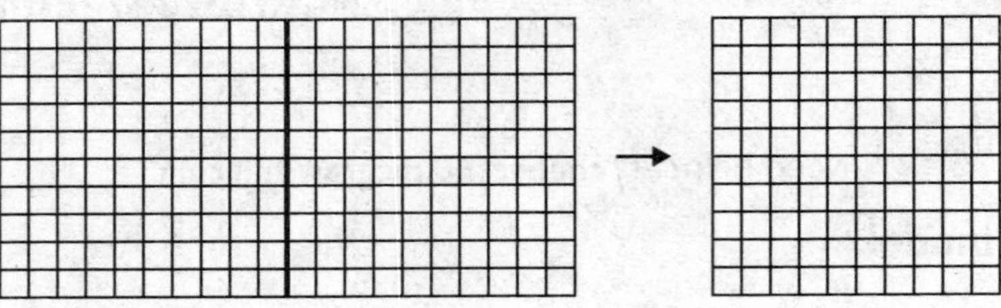

____

5. Abbey purchased a poster that had a height of 2 feet and a width of 0.7 foot. What is the area of Abbey's poster?

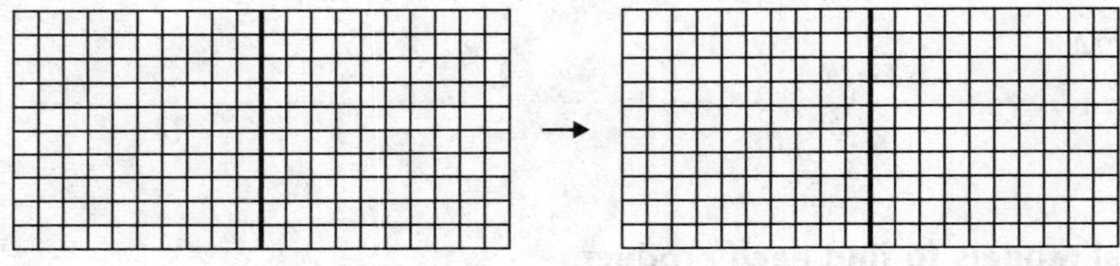

____

6. Mathematical PRACTICE 5 **Use Math Tools** Diana spends 0.8 hour per day at the pool. If she goes to the pool for 2 days one week, how many total hours does Diana spend at the pool that week?

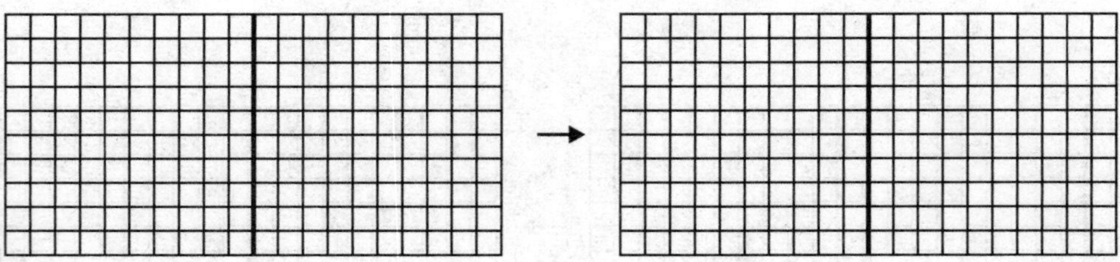

____

Name ....

Multiply Decimals

Lesson 5

ESSENTIAL QUESTION
How is multiplying and dividing decimals similar to multiplying and dividing whole numbers?

Math in My World

I am nuts about decimals!

Example 1

Nina is buying peanuts in bulk. The peanuts cost \$0.75 for each pound. She purchased 2.1 pounds of peanuts. How much will Nina pay? Round to the nearest cent.

Find \$0.75 × 2.1.

Estimate \$0.75 × 2.1 ⟶ ______ × ______ or ______

1. Multiply as with whole numbers.

2. Count the decimal places.

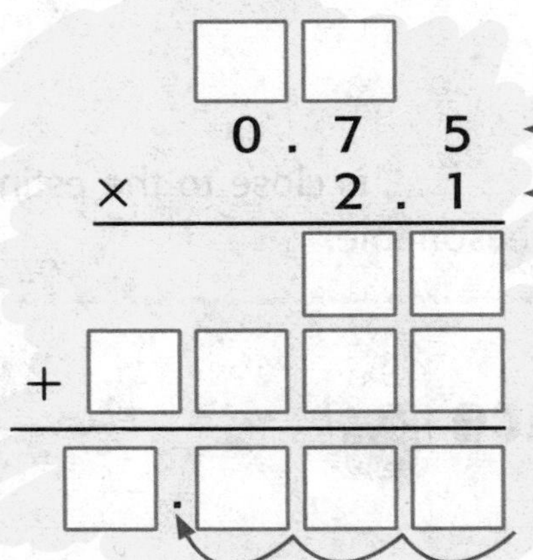

3. Since 2 + 1 = 3, count 3 decimal places from the right to place the decimal point.

4. Round to the nearest cent.

\$1.575 ⟶ ______

So, Nina will need to pay ______.

Check \$2 ≈ ______

Example 2 

Demarcus is buying a new guitar pick that costs \$0.50. The sales tax is found by multiplying the cost of the guitar pick by 0.06. What is the cost of the sales tax for the pick?

Find \$0.50 × 0.06.

Estimate \$0.50 × 0.06 ⟶ \$1 × 0 = \$0

1 Multiply as with whole numbers.

2 Count the decimal places.

3 Since 2 + 2 = 4, count 4 decimal places from the right. Annex a zero to place the decimal point.

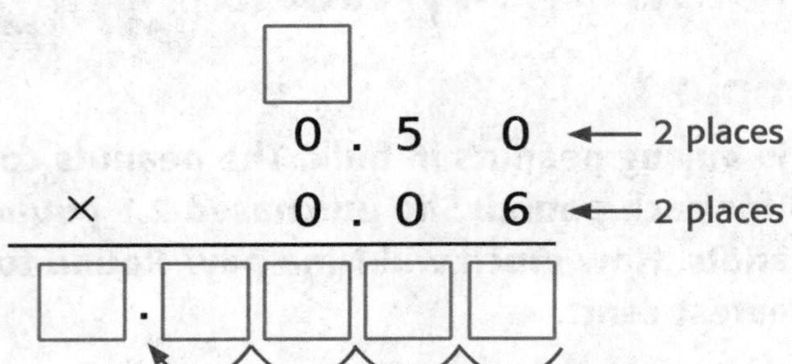

So, 0.50 × 0.06 = ________.

The sales tax is ________.

Check The product, ________, is close to the estimate, ________. The answer is reasonable.

Talk MATH

Describe a multiplication problem in which the product is between 0.005 and 1.

Guided Practice 

Multiply.

1. 0.6 × 0.8 = ________

2. 1.7 × 2.4 = ________

3. 0.9 × 3.8 = ________

Name

Independent Practice

Multiply. Check for reasonableness.

4. $\begin{array}{r} 0.96 \\ \times\ 7.1 \\ \hline \end{array}$

5. $\begin{array}{r} 3.65 \\ \times\ 2.6 \\ \hline \end{array}$

6. $\begin{array}{r} 0.07 \\ \times\ 5.2 \\ \hline \end{array}$

7. $\begin{array}{r} 2.78 \\ \times\ 0.8 \\ \hline \end{array}$

8. $\begin{array}{r} 0.35 \\ \times\ 0.15 \\ \hline \end{array}$

9. $\begin{array}{r} 3.24 \\ \times\ 6.4 \\ \hline \end{array}$

10. $0.9 \times 0.3 =$ ______

11. $1.6 \times 3.2 =$ ______

12. $0.5 \times 6.7 =$ ______

Algebra **Find each unknown.**

13. $0.81 \times 7.3 = b$

$b =$ ______

14. $5.6 \times 3.9 = p$

$p =$ ______

15. $1.2 \times 0.05 = g$

$g =$ ______

Problem Solving

16. A nutrition label says that one serving out of a bag of chips has 3.7 grams of fat. How many grams of fat are there in 2.5 servings?

17. Apples cost $0.98 per pound. How much would it cost to purchase 5.5 pounds of apples?

18. The average honeybee can travel 10.5 feet per second. How many feet can the honeybee travel in 3.7 seconds?

19. Mathematical PRACTICE 6 **Be Precise** Jocelyn's swimming pool is shown. Find the area of the swimming pool.

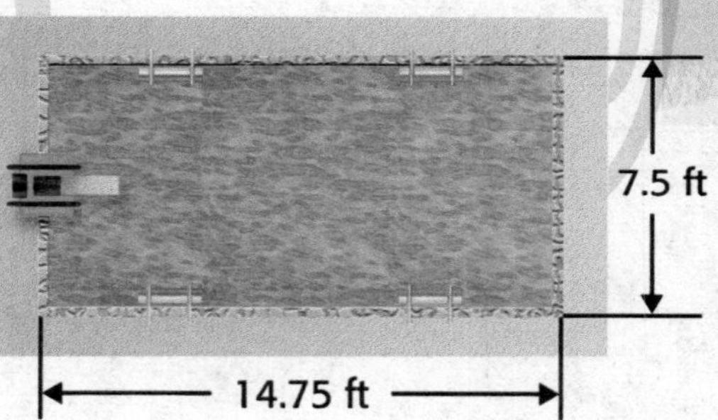

HOT Problems

20. Mathematical PRACTICE 4 **Model Math** Write a multiplication problem using two decimals that has a product greater than 0.1 but less than 0.2.

21. **Building on the Essential Question** How is multiplying decimals different from multiplying whole numbers?

Name ....

MY Homework

Lesson 5
Multiply Decimals

Homework Helper

Need help? connectED.mcgraw-hill.com

Find 0.32 × 0.3.

Estimate 0.32 × 0.3 ⟶ 0 × 0 = 0

1. Multiply as with whole numbers.

2. Count the decimal places.

$$\begin{array}{r} 0.32 \\ \times\ 0.3 \\ \hline 0.096 \end{array}$$

0.32 ← 2 decimal places
× 0.3 ← 1 decimal place
0.096 ← Annex a zero.

3. Since 2 + 1 = 3, count 3 decimal places from the right. Annex a zero to place the decimal point.

So, 0.32 × 0.3 = 0.096.

Check The product, 0.096, is close to the estimate, 0. The answer is reasonable.

Practice

Multiply. Check for reasonableness.

1. 0.28 × 0.03 = ______

2. 0.11 × 0.91 = ______

3. 5.14 × 0.4 = ______

4. 0.98 × 0.23 = ______

5. 0.52 × 0.48 = ______

6. 6.34 × 0.7 = ______

Algebra **Find each unknown.**

7. $0.47 \times 0.18 = t$

$t =$ ___________

8. $4.15 \times 6.3 = x$

$x =$ ___________

9. $0.9 \times 1.02 = w$

$w =$ ___________

Problem Solving

10. Tamara recycles 29.5 pounds of aluminum cans at a recycling center that pays $0.26 per pound. How much does Tamara receive for the cans?

____

11. Wes measures the floor area of his rectangular living room to replace the carpet with tiles. The room is 12.4 feet long and 9.8 feet wide. How many square feet of tile will Wes need?

____

12. Mathematical PRACTICE 2 **Use Number Sense** Lenora buys 3.7 pounds of potatoes that cost $0.29 per pound. How much does Lenora pay for the potatoes? Round to the nearest cent.

____

My Work!

1 potato, 2 potato, 3 potato, four

Test Practice

13. Mikayla is buying deli meat and cheese for sandwiches for a family reunion. How much will she spend on 4.5 pounds of ham and 3.5 pounds of cheese?

Ⓐ $7.39

Ⓑ $13.27

Ⓒ $20.66

Ⓓ $33.92

Check My Progress

Vocabulary Check

Draw lines to match each word(s) with its correct description or meaning.

1. compatible numbers • an approximate value

2. estimate • the result of a multiplication problem

3. product • numbers that are easy to compute mentally

Concept Check

Estimate each product.

4. \$1.80 × 8 ________

5. \$2.83 × 7 ________

Multiply. Check for reasonableness.

6. $\begin{array}{r} 1.9 \\ \times\ 8 \\ \hline \end{array}$

7. $\begin{array}{r} 3.4 \\ \times\ 7 \\ \hline \end{array}$

8. 2.3 × 2 = ________

9. 0.24 × 5 = ________

Problem Solving

10. Jorge bought 8 pounds of ground beef for $3.29 a pound. About how much did he pay altogether?

My Work!

11. The Marino family bought 4 tickets to the circus. What was the total cost of the tickets?

12. Wendy paints a rectangular wall that is 10.5 feet tall and 9.2 feet wide. What is the area of the wall that she paints?

13. One pound of tomatoes costs $1.59. How much do 5 pounds of tomatoes cost?

Test Practice

14. Mr. Garner's gas tank holds 17.5 gallons of gasoline. How much will it cost him to fill up his gas tank if gasoline costs $2.48 per gallon?

Ⓐ $34

Ⓑ $42.16

Ⓒ $43.40

Ⓓ $45.20

Name ______________________

Multiply Decimals by Powers of Ten

Lesson 6

ESSENTIAL QUESTION
How is multiplying and dividing decimals similar to multiplying and dividing whole numbers?

Math in My World

Example 1

Josh will need to make 10 payments of $32.25 to purchase a new skimboard. What is the total cost of the skimboard?

Totally worth it!

Find $32.25 × 10.

$$\begin{array}{r} 32.25 \\ \times\ 10 \\ \hline 0000 \\ +\ 32250 \\ \hline 322.50 \end{array}$$

The non-zero digits are the same. The decimal point in the product moved one place to the right.

Multiplying a decimal by 10 moves the decimal point one place to the right.

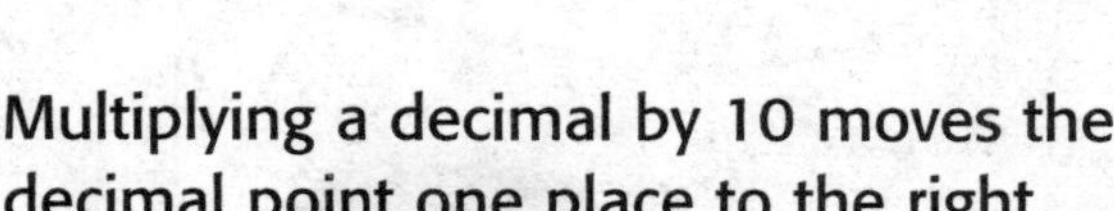

So, the skimboard costs ________.

Numbers like 10, 100, and 1,000 are powers of ten. They can be written with exponents with a base of ten.

There is 1 zero in 10. The exponent on 10^1 is 1.

There are 2 zeros in 100. The exponent on 10^2 is 2.

There are 3 zeros in 1,000. The exponent on 10^3 is 3.

Power of Ten	Written with Exponent
10	10^1
100	10^2
1,000	10^3

To multiply a decimal by a power of ten, move the decimal point to the right the same number of zeros in the power of ten. This is also the same number as the exponent on 10.

Example 2 

Find 24.7×10^2.

One Way **Multiply.**

Since $10^2 = 100$, find 24.7×100.

$$\begin{array}{r} 24.7 \\ \times\ 100 \\ \hline 000 \\ 0000 \\ +\ 24700 \\ \hline 2{,}470.0 \end{array}$$

The non-zero digits are the same. The decimal point in the product moved two places to the right.

Another Way **Move the decimal point.**

Multiplying a decimal by 100 moves the decimal point two places to the right.

Annex a zero.

24.70 → 2,470

So, $24.7 \times 10^2 =$ ________.

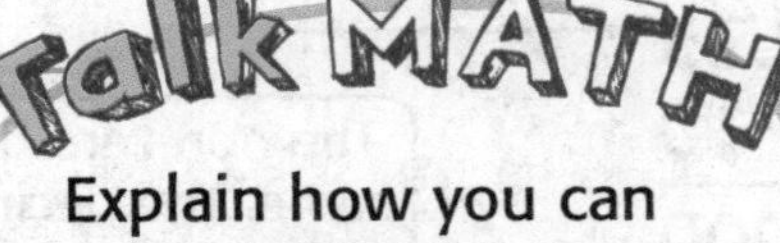

Explain how you can mentally find the cost of 10 text messages that each cost $0.25.

Guided Practice 

Multiply.

1. $0.54 \times 10 =$ ________

2. $8.32 \times 100 =$ ________

3. $7.46 \times 1{,}000 =$ ________

Name

Independent Practice

Multiply.

4. $1.63 \times 10 =$ ______ **5.** $0.853 \times 10^3 =$ ______ **6.** $0.397 \times 10^1 =$ ______

7. $1.76 \times 100 =$ ______ **8.** $0.78 \times 10^2 =$ ______ **9.** $76.5 \times 10^3 =$ ______

10. $0.81 \times 10 =$ ______ **11.** $1.23 \times 10^2 =$ ______ **12.** $0.48 \times 100 =$ ______

Algebra Find each unknown.

13. $0.93 \times 10 = a$

$a =$ ______

14. $22.94 \times 10^2 = n$

$n =$ ______

15. $0.05 \times 1{,}000 = w$

$w =$ ______

Problem Solving

16. A music store has 10 flutes. Each flute costs \$325.50. What is the cost of all 10 flutes?

17. Mathematical PRACTICE 1 **Make a Plan** One carton of milk costs \$0.99. What is the total cost of 10^2 cartons of milk?

18. Titus's and Cam's hourly charges for yard work are shown. Suppose Titus and Cam each worked 10 hours. How much money did they earn together?

Titus	Cam
\$8.25	\$5.58

19. The length of Jenny's dog is 10 times the length of her hamster. The length of her hamster is 8.4 centimeters. What is the length of her dog?

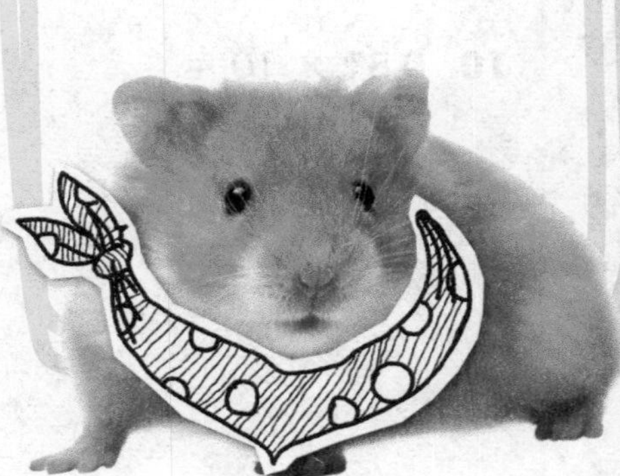

HOT Problems

20. Mathematical PRACTICE 2 **Use Number Sense** Find $0.346 \times 10^2 \times 10$.

21. **Building on the Essential Question** How does the exponent of each power of ten correspond with placing a decimal?

MY Homework

Lesson 6
Multiply Decimals by Powers of Ten

Homework Helper

Need help? connectED.mcgraw-hill.com

Find 0.35 × 100.

One Way **Multiply.**

$$\begin{array}{r} 0.35 \\ \times\ 100 \\ \hline 000 \\ 0000 \\ +\ 3500 \\ \hline 35.00 \end{array}$$

The non-zero digits are the same. The decimal point in the product moved two places to the right.

Another Way **Move the decimal point.**

Multiplying a decimal by 100 moves the decimal point two places to the right.

0.35 → 35

So, 0.35 × 100 = 35.

Practice

Multiply.

1. $0.13 \times 10 =$ ______

2. $1.4 \times 1{,}000 =$ ______

3. $4.81 \times 10^3 =$ ______

4. $0.72 \times 10^2 =$ ______

5. $0.179 \times 1{,}000 =$ ______

6. $67.2 \times 10^1 =$ ______

Problem Solving

For Exercises 7 and 8, use the table which shows a school store's prices.

Item	Price
Notebook	$1.25
Pencil	$0.50
Binder	$2.15
Pen	$0.80

7. How much would it cost to buy 10 pens from the school store?

8. How much will it cost for 10 pencils and one notebook?

9. The school purchased 100 trophies to give to the honor roll students. If one trophy costs $4.32, how much would it cost for all of the trophies?

10. Mathematical PRACTICE 3 **Find the Error** Rico is finding 7.5×100. Find and correct his mistake.

7.5×100

$07.5 = 0.075$

Test Practice

11. Elyse is cutting strips of paper for a scrapbook page. Each strip is 1.5 centimeters wide. How wide will 10 strips of paper be?

Ⓐ 1.5 centimeters
Ⓑ 15 centimeters
Ⓒ 150 centimeters
Ⓓ 1,500 centimeters

Name ....

Real World

Problem-Solving Investigation

STRATEGY: Look For a Pattern

Lesson 7

ESSENTIAL QUESTION
How is multiplying and dividing decimals similar to multiplying and dividing whole numbers?

Learn the Strategy

Watch Tutor

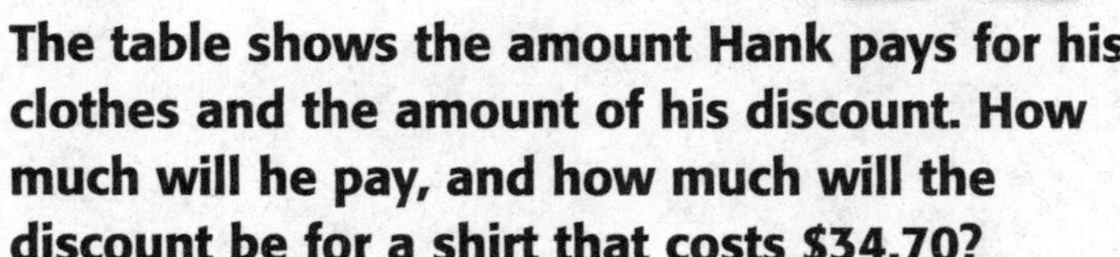

The table shows the amount Hank pays for his clothes and the amount of his discount. How much will he pay, and how much will the discount be for a shirt that costs $34.70?

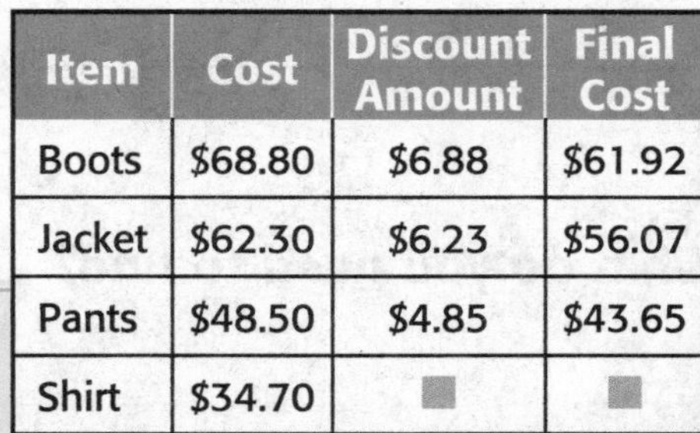

Item	Cost	Discount Amount	Final Cost
Boots	$68.80	$6.88	$61.92
Jacket	$62.30	$6.23	$56.07
Pants	$48.50	$4.85	$43.65
Shirt	$34.70	■	■

1 Understand

What facts do you know?

- Hank will receive a __________ on his __________.

What do you need to find?

- I need to find the amount Hank will pay for the __________.

2 Plan

Look for a pattern. Extend the pattern to find the final cost.

3 Solve

Multiply the cost of the item by __________ to find the discount.

Subtract the discount from the __________ to find the final cost.

$34.70 × __________ = __________ The discount is __________.

$34.70 − __________ = __________ Hank will pay __________.

4 Check

Does your answer make sense? Explain.

____

Practice the Strategy

Adrienne did 13 sit-ups the first day, 20 sit-ups the second day, and 27 sit-ups the third day. If the pattern continues, how many sit-ups will she do on the sixth day?

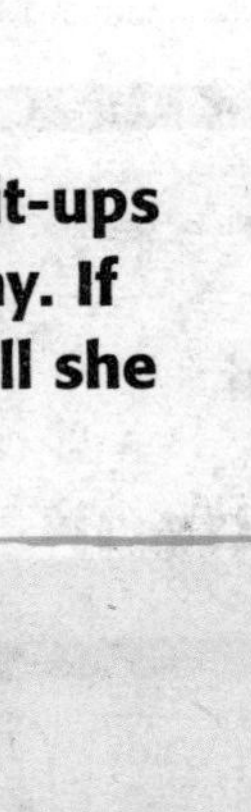

1 Understand

What facts do you know?

____

____

What do you need to find?

____

2 Plan

____

____

3 Solve

4 Check

Is my answer reasonable?

____

Name

Apply the Strategy

Solve each problem by looking for a pattern.

1. Every year, Victoria receives $30 for her birthday, plus $2 for each year of her age. Lacey receives $20 for her birthday and $4 for each year of her age. In 2013, Victoria is 10, and Lacey is 6. In what year will they both receive the same amount of money?

2. Trent lifts weights 7 days a week. He spends 18 minutes lifting weights on Monday, 29 minutes on Tuesday, 40 minutes on Wednesday, and 51 minutes on Thursday. If this pattern continues, how many minutes will Trent lift weights on Saturday?

3. Find the missing numbers in the table. Then describe the pattern.

Input	Output
5	9
10	19
15	■
20	39
■	49

4. Mathematical PRACTICE 8 **Look for a Pattern** Describe the pattern below. Then find the next three numbers.

0.03, 0.3, 3, 30, ______, ______, ______

5. Describe the pattern below. Then find the next two numbers.

784.5, 78.45, 7.845, ______, ______

My Work!

Review the Strategies

Use any strategy to solve each problem.

- Make a table.
- Choose an operation.
- Act it out.
- Draw a picture.

6. Mathematical PRACTICE 1 **Plan Your Solution** Samuel will arrive at the airport on the first plane after 10 A.M. Airplanes arrive every 50 minutes beginning at 6 A.M. When will Samuel's plane arrive?

7. On Serena's tenth birthday, her mom was 3 times Serena's age. How old will Serena and her mom be when her mom's age times 0.5 will equal Serena's age?

8. Mindy read 8 pages of her book the first day, 15 pages the second day, and 22 pages the third day. If the pattern continues, how many pages of her book will she read on the sixth day?

9. A 1-mile hiking path has signs placed every 240 feet. There are signs placed at the beginning and end of the mile. How many signs are there? (Hint: 1 mile = 5,280 feet)

10. Jada lives in a city that has an area of 344.6 square miles. Her friend lives in a town that is one-tenth, or 0.1, that size. What is the area of her friend's town?

11. Dennis has 9.5 weeks to prepare for a bike tour. If he rides 8.2 miles each week, how many miles will he ride before the bike tour?

My Work!

Name ....

MY Homework

Lesson 7
Problem Solving: Look for a Pattern

Homework Helper

Need help? connectED.mcgraw-hill.com

Deidra wants to buy new summer clothes. She needs to find the total cost in order to determine how much she can afford. The table shows the prices of some items. Based on the pattern, what is the sales tax and total cost for jeans that are $18.00? Round each amount to the nearest cent.

Item	Price	Sales Tax	Total Cost
T-shirt	$8.00	$0.56	$8.56
Shorts	$10.00	$0.70	$10.70
Sandals	$15.00	$1.05	$16.05
Jeans	$18.00	■	■

1 Understand

What facts do you know?

The price, sales tax, and total cost of the T-shirt, shorts, and sandals.

The price of the jeans.

What do you need to find?

The sales tax and total cost of the jeans.

2 Plan

Look for a pattern to solve the problem.

3 Solve

Multiply the cost of the item by 0.07 to find the sales tax, rounded to the nearest cent. Add the sales tax to the price of the item.

$18.00 × 0.07 = $1.26 — The sales tax is $1.26.

$18.00 + $1.26 = $19.26 — The total cost is $19.26.

4 Check

$19.26 − $1.26 = $18.00, so the answer is correct.

Problem Solving

Solve each problem by looking for a pattern.

1. Harold has 4 classes each school morning. Each class is 1 hour long, and there are 10 minutes between classes. The first class starts at 8:00 A.M. What time does the fourth class end?

____

2. Mr. Moore read 25 pages of a book on Monday. He read 32 pages on Tuesday, 39 pages on Wednesday, and 46 pages on Thursday. If this pattern continues, how many pages will Mr. Moore read on Sunday?

____

3. Mathematical **PRACTICE** 8 **Look for a Pattern** Describe the pattern below. Then find the next two numbers.

547, 54.7, 5.47, 0.547, ________, ________

____

4. Draw the next two figures in the pattern.

5. Marissa can buy 3 oranges for \$1.35, 4 oranges for \$1.80, or 5 oranges for \$2.25. Marissa can buy 10 oranges for a price that fits this pattern or a bag of 10 oranges for \$4.25. Which way costs less for 10 oranges? Explain.

____

____

____

____

Multiplication Properties

Lesson 8

ESSENTIAL QUESTION
How is multiplying and dividing decimals similar to multiplying and dividing whole numbers?

Math in My World

Example 1

A coach had 16 players in each of 2 groups. Each player scored 5 goals. Find the total number of goals scored.

Find $(16 \times 2) \times 5$.

Since you can easily multiply 2 and 5, change the way the numbers are grouped.

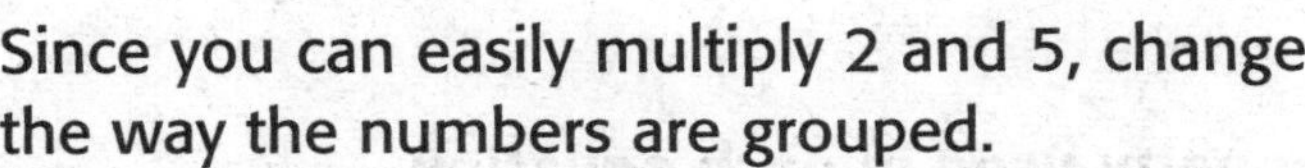

The **Associative Property of Multiplication** states that the way in which factors are grouped does not change the product.

It is easier to find 2×5 than 16×2. You can group the numbers differently to find 2×5 first.

$(16 \times 2) \times 5 = 16 \times (2 \times 5)$ Associative Property

$= 16 \times$ ____ Multiply. Parentheses tell you which factors to multiply first.

$=$ ____ Multiply.

So, the total number of goals scored is ____.

Example 2

Find the unknown in the equation
35.5 × ■ = 17 × 35.5.

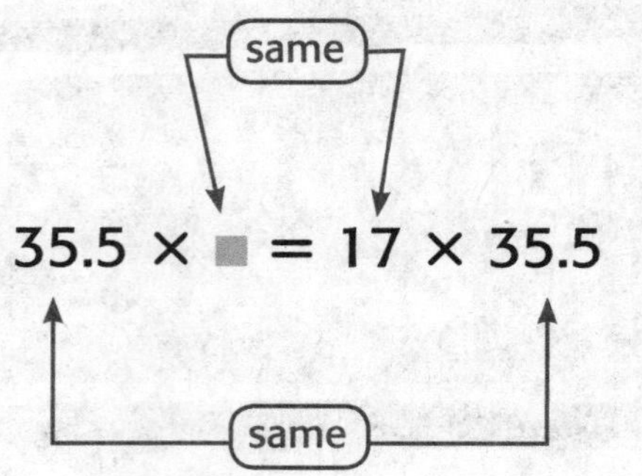

The **Commutative Property of Multiplication** shows that the order in which factors are multiplied does not change the product.

So, the unknown is ______.

Example 3

Find the unknown in the equation 17 × ■ = 17.

17 × ■ = 17

same

The **Identity Property of Multiplication** states that the product of any factor and 1 equals the factor.

So, the unknown is ______.

Guided Practice

Draw lines to match the multiplication property used in each equation.

1. 6.2 × 100 = 100 × 6.2 • Identity Property

2. (8 × 2) × 3 = 8 × (2 × 3) • Commutative Property

3. 78.56 × 1 = 78.56 • Associative Property

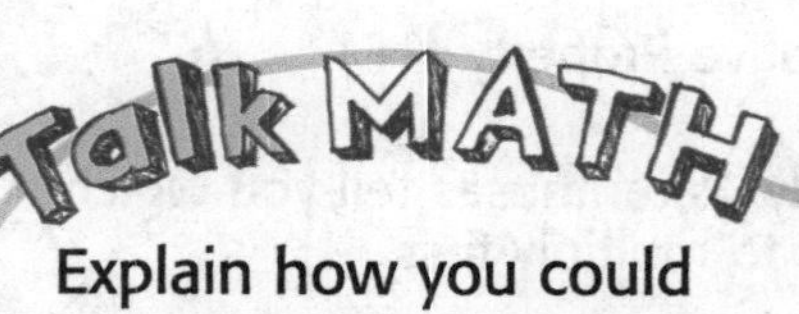

Explain how you could use mental math and multiplication properties to find (5.5 × 50) × 2.

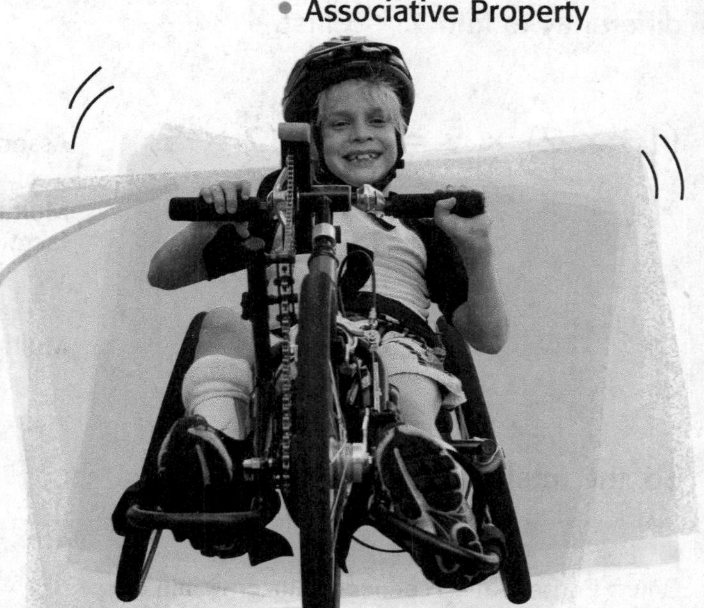

Name

Independent Practice

Use properties of multiplication to find each product mentally. Show your steps and identify the properties that you used.

4. $(5.1 \times 2) \times 50 =$ ______

5. $4 \times (2.5 \times 6) =$ ______

6. $(9.8 \times 500) \times 2 =$ ______

7. $(1.4 \times 50) \times 20 =$ ______

Mathematical PRACTICE 2 **Use Algebra Find the unknown in each equation. Circle which property you used.**

8. $19.5 \times$ ■ $= 19.5$

■ = ______

Commutative Property

Associative Property

Identity Property

9. $34 \times 65 = 65 \times$ ■

■ = ______

Commutative Property

Associative Property

Identity Property

10. $2.1 \times$ ■ $= 4.3 \times 2.1$

■ = ______

Commutative Property

Associative Property

Identity Property

11. $(17 \times 2) \times 5 = 17 \times ($■ $\times 5)$

■ = ______

Commutative Property

Associative Property

Identity Property

Problem Solving

12. Elijah and 2 of his friends are each paid $20 per afternoon for stuffing envelopes. If they work 5 afternoons, what is the total amount of their earnings?

____

13. Replace the ■ in (8.7 × ■) × 5 with a number greater than 10 so that the product is easy to find mentally. Explain.

____

14. Mathematical PRACTICE 3 **Draw a Conclusion** Each juice box contains 6.4 ounces. Each value pack of juice holds 10 juice boxes. If you have fifty value packs, how many ounces of juice do you have?

____

HOT Problems

15. Mathematical PRACTICE 2 **Use Number Sense** Which two properties can you use to find (3.4 × 4) × (25 × 1) mentally? Explain. Remember that the parentheses tell you which factors to multiply first.

____

16. **Building on the Essential Question** How do the multiplication properties help me to find products mentally?

____

Name ....

MY Homework

Lesson 8
Multiplication Properties

Homework Helper

Need help? connectED.mcgraw-hill.com

Use properties of multiplication to find (1.7 × 5) × 2.

$(1.7 \times 5) \times 2 = 1.7 \times (5 \times 2)$ Associative Property

$= 1.7 \times 10$ Multiply.

$= 17$ Multiply.

So, $(1.7 \times 5) \times 2 = 17$.

Practice

Use properties of multiplication to find each product mentally. Show your steps and identify the properties that you used.

1. $(1.6 \times 2) \times 5 =$ ______

2. $(27 \times 2.5) \times 4 =$ ______

Algebra **Find each unknown.**

3. ■ × 5.5 = 5.5

■ = ______

4. 49 × 201 = ■ × 49

■ = ______

Problem Solving

5. For a party, Shandra and James each bought 5 packages of hot dog buns that each cost \$1.50. How much did the hot dog buns cost altogether?

6. Eva needs to find: $(6 \times 1.5) \times 2$. Give the product and name the property she could use.

7. Write a multiplication sentence to show how the Associative Property can help you solve a problem mentally. Explain.

8. Mathematical PRACTICE 2 **Reason** Without calculating, is the equation $(1.8 \times 3) \times 2.1 = 3 \times (1.8 \times 2.1)$ true or false? Explain your reasoning.

Test Practice

9. Which multiplication property is shown in the equation below?

$$3.1 \times (2 \times 9) = (3.1 \times 2) \times 9$$

Ⓐ Identity Property

Ⓑ Commutative Property

Ⓒ Zero Property

Ⓓ Associative Property

Name ....

Estimate Quotients

Lesson 9
ESSENTIAL QUESTION
How is multiplying and dividing decimals similar to multiplying and dividing whole numbers?

Math in My World

Example 1

Ms. Glover buys 2 kites for a total of \$15.18. If each kite costs the same, about how much did each kite cost? Explain why your answer is reasonable.

Estimate the quotient of 15.18 and 2.

Use compatible numbers.

\$15.18 ÷ 2
↓ ↓
☐ ÷ 2
↓ ↓

Helpful Hint
Compatible numbers are numbers that are easy to compute mentally.

Divide. _____ ÷ 2 = _____

So, \$15.18 ÷ 2 is about _____.

Since 2 × 8 = _____ and _____ ≈ 15.18, the answer is reasonable.

The ≈ means *approximately equal to.*

Example 2

Three friends went out to dinner. The total cost of their meals was $32.57. If the friends split the bill evenly, about how much will each person pay? Explain why your answer is reasonable.

Estimate $32.57 ÷ 3 by rounding.

Round $32.57 to the nearest whole dollar because 33 and 3 are compatible numbers.

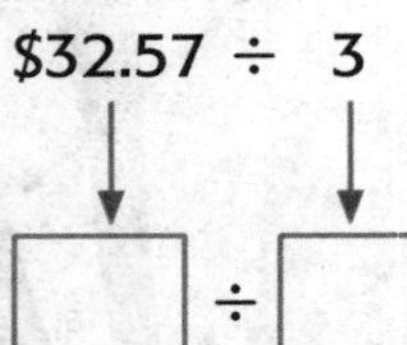

2 Divide.

____ ÷ ____ = ____

Each friend will pay about ____.

Since 3 × ____ = ____ and

____ ≈ 32.57, the answer is reasonable.

Guided Practice

Estimate each quotient.

1. $19.50 ÷ 5

☐ ÷ ☐ = ☐

2. $47.25 ÷ 25

☐ ÷ ☐ = ☐

3. 16.8 ÷ 4

☐ ÷ ☐ = ☐

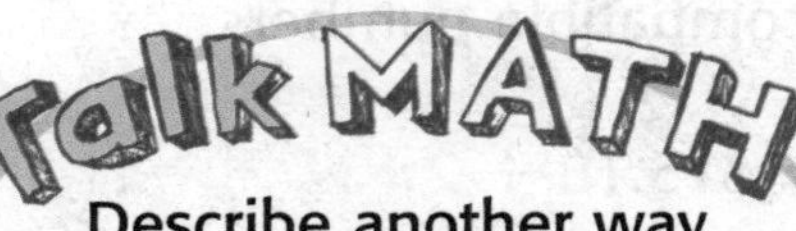

Describe another way you could estimate in Example 2. Are both estimates reasonable? Explain.

Name

Independent Practice

Estimate each quotient.

4. 87.3 ÷ 11

5. 44.7 ÷ 5

6. 195.8 ÷ 12

7. \$28.20 ÷ 6

8. \$7.92 ÷ 6

9. 88.3 ÷ 9

10. 128.9 ÷ 12

11. 576.4 ÷ 62

12. \$15.47 ÷ 7

13. 56.3 ÷ 18

Problem Solving

14. It costs $158.75 to purchase 15 tickets to the state fair. About how much does one ticket cost?

____

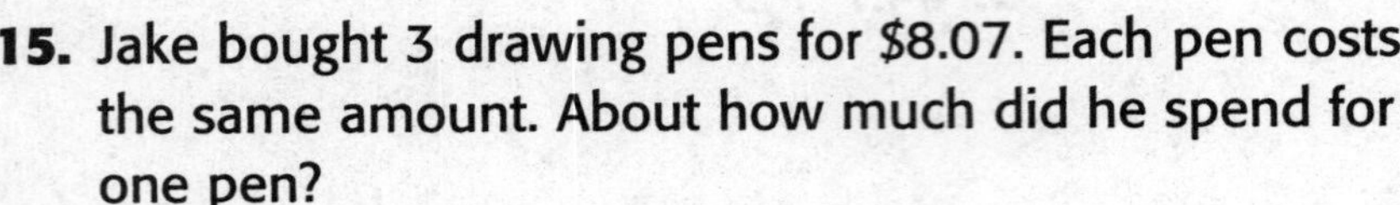

15. Jake bought 3 drawing pens for $8.07. Each pen costs the same amount. About how much did he spend for one pen?

____

16. A canoe rental company offers two trips along the river. Neil and Gabriela choose the longer trip. If they split the cost of the rental, about how much will each person pay?

Canoe Rental		
Trip	Distance (miles)	Cost
A	5.75	$12.98
B	8.5	$16.32

____

17. Mathematical PRACTICE 1 **Make Sense of Problems** It costs $88.50 for 6 adults to see an exhibit on Egyptian mummies. About how much does it cost one adult to see the exhibit?

____

My Work!

HOT Problems

18. Mathematical PRACTICE 4 **Model Math** Write a real-world estimation problem involving dividing a decimal by a whole number. Then estimate the quotient.

____

____

19. **Building on the Essential Question** How can I use compatible numbers to estimate the quotient of a decimal and whole number?

____

____

Name ____________________

MY Homework

Lesson 9
Estimate Quotients

Homework Helper

Need help? connectED.mcgraw-hill.com

Estimate 78.74 ÷ 42.

Estimate 78.74 ÷ 42 by rounding.

1 Round 78.74 and 42 to the nearest ten.

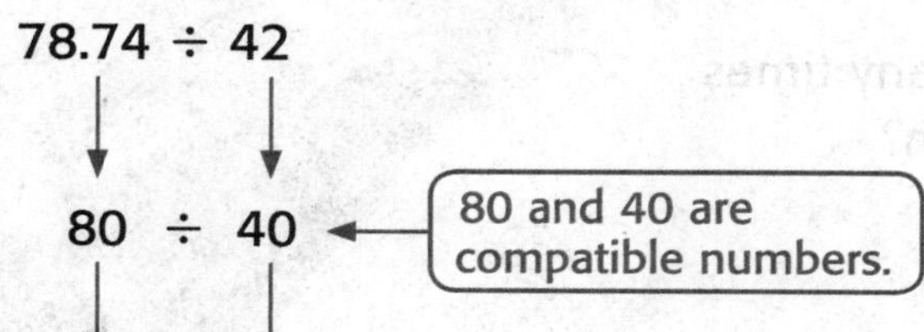

2 Divide. 80 ÷ 40 = 2

So, 78.74 ÷ 42 is about 2.

Practice

Estimate each quotient.

1. 32.17 ÷ 7

2. $175.32 ÷ 3

3. 21.9 ÷ 3

4. 36.3 ÷ 6

5. 17.5 ÷ 5

6. 120.6 ÷ 2

Problem Solving

Use the table to answer Exercises 7–9.

The information in the table can be used to determine the density of each object. Density describes how tightly the particles in an object are packed together. You can find density by dividing an object's mass by its volume.

Substance	Mass (g)	Volume (cm^3)
Aluminum	13.5	5
Gold	56.7	3
Mercury	121.5	9

7. Estimate the density of aluminum.

8. Is the density of gold greater than the density of mercury? Explain.

9. The density of gold is about how many times greater than the density of aluminum?

10. Harry's mother makes cakes for a local restaurant. She buys flour and sugar in large amounts. She has 157.86 pounds of flour and 82.69 pounds of sugar. If she uses 15 pounds of flour and 8 pounds of sugar each day, for about how many days will the flour last?

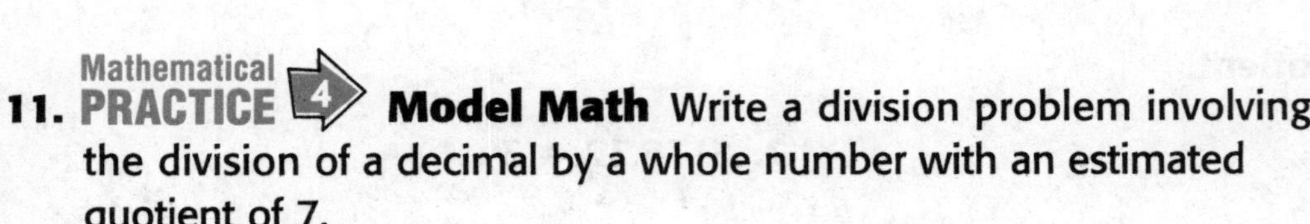

11. Mathematical PRACTICE 4 **Model Math** Write a division problem involving the division of a decimal by a whole number with an estimated quotient of 7.

Test Practice

12. Kendall measured the rainfall in her area for a year. Her readings totaled 34.56 inches. Which is the best estimate of the average rainfall per month?

Ⓐ about 1 inch per month

Ⓑ about 2 inches per month

Ⓒ about 3 inches per month

Ⓓ about 4 inches per month

Check My Progress

Vocabulary Check

Identify the multiplication property for each equation.

Associative Property	Commutative Property	Identity Property

1. $1.5 \times 2 = 2 \times 1.5$ __________

2. $5.5 \times (3 \times 10) = (5.5 \times 3) \times 10$ __________

3. $7.8 \times 1 = 7.8$ __________

Concept Check

Find each product.

4. $0.87 \times 10 =$ ______

5. $0.742 \times 100 =$ ______

6. $0.22 \times 10^2 =$ ______

7. $1.5 \times 10^3 =$ ______

Describe each pattern below. Then find the next two numbers.

8. 0.03, 0.3, 3, 30, ■, ■

9. 784.5, 78.45, 7.845, ■, ■

Estimate each quotient.

10. $12.5 \div 5$

11. $17.2 \div 8$

12. $23.7 \div 6$

Problem Solving

13. Neal is saving $4.25 a week for 100 weeks. How much will he have saved altogether?

14. Five friends went to the movies. The total cost was $55.91. If the friends split the cost equally, about how much did each person pay?

15. Wendy paints a wall that is 10.5 feet tall and 9.2 feet wide. What is the area of the wall that she paints?

16. Kevin swims 50 meters in 1.1 minutes on his first day of swimming class. He swims the same distance in 0.9 minute a week later. The next week, he swims the same distance in 0.7 minute. If the pattern continues, what will be Kevin's time one week later?

Test Practice

17. The Weston Laundry washes all the linens for local hotels. In 7 days, they washed 285.38 pounds of towels and 353.47 pounds of sheets. About how many pounds of laundry did they wash each day?

Ⓐ 80 pounds
Ⓑ 100 pounds
Ⓒ 120 pounds
Ⓓ 140 pounds

Name ....

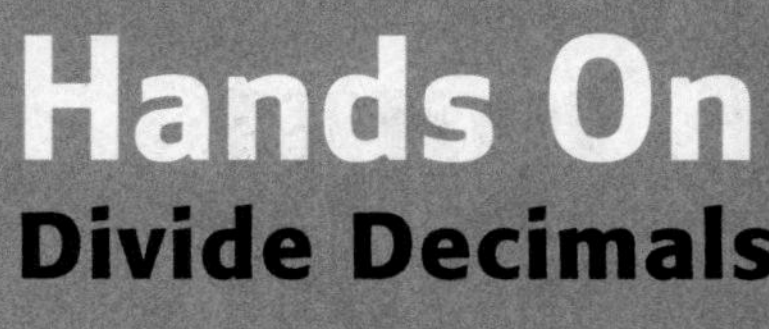

Hands On
Divide Decimals

Lesson 10

ESSENTIAL QUESTION
How is multiplying and dividing decimals similar to multiplying and dividing whole numbers?

You can use models to divide a decimal by a whole number.

Build It

Find 3.6 ÷ 3 using models.

1. The model shows 3.6 using three wholes and six tenths.

2. Divide the blocks into three equal groups. Draw a model that represents the equal groups.

How many whole blocks are in each group? ______

How many tenths are in each group? ______

So, 3.6 ÷ 3 = ______.

Check Use multiplication to check your answer.

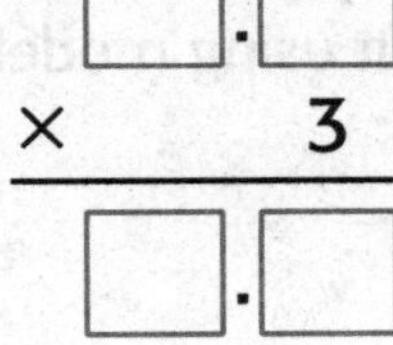

Try It

Find 4.8 ÷ 2 using models.

1 The model shows 4.8 using four wholes and eight tenths.

2 Divide the blocks into two equal groups. Draw a model to represent the equal groups.

How many whole blocks are in each group? ______

How many tenths are in each group? ______

So, 4.8 ÷ 2 = ______.

Check Use multiplication to check your answer.

$$\begin{array}{r} \square.\square \\ \times \quad 2 \\ \hline \square.\square \end{array}$$

Talk About It

1. The table shows some quotients. Study the table. Write a rule you can use to find the quotient of a decimal and a whole number without using models.

Decimals	Whole Numbers
3.6 ÷ 3 = 1.2	36 ÷ 3 = 12
4.8 ÷ 2 = 2.4	48 ÷ 2 = 24
2.4 ÷ 4 = 0.6	24 ÷ 4 = 6

2. **Mathematical PRACTICE 3** **Justify Conclusions** Use your rule from Exercise 1 to find 3.5 ÷ 7 without using models. Explain the process you used.

Name ________________

Practice It

Use models to find each quotient. Draw the equal groups.

3. 3.4 ÷ 2 = ________

4. 6.3 ÷ 3 = ________

5. 5.6 ÷ 4 = ________

6. 2.7 ÷ 3 = ________

Apply It

7. Marketa used 4.5 cups of sugar for 5 batches of cookies. If an equal amount was used for each batch, how many cups of sugar were used for each batch? Draw models to help you divide.

____

8. Alexander ran 3.2 miles over the past 2 days. If he ran an equal amount each day, how many miles did he run each day? Draw models to help you divide.

____

9. Mathematical PRACTICE 3 **Find the Error** Joseph used base-ten blocks to find 2.1 ÷ 3. He stated that 2.1 ÷ 3 = 0.6. Find his mistake and correct it.

____

____

____

Write About It

10. How can I use models to divide decimals by whole numbers?

____

____

MY Homework

Lesson 10
Hands On: Divide Decimals

Homework Helper

Need help? connectED.mcgraw-hill.com

Find 2.8 ÷ 4 using models.

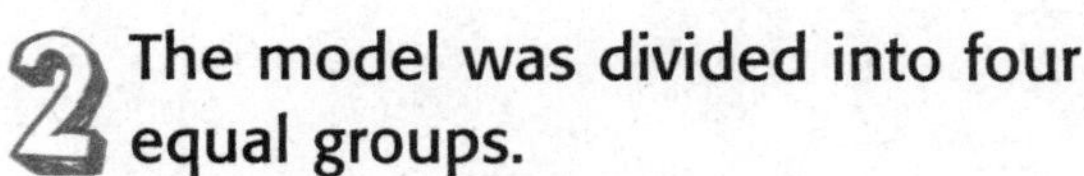

1 The model shows 2.8 using two wholes and eight tenths.

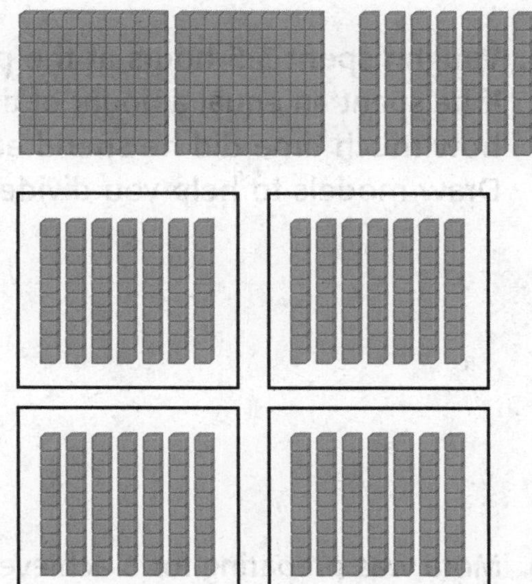

2 The model was divided into four equal groups.

There are no whole blocks in each group.

There are 7 tenths in each group.

So, 2.8 ÷ 4 = 0.7.

Check Use multiplication to check your answer.

$$\begin{array}{r} \overset{2}{0}.7 \\ \times\ 4 \\ \hline 2.8 \end{array}$$

Practice

Use models to find each quotient. Draw the equal groups.

1. 1.6 ÷ 2 = ______

2. 2.4 ÷ 4 = ______

My Drawing!

Real World Problem Solving

3. Elaine and her 3 friends purchased snacks after school for $3.60. If each person paid an equal amount, how much did each person pay? Draw models to help you divide.

My Work!

4. Vaughn spent 3.5 hours at the pool over the past week. If he spent an equal amount of time over the last 7 days, how much time did he spend each day at the pool? Draw models to help you divide.

5. Mora was preparing for 3 achievement assessments. She spent 3.6 hours studying over the weekend. If she spent an equal amount of time studying for each assessment, how much time did she spend on each assessment? Draw models to help you divide.

6. Mathematical PRACTICE 5 **Use Math Tools** Faith helped pull weeds, trim shrubs, and mow the lawn over the past 1.8 hours. If she spent an equal amount of time on each activity, how much time did she spend on each activity? Draw models to help you divide.

Name ....

Divide Decimals by Whole Numbers

Lesson 11

ESSENTIAL QUESTION
How is multiplying and dividing decimals similar to multiplying and dividing whole numbers?

Math in My World

Example 1

There are 8.4 meters left on a roll of ribbon. Nancy wants to cut the ribbon in half. What will be the size of each piece of ribbon?

Find 8.4 ÷ 2.

Estimate 8 ÷ 2 = _______

LET'S WRAP IT UP!

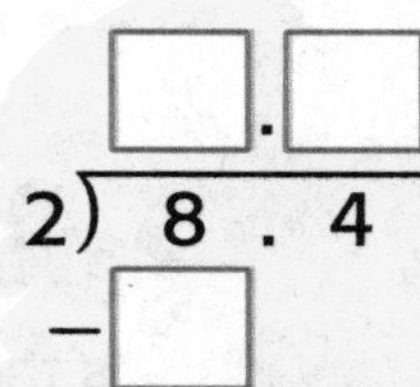

1. Place the decimal point directly above the decimal point in the dividend.

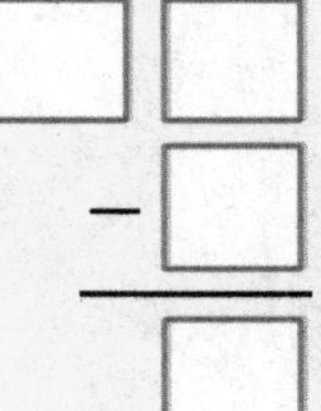

2. Divide as with whole numbers.

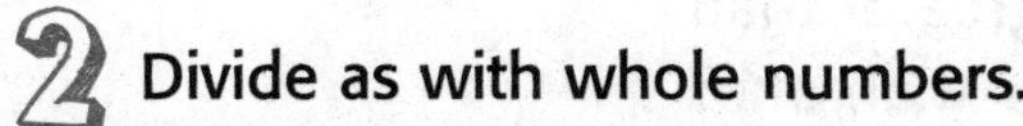

So, 8.4 ÷ 2 = _______. Each piece of ribbon will be _______ meters long.

Check Use multiplication. 4.2 × 2 = _______

Use models.

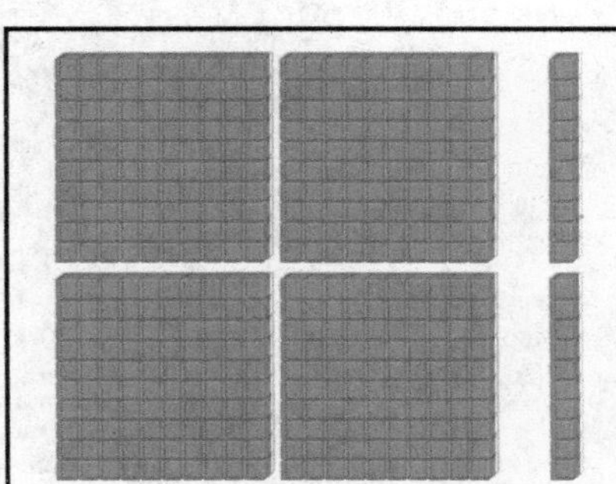

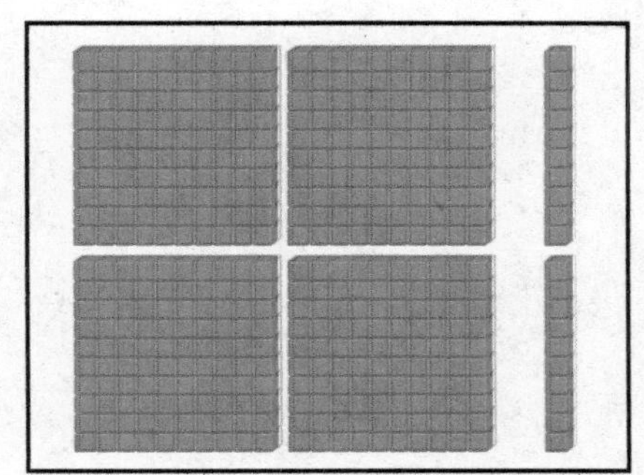

Example 2

Find $6.54 ÷ 12. Round to the nearest cent.

1. Place the decimal point directly above the decimal point the dividend.

2. Divide as with a whole numbers.

3. Annex a zero after 6.54 and continue dividing.

$$\begin{array}{r} 0.\square\square\square \\ 12\overline{)\,6.\ \ 5\ \ 4\ \ 0} \\ -\square\square\quad\quad \\ \hline \square\square\quad \\ -\square\square\quad \\ \hline \square\square \\ -\square\square \\ \hline \square \end{array}$$

So, $6.54 ÷ 12 = $________. Rounded to the nearest cent, this is $0.55.

Check Use multiplication.

$$\begin{array}{r} 0.545 \\ \times \quad 12 \\ \hline 1090 \\ +\,5450 \\ \hline 6.540 \end{array}$$

Talk MATH

Is the quotient of 9.3 ÷ 15 greater than one or less than one? Explain without calculating.

Guided Practice

Divide.

1. $3\overline{)48.33}$

2. $2\overline{)8.8}$

Name

Independent Practice

Divide. Check your answer using multiplication.

3. 145.8 ÷ 12 = ______ **4.** 22.11 ÷ 11 = ______ **5.** 38.4 ÷ 16 = ______

6. $8\overline{)12.4}$ **7.** $14\overline{)14.14}$ **8.** $11\overline{)55.44}$

Divide. Round to the nearest tenth.

9. 7.21 ÷ 7 = ______ **10.** 6.28 ÷ 4 = ______ **11.** $5\overline{)276.2}$

Divide. Round to the nearest hundredth.

12. 78.04 ÷ 8 = ______ **13.** $24\overline{)75.48}$ **14.** $25\overline{)4.60}$

Problem Solving

15. Mandy and 5 of her friends bought a package of bottled water for $4.98. How much will each friend pay to the nearest cent, if the cost is divided equally?

16. Four girls swam the 4-by-200 meter freestyle relay in a total of 540.4 seconds. If each girl swam her part of the race in the same amount of time, what was the time for one girl?

17. Mathematical PRACTICE 2 **Reason** The table shows the prices for two packs of batteries. Which pack is a better buy? Explain your reasoning.

Batteries		
Pack	Number of Batteries	Price
A	4	$2.44
B	6	$3.90

18. The total cost of four movie tickets is $33.40. What is the cost of one ticket?

My Work!

HOT Problems

19. Mathematical PRACTICE 4 **Model Math** Write a real-life problem that involves dividing a decimal by a whole number.

20. **Building on the Essential Question** Why is rounding important when dividing a decimal by a whole number?

Name ______________________

MY Homework

Lesson 11
Divide Decimals by Whole Numbers

Homework Helper

eHelp 

Need help? connectED.mcgraw-hill.com

Find 15.3 ÷ 3.

Estimate 15 ÷ 3 = 5

1. Place the decimal point directly above the decimal point in the dividend.

2. Divide as with whole numbers.

```
   5.1
3)15.3
 -15
   0 3
  -  3
     0
```

So, 15.3 ÷ 3 = 5.1.

Check Use multiplication to check your answer. 5.1 × 3 = 15.3

Practice

Divide. Check your answer using multiplication.

1. 223.6 ÷ 40 = ________ **2.** 8.14 ÷ 20 = ________ **3.** 361.5 ÷ 12 = ________

Divide. Check your answer using multiplication.

4. $2\overline{)26.48}$ **5.** $5\overline{)7.75}$ **6.** $3\overline{)6.9}$

Problem Solving

7. Four girls run a relay and finish the race in 6.48 minutes. If each girl ran at the same speed, how many minutes did each girl run? Round to the nearest tenth.

____

8. Mathematical PRACTICE 6 **Explain to a Friend** Is the quotient 3.6 ÷ 9 greater than or less than 1? Explain to a friend.

____

____

____

My Work!

9. A paper company ships 110.4 tons of paper in 4 trucks. If each truck carries the same weight, how many tons of paper are in each truck?

____

Test Practice

10. Marcel had drama practice 8.75 hours from Monday to Friday. If he practices the same amount each day, how many hours does Marcel spend in drama practice each day?

Ⓐ 1.75 hours Ⓒ 2.25 hours

Ⓑ 2 hours Ⓓ 2.50 hours

Lisa M. Macias/U.S. Air Force

Name ____________________

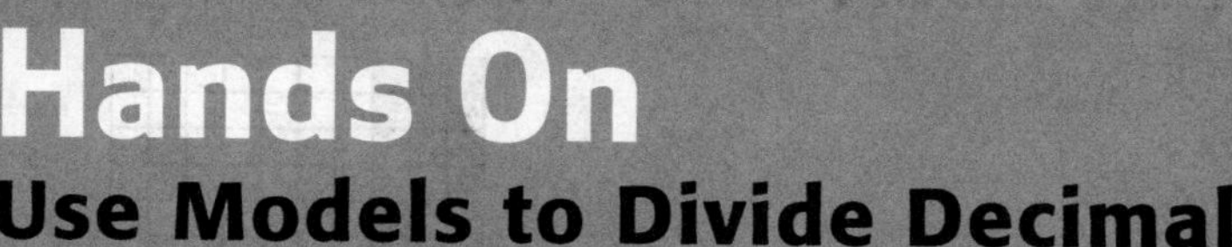

Hands On
Use Models to Divide Decimals

Lesson 12

ESSENTIAL QUESTION
How is multiplying and dividing decimals similar to multiplying and dividing whole numbers?

You can use models to divide decimals by decimals.

Build It

How many groups of 8 tenths are in 2.4? Find 2.4 ÷ 0.8.

1 The model shows 2.4.

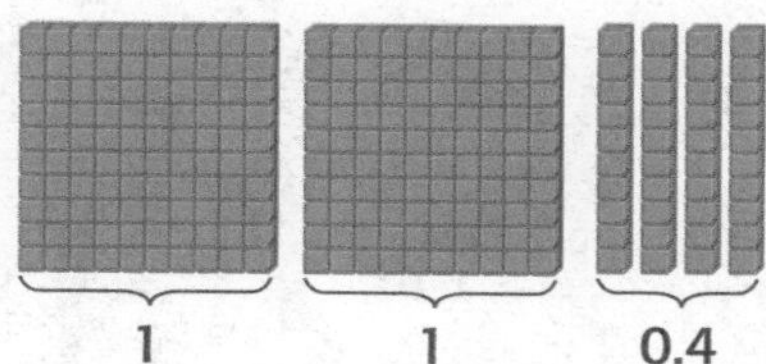

2 Since you are dividing by tenths, replace both of the wholes with tenths.

Each whole equals how many tenths?

____

How many tenths do you have altogether after replacing the two wholes with tenths? ________

3 Separate the tenths into groups of eight tenths to show dividing by 0.8.

How many groups will you have?

____

Are there any left over? ________
Draw the result at the right.

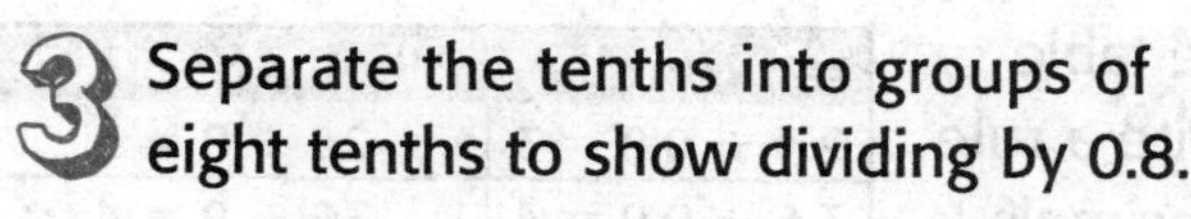

My Drawing!

So, 2.4 ÷ 0.8 = ________.

Check Use multiplication to check your answer.
0.8 × 3 = 2.4

Try It

How many groups of 6 hundredths are in 3 tenths? Find 0.3 ÷ 0.06.

1 The decimal model shows 0.3.

How many hundredths are in 3 tenths? ________

2 Separate these hundredths into groups of 6 hundredths. Draw the equal groups.

How many groups will you have? ________

Are there any left over? ________

So, 0.3 ÷ 0.06 = ________.

Talk About It

1. Mathematical PRACTICE 8 **Look for a Pattern** The table shows some quotients. Study the table. Write a rule you can use to find the quotient of two decimals, both to the tenths place, without using models.

Decimals	Whole Numbers
2.4 ÷ 0.8 = 3	24 ÷ 8 = 3
3.6 ÷ 0.9 = 4	36 ÷ 9 = 4
2.8 ÷ 0.4 = 7	28 ÷ 4 = 7

____

____

2. Mathematical PRACTICE 3 **Draw a Conclusion** Use your rule from Exercise 1 to find 4.8 ÷ 0.8 without using models. Explain the process you used.

____

Name

Practice It

Use models to find each quotient. Draw the equal groups.

3. 2.4 ÷ 0.3 = ______

4. 1.6 ÷ 0.4 = ______

5. 0.1 ÷ 0.05 = ______

6. 0.2 ÷ 0.04 = ______

7. 0.27 ÷ 0.03 = ______

Apply It

8. Mathematical PRACTICE 5 **Use Math Tools** Grace purchased some packages of gum for $1.50. If each package costs $0.50, how many packages did Grace purchase? Use models to help you divide.

9. Aidan purchased 1.5 feet of wood to make a napkin holder. If he used 0.5 foot of wood for each side, how many sides does the napkin holder have? Use models to help you divide.

10. Ryan used 2.5 cups of flour to make bread. If he used 0.5 cup of flour for each loaf, how many loaves of bread did Ryan make? Use models to help you divide.

11. Mathematical PRACTICE 6 **Explain to a Friend** Explain how to use models to divide 2.1 by 0.7.

Write About It

12. How can I use models to divide decimals by decimals?

Name

MY Homework

Lesson 12

Hands On: Use Models to Divide Decimals

Homework Helper

Need help? connectED.mcgraw-hill.com

How many groups of 7 tenths are in 1.4? Find 1.4 ÷ 0.7.

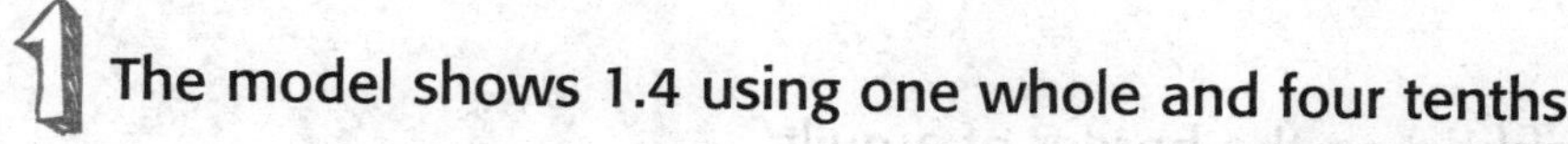

1 The model shows 1.4 using one whole and four tenths.

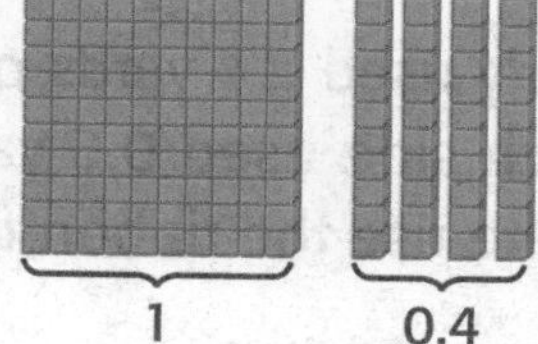

2 Since you are dividing by tenths, the whole block was replaced with tenths.

There are a total of 14 tenths altogether after replacing the one whole with ten tenths.

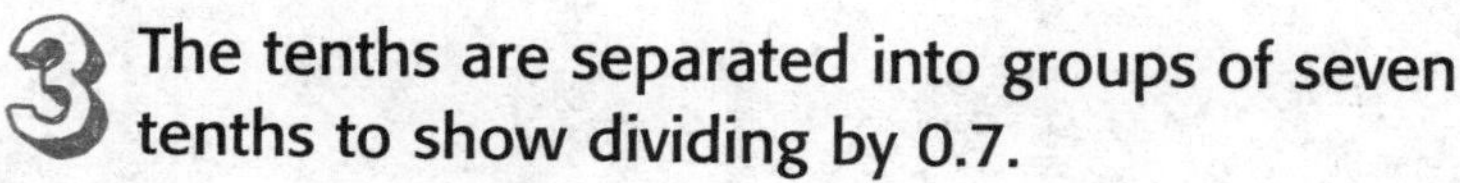

3 The tenths are separated into groups of seven tenths to show dividing by 0.7.

There are 2 groups with none left over.

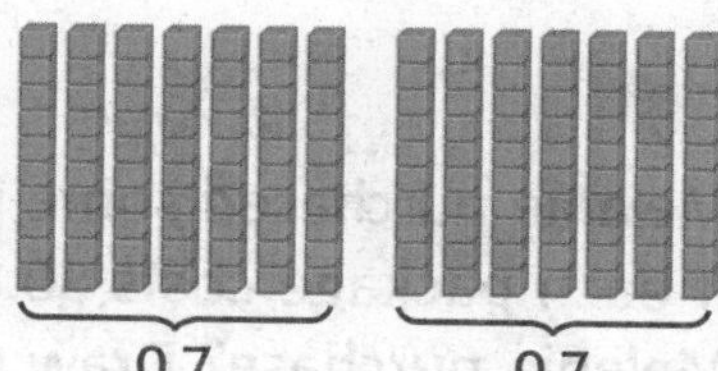

So, 1.4 ÷ 0.7 = 2.

Practice

Use models to find each quotient. Draw the equal groups.

1. 0.4 ÷ 0.05 = ______

2. 0.2 ÷ 0.02 = ______

Problem Solving

3. **Mathematical PRACTICE 5 Use Math Tools** Adam used 1.8 pieces of paper to make his project. He decorated 0.2 of each piece of paper with a different color. How many sections does Adam's art project have? Draw models to help you divide.

My Drawing!

4. Beverly used 2.4 yards of ribbon on the border of a quilt. If each side used 0.6 yard of ribbon, how many sides did she use the ribbon on? Draw models to help you divide.

5. Melanie purchased some packages of pencils for $1.60. If each package costs $0.80, how many packages did Melanie purchase? Draw models to help you divide.

6. Evan had 2.2 pounds of candy to give to his friends. If he gave 0.2 pound to each friend, how many friends did Evan give candy to? Draw models to help you divide.

Divide Decimals

Lesson 13

ESSENTIAL QUESTION

How is multiplying and dividing decimals similar to multiplying and dividing whole numbers?

When dividing by decimals, change the divisor into a whole number. To do this, multiply both the divisor and the dividend by the same power of 10. Then divide as with whole numbers.

Math in My World

Example 1

Kasey is 4.5 feet tall. Her brother, Jerome, is 6.75 feet tall. How many times taller is Jerome than Kasey?

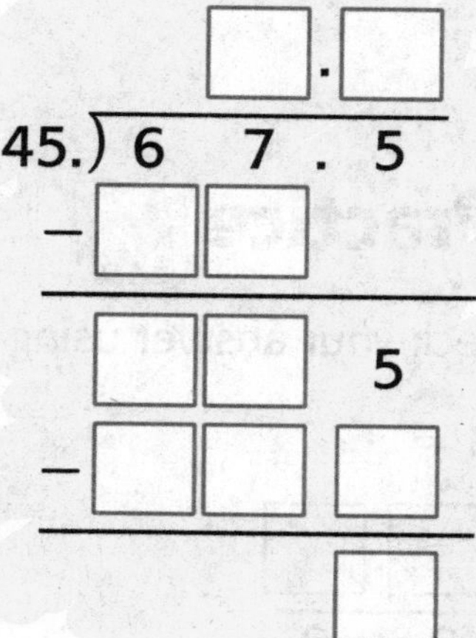

Find 6.75 ÷ 4.5.

Multiply 4.5 by 10 to make 45. Then, multiply 6.75 by the same number, 10, to make 67.5.

$4.5\overline{)6.75}$

Place the decimal point in the quotient. Divide as with whole numbers.

$$\begin{array}{r} \square.\square \\ 45.\overline{)\,6\ \ 7\,.\ 5} \\ -\ \square\ \square\ \ \ \ \\ \hline \square\ \square\ \ 5 \\ -\ \square\ \square\ \square \\ \hline \square \end{array}$$

So, Jerome is ______ times taller than Kasey.

Check Multiply to check your answer.

$$\begin{array}{r} 1.\ \ 5 \\ \times\ 4.\ \ 5 \\ \hline \square.\square\square \end{array}$$

Example 2

Find 0.06 ÷ 1.5.

1 Multiply each number by 10.

1.5)0.06

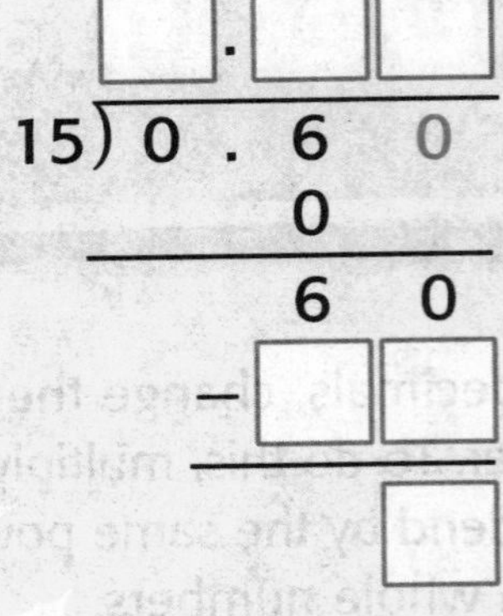

2 Place the decimal point in the quotient. Divide as with whole numbers. Annex zeros in the dividend as needed.

15)0.60

15 does not go into 6, so write a 0 in the hundredths place.

Check Multiply to check your answer.

$$\begin{array}{r} 0.04 \\ \times\ 1.5 \\ \hline 0.06 \end{array}$$

Talk MATH

When finding 0.808 ÷ 0.4, by what number should you multiply the divisor? Explain.

Guided Practice

1. Divide. Check your answer using multiplication.

6.89 ÷ 1.3 = ______

1.3) 6.8 9

Name

Independent Practice

Divide. Check your answer using multiplication.

2. $0.66 \div 0.3 =$ _______ **3.** $16.5 \div 0.03 =$ _______ **4.** $0.462 \div 0.2 =$ _______

5. $0.12\overline{)18.6}$ **6.** $1.4\overline{)0.07}$ **7.** $0.04\overline{)3.822}$

8. $14.4 \div 0.4 =$ _______ **9.** $0.78\overline{)3.51}$ **10.** $0.15\overline{)84.78}$

Algebra Find each unknown.

11. $0.08 \div 1.6 = z$

$z =$ _______

12. $13.2 \div 0.3 = d$

$d =$ _______

13. $0.92 \div 0.4 = q$

$q =$ _______

Problem Solving

14. Mathematical PRACTICE 4 **Model Math** Mrs. Chibas is making chocolate chip cookies for her daughter's class. She bought a tub of chocolate chip cookie dough that contained 57.6 ounces of dough. If each cookie needs 1.2 ounces of dough, how many cookies can she make?

____

15. Elliot ran 4 laps in 209.2 seconds. If he ran at the same speed for each lap, what was his time for each lap?

____

16. Kaya works for a T-shirt company and has 49.5 yards of fabric to make specialty T-shirts. If each T-shirt needs 4.5 yards of fabric, how many T-shirts can she make?

____

HOT Problems

17. Mathematical PRACTICE 6 **Be Precise** Without solving, would $1.98 \div 0.51$ be closer to 4 or 5? Explain.

____

18. **Building on the Essential Question** How is dividing decimals different from dividing whole numbers?

____

____

____

____

Name ....

MY Homework

Lesson 13
Divide Decimals

Homework Helper

Need help? connectED.mcgraw-hill.com

The Prease family drove 213.9 miles to the beach. Their car used 9.3 gallons of gas. How many miles did they drive per gallon of gas?

Find 213.9 ÷ 9.3.

1. Multiply 9.3 by 10 to make 93. Then, multiply 213.9 by the same number, 10, to make 2,139.

$9.3\overline{)213.9}$

2. Place the decimal point in the quotient. Divide as with whole numbers.

$$\begin{array}{r} 23. \\ 93\overline{)2{,}139.} \\ -\,186 \\ \hline 279 \\ -\,279 \\ \hline 0 \end{array}$$

So, they drove 23 miles per gallon of gas.

Check Multiply to check your answer.

$$\begin{array}{r} 23 \\ \times\ 9.3 \\ \hline 213.9 \end{array}$$

Practice

Divide. Check your answer using multiplication.

1. 12.42 ÷ 4.6 = ______

2. $0.4\overline{)0.242}$

3. 1.404 ÷ 0.45 = ______

Problem Solving

My Work!

4. Lamar spent $16.25 on peanuts. If he bought 2.6 pounds of peanuts, how much does one pound of peanuts cost?

5. Sei paid $10.48 for a number of washcloths. If each washcloth is $1.31, how many washcloths did Sei get?

6. Mathematical PRACTICE 6 **Be Precise** Write and solve a real-world problem that involves dividing a decimal by a decimal.

7. Patricia has 3.75 pounds of hamburger. She is making hamburgers with 0.25 pound of hamburger each. How many hamburgers can she make?

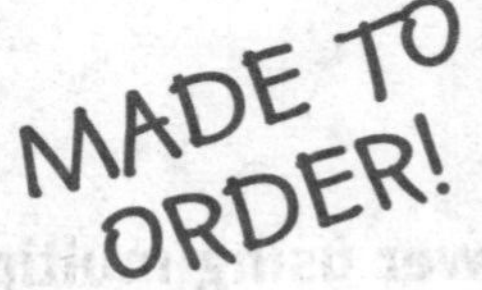

Test Practice

8. Mason has 1.71 meters of string to decorate his locker. He cut the string into 0.19-meter lengths. How many pieces does he have?

Ⓐ 6 pieces

Ⓑ 7 pieces

Ⓒ 8 pieces

Ⓓ 9 pieces

Name ....

Divide Decimals by Powers of Ten

Lesson 14

ESSENTIAL QUESTION
How is multiplying and dividing decimals similar to multiplying and dividing whole numbers?

You can divide decimals by powers of ten, such as 10, 100, and 1,000. Powers of ten can be written with exponents.

Math in My World

ZOOM!

Example 1

Laura needs to make 10 payments to pay for her new motorized scooter. How much will each payment be if the scooter costs $219.50?

One Way Divide.

Find $219.50 ÷ 10.

```
     21.95
10)219.50
  - 20
    19
  - 10
     95
   - 90
     50
   - 50
      0
```

The non-zero digits are the same. The decimal point in the quotient moved one place to the left.

Another Way Move the decimal point.

Dividing a decimal by 10 moves the decimal point one place to the left.

219.5 → 21.95

So, each payment will be ________.

Check Use multiplication. 21.95 × 10 = 219.5

To divide a decimal by a power of ten, move the decimal point to the left same number of zeros in the power of ten. This is also the same number as the exponent on 10.

Example 2

In the past 10^2 years, scientists measured the movement of the continents to be 190.5 centimeters. If the continents moved the same amount each year, how much did the continents move in one year?

The quotient of a number and a power of ten that is greater than 1 will always be less than the original number.

Since $10^2 = 100$, find $190.5 \div 100$.

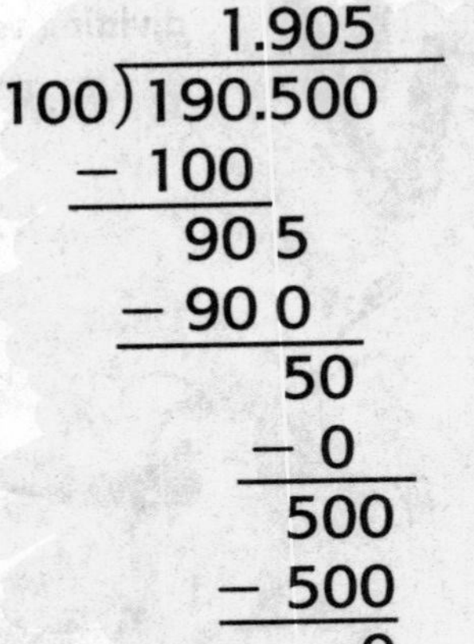

$$
\begin{array}{r}
1.905 \\
100\overline{)190.500} \\
-\,100 \\
\hline
90\,5 \\
-\,90\,0 \\
\hline
50 \\
-\,0 \\
\hline
500 \\
-\,500 \\
\hline
0
\end{array}
$$

The non-zero digits are the same. The decimal point in the quotient moved two places to the left.

Dividing a decimal by 100 moved the decimal point two places to the left.

$190.5 \rightarrow 1.905$

So, $190.5 \div 10^2 =$ ________. The continents moved ________ centimeters in one year.

Check Use multiplication. $1.905 \times 100 = 190.5$

Talk MATH

Use the number of zeros in the number 10 to explain why $4.5 \div 10 = 0.45$.

Guided Practice

Divide. Check your answer using multiplication.

1. $7.2 \div 10 =$ ________

2. $21.5 \div 10^2 =$ ________

3. $19.2 \div 10^3 =$ ________

Name

Independent Practice

Divide. Check your answer using multiplication.

4. $5.62 \div 100 =$ ______

5. $18.7 \div 100 =$ ______

6. $6.3 \div 10^3 =$ ______

7. $0.05 \div 1 =$ ______

8. $0.012 \div 10^2 =$ ______

9. $2.46 \div 10^1 =$ ______

10. $8.72 \div 100 =$ ______

11. $98.6 \div 10^2 =$ ______

12. $5.71 \div 1 =$ ______

13. $437 \div 1{,}000 =$ ______

Problem Solving

14. Herbert bought red beans to make chili to sell at a fundraiser. The total price of the red beans was $13.90. How much did it cost for each can?

Items Purchased	
Item	**Quantity**
Red beans	10
Shredded cheese	6

15. Mathematical PRACTICE 4 **Model Math** Sarah was trying to decide what kind of cat food to buy at the store. Smiley Cat costs $7.90 for 10 pounds. Purr Plus costs $0.85 per pound. Which brand is the better buy? Explain your reasoning.

16. In Alaska, scientists are studying the melting of glaciers. They found that in the past 10 years, the glaciers receded about 59.05 feet. If the glaciers receded at the same rate each year, how much did they recede in one year?

HOT Problems

17. Mathematical PRACTICE 3 **Which One Doesn't Belong?** Circle the expression that does not belong with the other three. Explain your reasoning.

$6.7 \div 10$	$4 \div 1{,}000$	$0.2 \div 1$	$52.1 \div 100$

18. **Building on the Essential Question** Describe the relationship between the number of places a decimal point is moved to the left and the change in the value of the number.

Name

MY Homework

Lesson 14
Divide Decimals by Powers of Ten

Homework Helper

Need help? connectED.mcgraw-hill.com

Find 36.4 ÷ 100.

One Way **Divide.**

$$\begin{array}{r} 0.364 \\ 100\overline{)36.4} \\ -\ 30\ 0 \\ \hline 6\ 40 \\ -\ 6\ 00 \\ \hline 400 \\ -\ 400 \\ \hline 0 \end{array}$$

The non-zero digits are the same. The decimal point in the quotient moved two places to the left.

Another Way **Move the decimal point.**

Dividing a decimal by 100 moves the decimal point two places to the left.

36.4 → 0.364

So, 36.4 ÷ 100 = 0.364.

Check Use multiplication. 0.364 × 100 = 36.4

Practice

Divide. Check your answer using multiplication.

1. 8.3 ÷ 100 = ________

2. 208 ÷ 10^2 = ________

3. 0.07 ÷ 1 = ________

4. 32.7 ÷ 1,000 = ________

Real World Problem Solving

Use the table to answer Exercises 5–7.

Location	Number of Parking Spots per Row	Combined Width (m)
Grocery Store	10	31.92
Hardware Store	10	31.96
Mall	10	31.30

5. Each row of the grocery store parking lot has 10 parking spots of equal width. What is the width for each spot?

____

6. How much space is given for each parking spot at the Mall parking lot if each spot has an equal width?

____

7. Which location gives the greatest space for each parking spot?

____

8. Mathematical PRACTICE 4 **Model Math** Christy purchased 6.75 pounds of licorice. How much licorice does she need to put in each bag if she divides the total amount into 10 equal-sized bags?

____

9. Mathematical PRACTICE 6 **Explain to a Friend** Tell what number you must divide 180 by to get 1.8.

____

Test Practice

10. Tomas buys a trumpet for $108.90. He will make 10 equal payments to pay for the trumpet. What is the amount of each payment that Tomas will make?

Ⓐ $1.09 Ⓒ $19.90

Ⓑ $10.89 Ⓓ $108.90

Review

Chapter 6

Multiply and Divide Decimals

Vocabulary Check

Match each word to its definition. Write your answers on the lines provided.

1. **Commutative Property** ______
2. **compatible numbers** ______
3. **Identity Property** ______
4. **Associative Property** ______
5. **powers of 10** ______
6. **rounding** ______

A. The product of any factor and 1 is equal to the factor.

B. Finding the approximate value of a number.

C. The order in which factors are multiplied does not change the product.

D. Numbers like 10, 100, and 1,000, because they can be written as 10^1, 10^2, and 10^3.

E. Numbers that are easy to multiply mentally.

F. The grouping of the factors does not change the product.

Concept Check

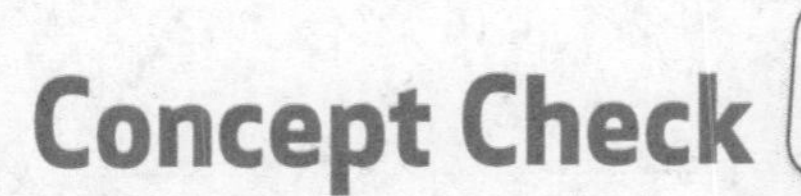

Estimate each product.

7. $4.80 × 5

8. 3.36 × 4

9. 10.8 × 7

Multiply. Check for reasonableness.

10. 6 × 1.8 = ______

11. $\begin{array}{r} 4.6 \\ \times\ 5 \\ \hline \end{array}$

12. $\begin{array}{r} 7.16 \\ \times\ 2.1 \\ \hline \end{array}$

Multiply. Tell which property you used.

13. 50 × (20 × 1.3) = ______

14. 2 × (50 × 5.6) = ______

Estimate each quotient.

15. 20.6 ÷ 4

16. 24.3 ÷ 8

17. 118.1 ÷ 10

Divide. Check your answer using multiplication.

18. $102.6 \div 10^2 =$ ______

19. 1.8 ÷ 0.08 = ______

20. $6.2\overline{)12.71}$

Name

Problem Solving

21. Kasi bought 7 pounds of mozzarella cheese. Each pound costs $4.29. About how much did he spend altogether?

22. Ariel buys 2 fish tanks at the pet store. Each tank holds 9.7 liters of water. How many liters of water will she need to fill both tanks?

23. It costs $4.25 for one pound of roast beef. How much will it cost to purchase 2.5 pounds of roast beef? Round to the nearest cent.

24. Describe the pattern below. What are the next two numbers?

0.41, 0.82, 1.64, 3.28

25. The total bill for a dinner party was $113.58. If each person pays $12.62, how many people attended the dinner party?

Test Practice

26. Sofia paints a rectangular mural on the wall. The area of the mural is 128.52 square feet. The mural is 12.6 feet wide. What is the height of the mural?

Ⓐ 115.92 ft　　Ⓒ 10.2 ft

Ⓑ 105.2 ft　　Ⓓ 8.6 ft

Reflect

Chapter 6

Answering the ESSENTIAL QUESTION

Use what you learned about multiplying and dividing decimals to complete the graphic organizer.

Whole Numbers

Decimals

ESSENTIAL QUESTION

How is multiplying and dividing decimals similar to multiplying and dividing whole numbers?

Vocabulary

Now reflect on the ESSENTIAL QUESTION  **Write your answer below.**

Chapter

Expressions and Patterns

ESSENTIAL QUESTION

How are patterns used to solve problems?

Fun with My Friends

Watch

Watch a video!

MY Common Core State Standards

Operations and Algebraic Thinking

5.OA.1 Use parentheses, brackets, or braces in numerical expressions, and evaluate expressions with these symbols.

5.OA.2 Write simple expressions that record calculations with numbers, and interpret numerical expressions without evaluating them.

5.OA.3 Generate two numerical patterns using two given rules. Identify apparent relationships between corresponding terms. Form ordered pairs consisting of corresponding terms from the two patterns, and graph the ordered pairs on a coordinate plane.

Geometry *This chapter also addresses these standards:*

5.G.1 Use a pair of perpendicular number lines, called axes, to define a coordinate system, with the intersection of the lines (the origin) arranged to coincide with the 0 on each line and a given point in the plane located by using an ordered pair of numbers, called its coordinates. Understand that the first number indicates how far to travel from the origin in the direction of one axis, and the second number indicates how far to travel in the direction of the second axis, with the convention that the names of the two axes and the coordinates correspond (e.g., *x*-axis and *x*-coordinate, *y*-axis and *y*-coordinate).

5.G.2 Represent real world and mathematical problems by graphing points in the first quadrant of the coordinate plane, and interpret coordinate values of points in the context of the situation.

Cool! This is what I'm going to be doing!

Standards for Mathematical PRACTICE

1. Make sense of problems and persevere in solving them.
2. Reason abstractly and quantitatively.
3. Construct viable arguments and critique the reasoning of others.
4. Model with mathematics.
5. Use appropriate tools strategically.
6. Attend to precision.
7. Look for and make use of structure.
8. Look for and express regularity in repeated reasoning.

= focused on in this chapter

Name ....

Am I Ready?

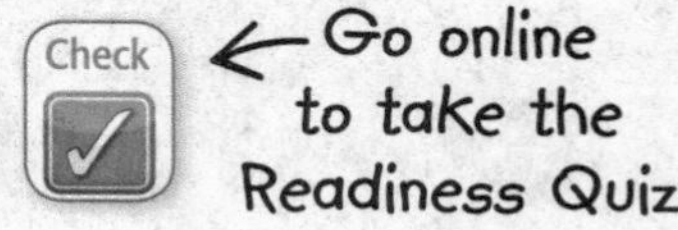

Find each missing number.

1. 5 + ______ = 7

2. ______ − 4 = 9

3. 27 − ______ = 18

4. Devry needed to read 36 pages by Friday. He read 10 on Tuesday and 15 on Wednesday. How many pages does he need to read on Thursday in order to read 36 pages by Friday?

____

5. Quinton struck out 12 batters in his first baseball game, 5 batters in the second game, and 8 batters in the third game. How many batters did he strike out in all three games?

____

Identify each pattern.

6. 2, 5, 8, 11 . . .

____

7. 4, 9, 14, 19 . . .

____

8. 27, 21, 15, 9 . . .

____

9. Antoinette runs each day. What is the rule for the pattern shown in her running log?

____

Running Log					
Day	1	2	3	4	5
Distance (mi)	2	4	6	8	10

Shade the boxes to show the problems you answered correctly.

How Did I Do?

1	2	3	4	5	6	7	8	9

Name

MY Math Words

Vocab

Review Vocabulary

perpendicular

Making Connections

Use the review vocabulary to complete each section of the bubble map.

Definition

Real-World Example

Perpendicular

Example

Non-example

MY Vocabulary Cards

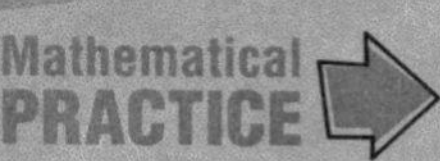

Lesson 7-8

coordinate plane

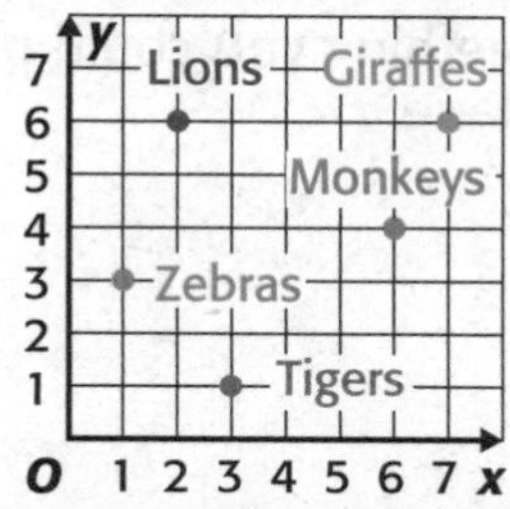

Lesson 7-1

evaluate

$5 + 7 = 12$

$15 - 10 = 5$

Lesson 7-1

numerical expression

$25 \div 5 - 3$

$15 + 7 - 8$

Lesson 7-8

ordered pair

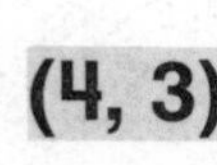

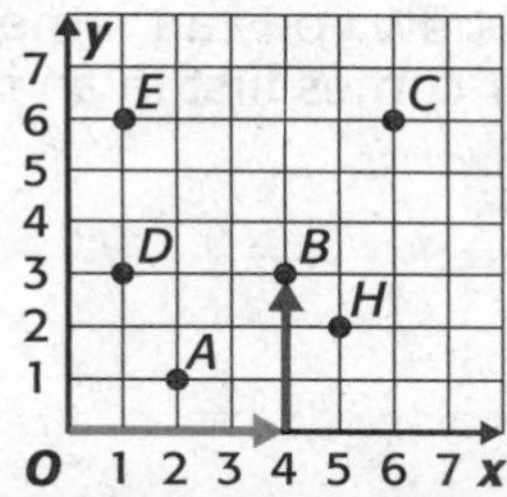

Lesson 7-2

order of operations

$[20 + 2 \times (7 - 5)] - 8$

$[20 + 2 \times 2] - 8$

$[20 + 4] - 8$

$24 - 8 = 16$

Lesson 7-8

origin

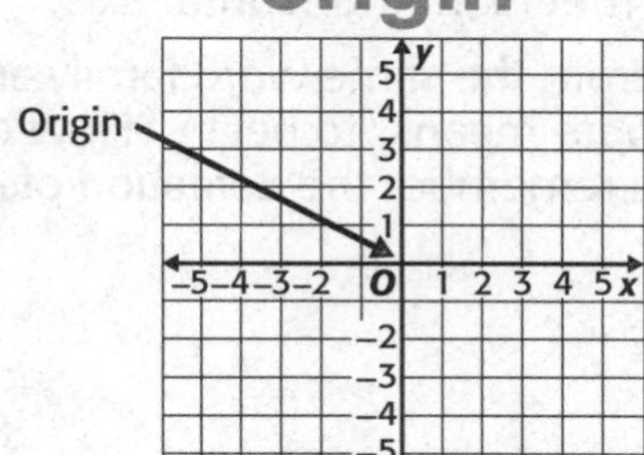

Lesson 7-6

sequence

2, 4, 6, 8, . . .

Lesson 7-6

term

2, 4, 6, 8, . . .

Ideas for Use

- Group 2 or 3 common words. Add a word that is unrelated to the group. Then work with a friend to name the unrelated word.
- Design a crossword puzzle. Use the definition for each word as the clues.

To find the value of a numerical expression by completing each operation.

When evaluating $5 + 3 + 6$, does the order in which you add matter? Explain.

A plane that is formed when two number lines intersect at a right angle.

Describe something you could plot on a coordinate plane.

A pair of numbers that is used to name a point on a coordinate plane.

Explain how you can remember which number comes first in an ordered pair.

A combination of numbers and at least one operation, such as $9 - 4$.

What is the difference between a numerical expression and an equation?

The point on a coordinate plane where the vertical axis meets the horizontal axis.

Originate* is from the same word family as *origin. Originate* means "to begin." How can this help you remember the definition of *origin?

The order in which operations on numbers should be done: parentheses, exponents, multiply and divide, add and subtract.

Write a sentence you could use to help you remember the correct order of operations.

A number in a pattern or sequence.

Term* is a multiple-meaning word. Write a sentence using another meaning of *term.

A list of numbers that follows a specific pattern.

Write a real-world example that is an example of a *sequence*.

MY Vocabulary Cards

Mathematical PRACTICE

Lesson 7-8

x-coordinate

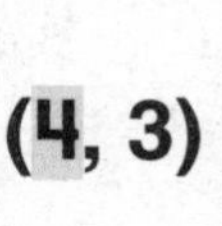

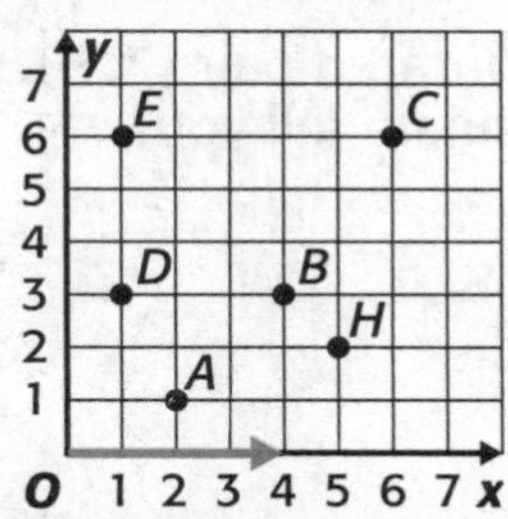

Lesson 7-8

y-coordinate

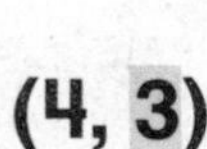

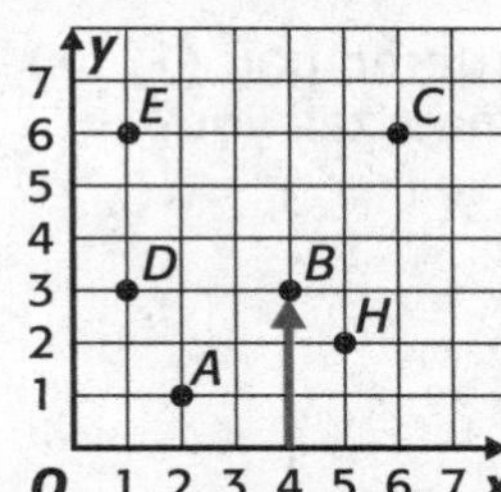

Ideas for Use

- Write the names of lessons on the front of blank cards. Write a few study tips on the back of each card.
- Use the blank cards to write your own vocabulary cards.

The second part of an ordered pair that shows how far away from the *x*-axis the point is.

In the ordered pair (1, 12), what does the *y*-coordinate tell you?

____

____

The first part of an ordered pair that shows how far away from the *y*-axis the point is.

In the ordered pair (7, 9), what does the *x*-coordinate tell you?

____

____

MY Foldable

FOLDABLES® Follow the steps on the back to make your Foldable.

Parentheses

( )

1st

Exponents

2^2

2nd

Multiply and Divide

× ÷

3rd

Add and Subtract

+ −

4th

$(15 \div 3) \times 5^2 - 5$

$5 \times 5^2 - 5$

$5 \times 5^2 - 5$

$5 \times 25 - 5$

$5 \times 25 - 5$

$125 - 5$

$125 - 5$

120

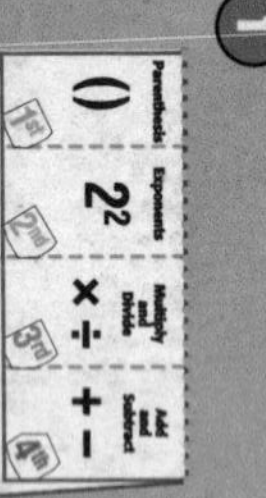

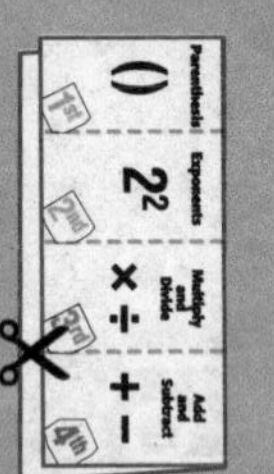

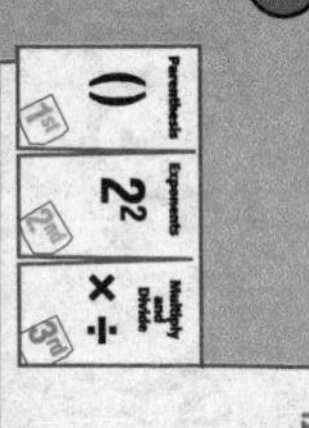

Name ....

Hands On
Numerical Expressions

Lesson 1

ESSENTIAL QUESTION
How are patterns used to solve problems?

A **numerical expression**, such as 8 + 7, is a combination of numbers and at least one operation. You can find the value, or **evaluate**, the numerical expression by completing each operation.

Draw It

Gregory and his family went hiking over the weekend. On Saturday, they hiked 5 miles and on Sunday, they hiked 5 miles. Use the bar diagram to write and evaluate two numerical expressions to represent the total number of miles hiked.

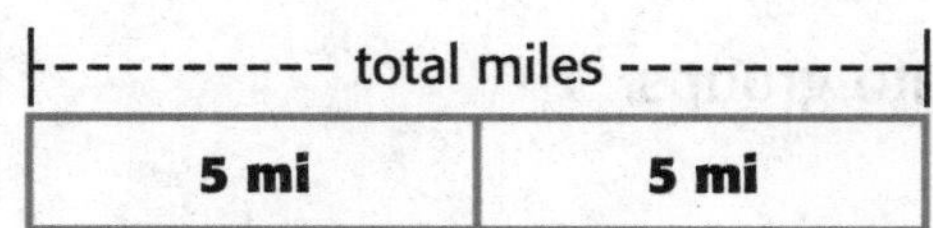

Use the bar diagram to write an addition expression.

5 + ________

Evaluate the expression.

5 + ________ = ________

Use the bar diagram to write a multiplication expression.

________ × 5

Evaluate the expression.

________ × 5 = ________

So, they hiked a total of ________ miles.

Try It

Mrs. Yearling has two groups of 5 students and two groups of 4 students. Use the bar diagram to write and evaluate two numerical expressions to represent the total number of students.

total students

5 students	5 students	4 students	4 students

Use the bar diagram to write an expression using only addition.

5 + ______ + 4 + ______

Evaluate the expression.

5 + ______ + 4 + ______ = ______

2 Use the bar diagram to write an expression using multiplication and addition.

(______ × 5) + (______ × 4)

Evaluate the expression.

(______ × 5) + (______ × 4) = ______

Helpful Hint

Parentheses tell you which numbers to group together. Perform operations inside parentheses first.

So, there are ______ students that are divided into groups.

Talk About It

1. Evaluate the addition expressions to find the sum. Does the order in which the expression is written change the sum? Explain.

addition expression	addition expression
7 + 7 + 5 + 5	7 + 5 + 7 + 5

____

____

2. Mathematical PRACTICE 4 **Model Math** Suppose Mrs. Yearling also had another group of 4 students. Write two new numerical expressions to represent the total number of students.

Expression 1	Expression 2
5 + ______ + 4 + 4 + ______	(2 × ______) + (______ × 4)

Name ....

Practice It

3. Caleb's music class is divided into 5 groups of 4 students for a project. Use the bar diagram to write and evaluate two numerical expressions to represent the total number of students in his music class.

total students

4 students	4 students	4 students	4 students	4 students

Write and evaluate an expression using only addition.

____

Write and evaluate an expression using multiplication.

____

So, there are ________ students in his music class.

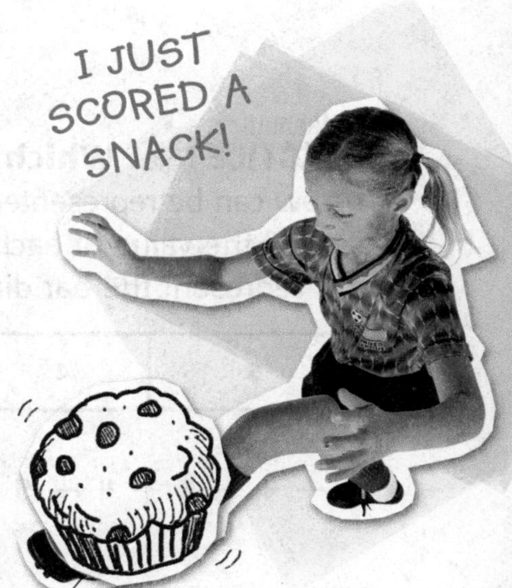

4. Mathematical PRACTICE 7 **Identify Structure** Bailey's soccer team had snacks after the game that included 12 granola bars, 12 mini muffins, and 14 bananas. Use the bar diagram to write and evaluate two numerical expressions to represent the total number of snacks after the soccer game.

total snacks

12 granola bars	12 mini muffins	14 bananas

Write and evaluate an expression using only addition.

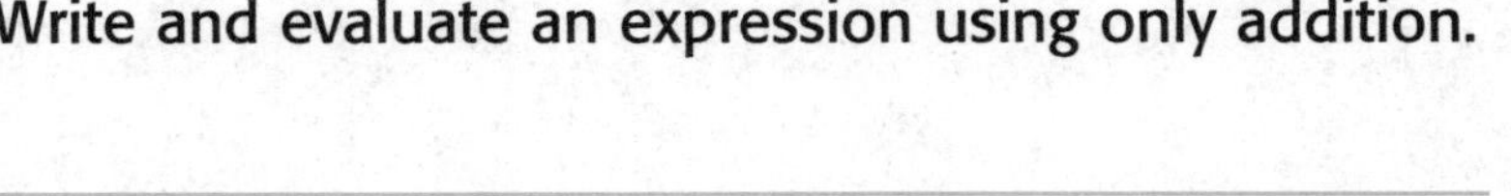

Write and evaluate an expression using multiplication and addition.

____

So, there are ________ total snacks after the game.

Apply It

5. Wasah went bird watching and spotted 6 robins, 5 sparrows, 6 cardinals, and 6 doves. Use the bar diagram to write and evaluate two numerical expressions to represent the total number of birds Wasah spotted.

total number of birds

6 robins	5 sparrows	6 cardinals	6 doves

Write and evaluate an expression using only addition.

____

Write and evaluate an expression using multiplication and addition.

____

So, Wasah spotted a total of _______ birds.

6. **Mathematical PRACTICE 3** **Which One Doesn't Belong?** The bar diagram below can be represented by three of the four expressions underneath it. Find the value of each expression and circle the one that does not represent the bar diagram.

5	4	5	4	5	4

$5 + 4 + 5 + 4 + 5 + 4$	$(3 \times 5) + (3 \times 4)$	$3 + 5 + 3 + 4$	$3 \times (5 + 4)$
_______	_______	_______	_______

Write About It

7. How can bar diagrams be used to model numerical expressions?

____

____

Name

MY Homework

Lesson 1
Hands On: Numerical Expressions

Homework Helper

Need help? connectED.mcgraw-hill.com

Paul has a fruit basket that includes 3 oranges, 4 apples, 3 bananas, and 2 grapefruits. Use the bar diagram to write and evaluate two numerical expressions to represent the total number of fruit in the basket.

total number of fruit

3 oranges	4 apples	3 bananas	2 grapefruits

1 Use the bar diagram to write an expression using only addition.

$3 + 4 + 3 + 2$

Evaluate the expression.

$3 + 4 + 3 + 2 = 12$

2 Use the bar diagram to write an expression using multiplication and addition.

$(2 \times 3) + 4 + 2$

Evaluate the expression.

$(2 \times 3) + 4 + 2 = 12$

So, there are 12 total pieces of fruit in the basket.

Practice

1. Refer to the Homework Helper. Suppose Paul went to the store and bought 4 peaches to add to the basket. Write two new numerical expressions to represent the total number of fruit in the basket.

Expression 1

3 + _____ + 3 + 2 + _____

Expression 2

(2 × _____) + (_____ × 4) + 2

Problem Solving

2. Deborah's playlist on her MP3 player had a variety of songs on it. She had 21 country songs, 18 alternative songs, 16 pop songs, and 18 rock songs. Use the bar diagram to write and evaluate two numerical expressions to represent the total number of songs on Deborah's MP3 player.

total songs

21 country songs	18 alternative songs	16 pop songs	18 rock songs

Write and evaluate an expression using only addition.

____

Write and evaluate an expression using multiplication and addition.

____

So, there are _______ total songs on Deborah's MP3 player.

3. Mathematical PRACTICE 7 **Identify Structure** Mrs. Conrad's art class was painting murals. She had 4 bottles of red paint, 5 bottles of green paint, 4 bottles of yellow paint, 6 bottles of blue paint, and 5 bottles of orange paint. Use the bar diagram to write and evaluate two numerical expressions to represent the total number of bottles of paint.

total number of bottles

4 red	5 green	4 yellow	6 blue	5 orange

Write and evaluate an expression using only addition.

____

Write and evaluate an expression using multiplication and addition.

____

So, Mrs. Conrad has a total of _______ bottles of paint.

Name ....

Order of Operations

Lesson 2

ESSENTIAL QUESTION
How are patterns used to solve problems?

Math in My World

Example 1

The table shows the number of Calories burned in one minute for two different activities. Nathan swims for 4 minutes and then runs for 8 minutes. How many Calories has Nathan burned in all?

Activity	Calories Burned per Minute
Swimming	12
Running	10

Evaluate the expression $12 \times 4 + 10 \times 8$.

Write the expression. $12 \times 4 + 10 \times 8$

Multiply 12 by 4. ______ $+ 10 \times 8$

Multiply 10 by 8. ______ + ______

Add. ______

So, Nathan has burned ______ Calories.

The **order of operations** is a set of rules to follow when more than one operation is used in an expression.

Key Concept Order of Operations

1. Perform operations in parentheses.
2. Find the value of exponents.
3. Multiply and divide in order from left to right.
4. Add and subtract in order from left to right.

Parentheses include brackets [ ] as well as braces { }. Perform operations inside parentheses first, then perform operations inside brackets, and finally, perform operations inside braces.

Example 2

Evaluate $20 - \{4 + [4 + (10 \div 2)]\}$.

Write the expression.	$\_\_\_\_ - \{4 + [\_\_\_\_ + (10 \div \_\_\_\_)]\}$	
Divide 10 by 2.	$20 - \{4 + [4 + \_\_\_\_]\}$	parentheses 1st
Add.	$20 - \{4 + \_\_\_\_\}$	brackets 2nd
Add.	$20 - \_\_\_\_$	braces 3rd
Subtract.		

So, $20 - \{4 + [4 + (10 \div 2)]\} = \_\_\_\_$.

Guided Practice

1. Evaluate $\{28 + [(2 \times 4^2) \div 8]\}$.

Write the expression.	$\{\_\_\_\_ + [(2 \times 4^2) \div \_\_\_\_]\}$	
Find 4^2.	$\{28 + [(2 \times \_\_\_\_) \div 8]\}$	
Multiply.	$\{28 + [\_\_\_\_ \div 8]\}$	parentheses 1st
Divide.	$\{28 + \_\_\_\_\}$	brackets 2nd
Add.	$\_\_\_\_$	braces 3rd

So, $\{28 + [(2 \times 4^2) \div 8]\} = \_\_\_\_$.

Talk MATH

Explain why it is important to follow the order of operations when evaluating $15 + 3 \times 4$.

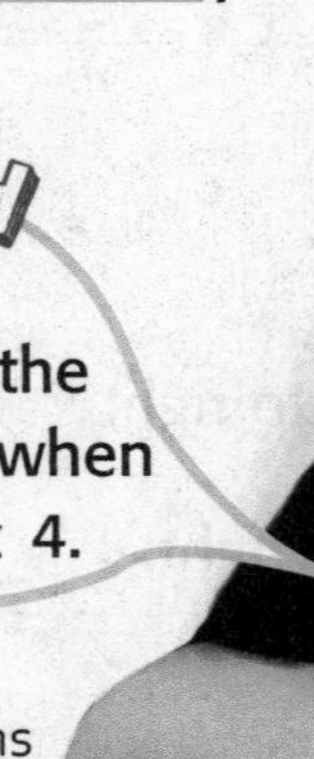

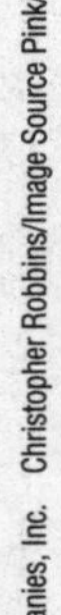

Name

Independent Practice

Evaluate each expression.

2. $5 \times (92 - 18) =$ ______

3. $12 + (4^2) - 11 =$ ______

4. $(15 - 5) \times [(9 \times 3) + 3] =$ ______

5. $58 - 6 \times 7 =$ ______

6. $55 - [(5^2 \times 3) - 5^2] =$ ______

7. $7 \times 10 + 3 \times 30 =$ ______

8. $2^2 + \{[1 \times (5 - 2)] \times 3\} =$ ______

9. $\{2 \times [4 - (6 \div 2)]\} \times 3 =$ ______

Algebra Find each unknown.

10. $3^3 + 3 \times 5 = k$

$k =$ ______

11. $12 - [(3^2 \times 4) - 30] = b$

$b =$ ______

Problem Solving

12. Three students are on the same team for a relay race. They finish the race in 54.3 seconds. The runners' times are shown in the table. Evaluate $54.3 - (18.8 + 17.7)$ to find the time of the third runner. Record your answer in the table.

Relay Times	
Runner	Time (seconds)
1	18.8
2	17.7
3	

My Work!

13. You can find the temperature in degrees Celsius by using the expression $5 \times (°F - 32) \div 9$. If the temperature of a cup of hot chocolate is 104°F, what is the temperature of the cup of hot chocolate in degrees Celsius?

____

14. Ryan and Maggie split the cost of a $12 pizza. They also have a coupon for $2 off. Evaluate $(12 - 2) \div 2$ to find the cost each person will pay.

____

HOT Problems

15. Mathematical PRACTICE 1 **Plan Your Solution** Write an expression using only multiplication and subtraction so that its value is 25.

____

16. **Building on the Essential Question** When and why does order matter?

____

____

MY Homework

Lesson 2
Order of Operations

Homework Helper

Need help? connectED.mcgraw-hill.com

Evaluate $\{5^3 \div [1 \times (10 - 5)]\} - 20$.

Write the expression.	$\{5^3 \div [1 \times (10 - 5)]\} - 20$	parentheses 1st
Subtract 5 from 10.	$\{5^3 \div [1 \times 5]\} - 20$	brackets 2nd
Multiply.	$\{5^3 \div 5\} - 20$	
Find 5^3.	$\{125 \div 5\} - 20$	braces 3rd
Divide.	$25 - 20$	
Subtract.	5	

So, $\{5^3 \div [1 \times (10 - 5)]\} - 20 = 5$.

Practice

1. Evaluate $64 \div [4 \times (27 - 5^2)]$.

Write the expression.	______ $\div [4 \times ($ ______ $- 5^2)]$	
Find 5^2.	$64 \div [4 \times (27 -$ ______ $)]$	parentheses 1st
Subtract.	$64 \div [4 \times$ ______ $]$	brackets 2nd
Multiply.	$64 \div$ ______	
Divide.	______	

So, $64 \div [4 \times (27 - 5^2)] =$ ______.

Problem Solving

2. Mathematical PRACTICE 4 **Model Math** Kishauna rode her bike for 35 minutes each on Monday, Wednesday, and Saturday and 55 minutes each on Tuesday and Thursday. Write an expression that shows the total amount of time she spent riding her bike. Then evaluate the expression.

____

3. Sarah evaluated the expression $[(2^3 \times 4) \div 2] + 2$. What was her answer?

____

4. Kylie and her three friends equally divided the cost to rent a movie for \$4 and order sandwiches for a total of \$15. They also have a coupon for \$3 off the sandwiches. Evaluate $[(4 + 15) - 3] \div 4$ to find the cost each person will pay.

____

Vocabulary Check

5. Fill in each blank with the correct word to complete the sentence. The rules of the order of operations tells you to multiply and divide in order from ________ to ________.

Test Practice

6. Keiko's class collected money to donate to charity. When Keiko counted the money, there were 140 five-dollar bills, and 255 one-dollar bills. What expression could he use to find out how much money was collected?

Ⓐ (140 × \$5) + (255 × \$1)

Ⓑ (140 × \$1) + (255 × \$5)

Ⓒ (140 + \$5 × 255 + \$1)

Ⓓ 140 + \$5 + 255 + \$1

Write Numerical Expressions

Lesson 3

ESSENTIAL QUESTION
How are patterns used to solve problems?

Math in My World

CHECK PLEASE!

Example 1

Terrell went to dinner with his friends and ordered 3 tacos. Each taco costs $2 and he has a coupon for a dollar off his purchase. The total cost in dollars of Terrell's purchase is represented by the phrase multiply three by two, then subtract one. Write the total cost as a numerical expression.

1 Write the phrase in parts.

Part 1 multiply three by ______

Part 2 then subtract ______

2 Write each part as a numerical expression.

Part 1 multiply three by two ——→ ______

Part 2 then subtract one ——→ ______

3 Combine the numerical expressions to represent the total cost in dollars. Add parentheses if needed.

______

Example 2

A ticket to a baseball game costs $25 and popcorn is $8. Three friends bought tickets and popcorn. The expressions below give the cost for one friend and for three friends. Compare the two expressions without evaluating them.

One Friend	**Three Friends**
$25 + 8$	$(25 + 8) \times 3$

Both expressions contain the same addition expression. Write the addition expression. ________

For three friends, the addition expression is multiplied by ________.

So, the second expression is ________ times as large as the first expression.

Guided Practice 

1. Write the phrase *add 7 and 11, then divide by 2* as a numerical expression.

   Write the phrase in parts.

   Part 1 ____________________

   Part 2 ____________________

   Write each part as a numerical expression.

   Part 1 add 7 and 11 ⟶ ________

   Part 2 then divide by 2 ⟶ ________

   Combine the numerical expressions. Add parentheses if needed.

   ____

Write a real-world problem that could be represented by a numerical expression.

Name ______________________________

Independent Practice

Write each phrase as a numerical expression.

2. divide 15 by 3, then add 13 ______________

3. subtract 4 from 20, then divide by 2 ______________

4. add 9 and 4, then multiply by 2 ______________

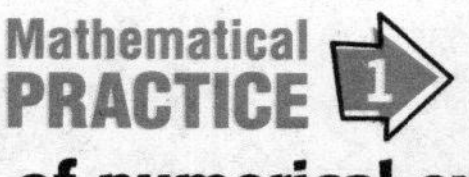

Make Sense of Problems **Compare each pair of numerical expressions without evaluating them.**

5. **Expression 1** $(7 \times 4) \div 2$ **Expression 2** 7×4

Both expressions contain the same multiplication expression.

Write the multiplication expression. __________

In Expression 1, the product is divided by ______.

So, Expression 1 is ______ as large as Expression 2.

6. **Expression 1** $2 + 5 + 8$ **Expression 2** $4 \times (2 + 5 + 8)$

Both expressions contain the same addition expression. Write the addition expression. ______________

In Expression 2, the addition expression is multiplied by ______.

So, Expression 2 is ______ times as large as Expression 1.

Problem Solving

7. Robin wants to find the area of the triangle below. To find the area of a triangle, multiply the base times the height and then divide by 2. The base and height of the triangle are shown. Represent the area of the triangle with a numerical expression.

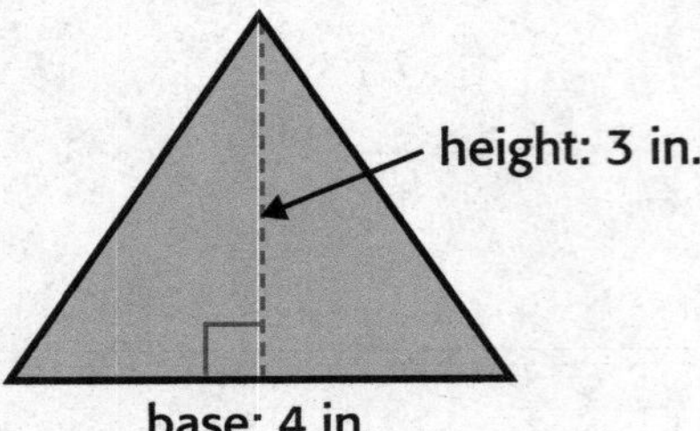

____

8. Deirdre doubled her savings account balance of $100. Then she withdrew $30 to buy some new clothes. Represent this situation with a numerical expression.

____

HOT Problems

9. Mathematical PRACTICE 2 **Use Number Sense** Explain why the numerical expression 3 less than 16 is written as 16 − 3 and not 3 − 16.

____

10. Mathematical PRACTICE 7 **Identify Structure** Circle the numerical expression that is four times as large as 52 − 9.

52 − (9 × 4)	(52 − 9) + 4	(52 − 9) × 4	(52 − 9) ÷ 4

11. **Building on the Essential Question** How do I compare numerical expressions without calculating them?

____

____

____

Name ....

MY Homework

Lesson 3
Write Numerical Expressions

Homework Helper

Need help? connectED.mcgraw-hill.com

Admission to a county fair is $10 for adults and $6 for children. The total cost in dollars of admission for 1 adult ticket and 4 children's tickets is represented by the phrase four multiplied by six, then add ten. Write the total cost of admission as a numerical expression.

1 Write the phrase in parts.

Part 1 four multiplied by six

Part 2 then add ten

2 Write each part as a numerical expression.

Part 1 four multiplied by six ⟶ 4×6

Part 2 then add ten ⟶ $+ 10$

3 Combine the numerical expressions to represent the total cost in dollars. Add parentheses if needed.

$4 \times 6 + 10$

Practice

1. Compare the two numerical expressions without evaluating them.

Expression 1	**Expression 2**
$8 - 3$	$(8 - 3) \times 4$

Both expressions contain the same subtraction expression.

Write the subtraction expression. ______________

In Expression 2, the difference is multiplied by ________.

So, Expression 2 is __________ times as large as Expression 1.

Problem Solving

2. Jeffrey purchased and downloaded 12 songs on Monday. He purchased an additional 3 songs on Tuesday. The cost to download each song is $2. Write a numerical expression to represent this situation.

____

3. Mora bought 3 bags of apples for her class. One full bag has 8 apples, and each apple weighs 6 ounces. Write a numerical expression to represent this situation.

____

4. Mathematical PRACTICE 2 **Use Number Sense** Jane wants to find the area of the trapezoid. To find the area of a trapezoid, add the two bases, multiply by the height, then divide by 2. The bases and height of the trapezoid are shown. Represent the area of the trapezoid with a numerical expression.

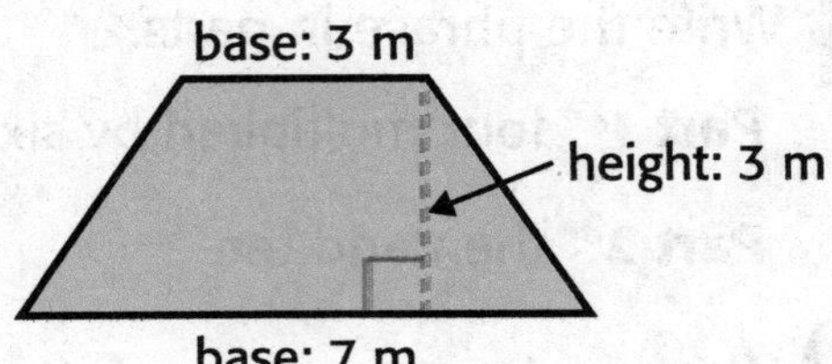

____

Vocabulary Check

5. Fill in the blank with the correct term or number to complete the sentence.

A ______________ expression like (3 + 5) × (4 − 1) is a combination of numbers and at least one operation.

Test Practice

6. Denzel and three friends go to the movies. Each person buys a movie ticket for $8, a snack for $4, and a drink for $2. Which numerical expression represents the total cost of the trip to the movies for Denzel and his friends?

Ⓐ 4 + ($8 × $4 × $2)

Ⓑ 4 × ($8 + $4 + $2)

Ⓒ (4 × $8) + ($4 × $2)

Ⓓ (4 × $8 + $4) + (4 × $4 + $2)

Name ....

Real World

Problem-Solving Investigation

STRATEGY: Work Backward

Lesson 4

ESSENTIAL QUESTION
How are patterns used to solve problems?

Learn the Strategy

The Nature Club raised $125 to buy and install bird houses at a wildlife site. Each house costs $5. It costs $75 to rent a bus so the members can travel to the site. How many boxes can the club buy?

1 Understand

What facts do you know?

_______ is available to buy and install the nesting boxes.

Each box costs _______ and the bus rental costs _______.

What do you need to find?

How many _______ can the club buy?

2 Plan

I can work backward to solve the problem.

3 Solve

Subtract the cost of the bus. Then divide by the cost for each box.

$125 − $75 = _______   _______ ÷ $5 = _______

So, _______ boxes can be bought.

4 Check

Is my answer reasonable? Explain.

Multiply. _______ × $5 = _______   Add. _______ + $75 = $125

Practice the Strategy

Mr. Evans bought the items listed. He had $5 left over. How much did Mr. Evans have to start with?

Items Purchased	
Toothpaste	$4
Toothbrush	$2
Floss	$1

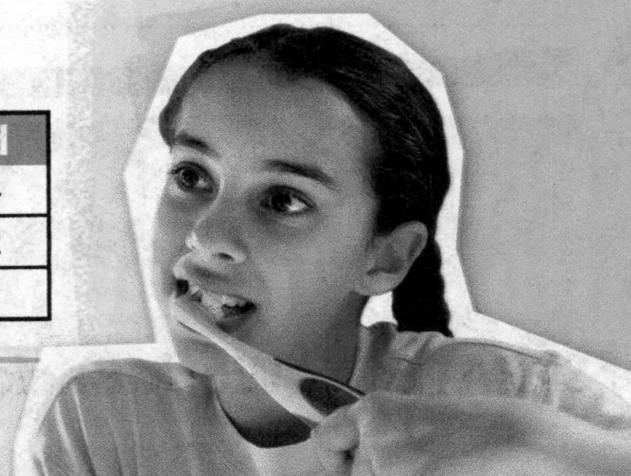

1 Understand

What facts do you know?

What do you need to find?

2 Plan

3 Solve

4 Check

Is my answer reasonable? Explain.

Name

Apply the Strategy

Solve each problem by working backward.

1. Seth bought a movie ticket, popcorn, and a drink. After the movie, he played 4 video games that each cost the same. He spent a total of $19. How much did it cost to play each video game?

Movie Costs	
Popcorn	$4
Drink	$3
Ticket	$8

2. Students sold raffle tickets to raise money for a field trip. The first 20 tickets sold cost $4 each. To sell more tickets, they lowered the price to $2 each. If they raise $216, how many tickets did they sell in all?

3. Jeanette's sister charges $5.50 per hour before 9:00 P.M. for babysitting and $8 per hour after 9:00 P.M. She finished babysitting at 11:00 P.M. and earned $38. At what time did she begin babysitting?

4. Mathematical PRACTICE 2 **Use Algebra** Work backward to find the value of the variable in the equation below.

$$d + 4 = 19$$

5. Allie collected 15 more cans of food than Peyton. Ling collected 8 more than Allie. Ling collected 72 cans of food. How many cans of food did Peyton collect?

My Work!

Review the Strategies

Use any strategy to solve each problem.

- Work backward.
- Make a table.
- Solve a simpler problem.
- Determine extra or missing information.

6. Rebecca sold 11 more magazine subscriptions than Chad. Laura sold 4 more than Rebecca. Laura sold 45 magazine subscriptions. How many magazine subscriptions did Chad sell?

7. Mathematical **PRACTICE** 8 **Look for a Pattern** Frankie is planning to buy a new MP3 player for $90. Each month he doubles the amount he saved the previous month. If he saves $3 the first month, in how many months will Frankie have enough money to buy the MP3 player?

8. The table shows the number of miles Michael ran each day over the past four days. How many more miles did he run on day 3 than on day 2? Determine if there is extra or missing information.

Day	Miles
1	5
2	2
3	7
4	3

9. Mrs. Stevens is delivering flowers to a local flower shop. She delivers the same number of flowers with each delivery. The flower shop has ordered 2,050 flowers and it will take 5 trips to deliver all the flowers. How many flowers will Mrs. Stevens have delivered after 4 trips?

10. Admission to a car show costs $5 for each ticket. After selling only 20 tickets, they decided to lower the price to $3 each. If they raise $217, how many tickets did they sell in all?

My Work!

Name

MY Homework

Lesson 4
Problem Solving: Work Backward

Homework Helper

Need help? connectED.mcgraw-hill.com

Peyton and his friends built an outdoor game board in the shape of a rectangle that has a length of 4 feet and a width of 2 feet. If they cut a circular hole that has an area of 1 square foot, what is the area of the game board that does not include the hole?

1 Understand

What facts do you know?

the length and width of the game board

the area the hole takes up

What do you need to find?

the area of the game board, not including the hole

2 Plan

I can work backward to solve the problem.

3 Solve

Find the area of the game board. (Hint: Area = length × width)

$4 \times 2 = 8$ square feet

Subtract the area of the hole from the area of the game board.

$8 - 1 = 7$ square feet

So, the game board that Peyton and his friends built has an area of 7 square feet, not including the hole.

4 Check

Is my answer reasonable? Explain.

area of game board + area of hole = total area

7 square feet + 1 square foot = 8 square feet

Problem Solving

Solve each problem by working backward.

1. The science club raised money to clean the beach. They spent $29 on trash bags and $74 on waterproof boots. They still have $47 left. How much did they raise?

____

2. Mr. Charles cut fresh roses from his garden and gave 10 roses to his neighbor. Then he gave half of what was left to his niece. He kept the remaining 14 roses. How many roses did he cut?

____

3. Mathematical PRACTICE 2 **Use Number Sense** A number is divided by 6. Then 8 is added to the quotient. Next 3 is subtracted from the sum. The result is 7. What is the number?

____

4. Hoover Dam, in the United States, is 223 meters high. Ertan Dam, in China, is 240 meters high. Write and evaluate a numerical expression to find the difference between the heights of the two dams.

____

DINO-MITE!

5. Ms. Houston's fifth-grade class is going to a museum. The class raises $68 for the trip. Transportation to the museum costs $40. The museum sells small fossils for $4 each. How many fossils can they buy with the money they have left?

____

Check My Progress

Vocabulary Check 

State whether each sentence is *true* or *false*. If *false*, replace the underlined word or number to make a true sentence.

1. A combination of numbers and operations is called a **formula**.

____

2. The **numerical expression** of (2 × 4) + (3 × 3) has a value of 33.

____

3. The **order of operations** is a set of rules to follow when more than one operation is used in an expression.

____

Concept Check 

4. Find the value of 2 × {15 − [(12 ÷ 3) × 2]}.

Write the expression.	___ × {15 − [(___ ÷ 3) × ___]}	
Divide 12 by 3.	2 × {15 − [___ × 2]}	parentheses 1st
Multiply.	2 × {15 − ___}	brackets 2nd
Subtract.	2 × ___	braces 3rd
Multiply.	___	

So, 2 × {15 − [(12 ÷ 3) × 2]} = ___.

Write each phrase as a numerical expression.

5. multiply 4 and 7, then subtract 5 ______________

6. add 3 to the product of 10 and 4 ______________

7. subtract 8 from the quotient of 15 and 3 ______________

8. subtract 9 from 13, then multiply the result by 2 ______________

Real World Problem Solving

9. Tia and her five friends are going to the ice skating rink. Each person pays \$5 for admission and \$5 for food. Write and evaluate a numerical expression to find the total cost for admission and food.

____

10. Cameron has 2 video game holder stands. Each can hold 2 rows of 20 games. Write and evaluate a numerical expression to find the total number of games Cameron's video game holder stands can hold.

____

Test Practice

11. Arturo buys 3 containers of ice cream for \$5 each and a cake that costs \$8 to take to his friend's party. Which expression will allow you to find how much money Arturo spent on ice cream and cake?

Ⓐ \$8 × 3 × \$5 Ⓒ (3 × \$8) + \$5

Ⓑ (3 × \$5) + \$8 Ⓓ 3 × (\$5 + \$8)

Name ....

Hands On
Generate Patterns

Lesson 5

ESSENTIAL QUESTION
How are patterns used to solve problems?

Build It

The pattern below is made from toothpicks. The first figure uses 4 toothpicks, the second figure uses 7 toothpicks, and the third figure uses 10 toothpicks. Assume the pattern continues.

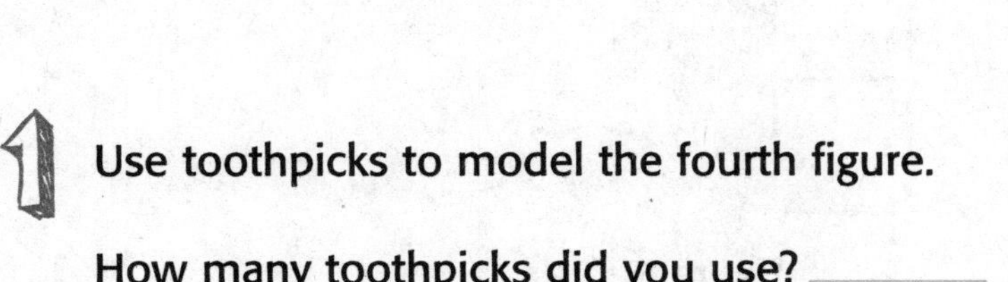

Figure 1 Figure 2 Figure 3

1 Use toothpicks to model the fourth figure.

How many toothpicks did you use? ________

2 Use toothpicks to model the fifth figure. Draw the result below.

How many toothpicks did you use? ________

My Drawing!

3 Complete the table to show the number of toothpicks needed if the pattern continues.

Figure Number	1	2	3	4	5	6	7
Number of Toothpicks	4	7	10				

What do you notice about the number of toothpicks needed for each new figure? ______

Talk About It

1. Using your rule, how many toothpicks would be needed for the eighth figure? ninth figure?

Figure 8 ______ Figure 9 ______

2. Mathematical PRACTICE 5 **Use Math Tools** Create a new pattern. Start with Figure 1 again. Add 6 toothpicks for each new figure as shown. Complete the table to show the number of toothpicks used for each figure.

Figure 1 Figure 2

Figure 3

Figure Number	1	2	3	4	5	6	7
Number of Toothpicks	4	10	16				

3. Compare the pattern in Exercise 2 to the pattern in the activity on the previous page.

The number of toothpicks in Figure 2 for the pattern on the previous page is ______ less than the number of toothpicks in Figure 2 for the pattern in Exercise 2.

The number of toothpicks in Figure 3 for the pattern on the previous page is ______ less than the number of toothpicks in Figure 3 for the pattern in Exercise 2.

Name

Practice It

Mathematical PRACTICE 8 Look for a Pattern **For each pattern, find the number of toothpicks needed for the next figure.**

4.

Figure 1 Figure 2 Figure 3

Figure 4 needs ______ toothpicks.

How is this pattern different than the pattern in the activity on the first page of this lesson?

____

____

5.

Figure 1 Figure 2 Figure 3

Figure 4 needs ______ toothpicks.

6.

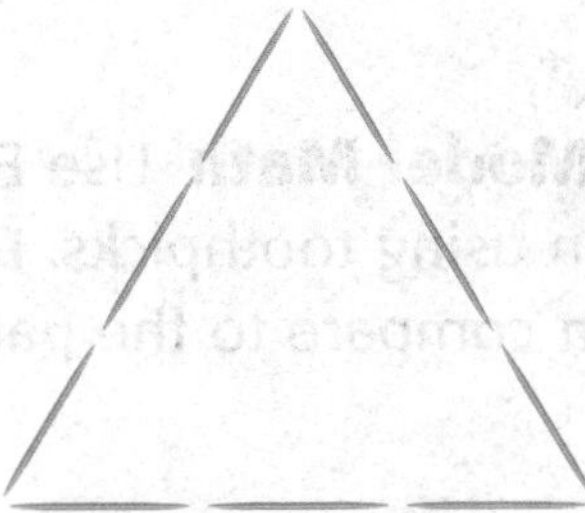

Figure 1 Figure 2 Figure 3

Figure 4 needs ______ toothpicks.

How does this pattern compare to the pattern for Exercise 5?

____

____

Apply It

7. The tables show the number of laps Tammi and Kelly swim each day. Complete each table if the patterns continue.

Compare the number of laps swam by each person on each day.

Tammi's Swimming Log	
Day	Number of Laps
1	0
2	6
3	12
4	
5	
6	

Kelly's Swimming Log	
Day	Number of Laps
1	0
2	3
3	6
4	
5	
6	

8. Two stores sell scented candles. Assume the pattern in the table below continues. Compare the price of candles sold by each store.

Store 1 Candles	
Number of Candles	Cost ($)
2	8
3	12
4	16

Store 2 Candles	
Number of Candles	Cost ($)
2	4
3	6
4	8

9. Mathematical PRACTICE 4 **Model Math** Use Figure 1 of Exercise 5 to create a different pattern using toothpicks. Draw the pattern below. How does your pattern compare to the pattern in Exercise 5?

Write About It

10. How can models be used to generate and analyze patterns?

Name ....

MY Homework

Lesson 5
Hands On: Generate Patterns

Homework Helper

Need help? connectED.mcgraw-hill.com

The pattern below is made from toothpicks. Figure 1 uses 4 toothpicks, Figure 2 uses 8 toothpicks, and Figure 3 uses 12 toothpicks. How many toothpicks will be needed for Figures 4, 5, 6, 7 and 8?

Figure 1

Figure 2

Figure 3

1 Use toothpicks to model Figure 4.

Sixteen toothpicks were used.

Figure 4

2 Use toothpicks to model Figure 5.

Twenty toothpicks were used.

Figure 5

3 Complete the table. The number of toothpicks increases by 4.

Figure Number	1	2	3	4	5	6	7	8
Number of Toothpicks	4	8	12	16	20	24	28	32

So, Figure 4 uses 16 toothpicks, Figure 5 uses 20 toothpicks, Figure 6 uses 24 toothpicks, Figure 7 uses 28 toothpicks, and Figure 8 uses 32 toothpicks.

Practice

Mathematical PRACTICE 8 **Look for a Pattern For each pattern, draw toothpicks to find the number of toothpicks needed for the next figure.**

1.

Figure 1 Figure 2 Figure 3

Figure 4 uses ______ toothpicks.

2.

Figure 1 Figure 2 Figure 3

Figure 4 uses ______ toothpicks.
How does this pattern compare to the pattern for Exercise 1?

____

____

Problem Solving

3. The tables show the height in centimeters each plant grew during a week. Assume the patterns continue. Compare the growth in height of each plant.

____

____

____

Plant A	
Day	Height (cm)
1	0
2	2
3	4
4	6
5	8
6	10
7	12

Plant B	
Day	Height (cm)
1	0
2	6
3	12
4	18
5	24
6	30
7	36

Name ______________________________

Patterns

Lesson 6

ESSENTIAL QUESTION
How are patterns used to solve problems?

A **sequence** is a list of numbers that follow a specific pattern. Each number in the list is called a **term.**

Math in My World

Example 1

Mary and her friends find a four-leaf clover during lunch. A four-leaf clover has four leaves. The table shows the total number of leaves for several four-leaf clovers. Extend the pattern to find the next three terms.

Number of Four-Leaf Clovers	1	2	3	4
Number of Leaves	4	8	12	16

4, 8, 12, 16, . . .
+ 4 + 4 + 4

Each term in the sequence can be found by adding ________ to the previous term.

16 + 4 = ________ ________ + 4 = ________ ________ + 4 = ________

The next three terms are ________________.

Example 2 

Maria and Jeong are training to run a half-marathon. A half-marathon is about 13 miles. Their weekly training plans are shown in the table. Use the information to write a sequence to represent each person's weekly training plan. Then compare the plans.

Runner	Starting Miles	Training Plan
Maria	2	Add 2 miles per week for each of the next 4 weeks.
Jeong	4	Add 4 miles per week for each of the next 4 weeks.

Each week, Maria will run two more miles and Jeong will run four more miles than the previous week.

Write a sequence with 5 terms for Maria's training plan. ________

Write a sequence with 5 terms for Jeong's training plan. ________

Compare the training plans.
Each week, Jeong plans to run ________ as many miles as Maria.

Talk MATH

How are the sequences 2, 5, 8, 11, . . . and 2, 6, 18, 54, . . . alike? How are they different?

Guided Practice 

1. Write the next three terms in the sequence 1, 4, 7, 10, ...

Each term in the sequence can be found

by adding ________ to the previous term.

$10 + 3 =$ ________

________ $+ 3 =$ ________

________ $+ 3 =$ ________

The next three terms are ________________.

Name

Independent Practice

Algebra Identify the pattern. Then write the next three terms in each sequence.

2. 0, 7, 14, 21, . . .

3. 1,458, 486, 162, 54, . . .

4. 72, 66, 60, 54 . . .

5. 1, 3, 9, 27 . . .

6. 2, 4, 8, 16, . . .

7. 94, 88, 82, 76, . . .

8. 12, 24, 36, 48, . . .

9. 512, 256, 128, 64, . . .

10. 8, 13, 18, 23, . . .

11. 11, 24, 37, 50, . . .

12. 83, 75, 67, 59, . . .

13. 2, 8, 32, 128, . . .

Problem Solving

14. An amusement park offers discounted tickets after 4 P.M. Both ticket prices are shown to the right. Write the total cost of 1, 2, 3, and 4 tickets for each time period. Compare the cost of 4 tickets before 4 P.M. to 4 tickets after 4 P.M.

Admission Tickets	
Time	**Cost ($)**
Before 4 P.M.	45
After 4 P.M.	15

HOT Problems

15. Mathematical PRACTICE 3 **Which One Doesn't Belong?** Circle the sequence that does not belong with the other three. Explain your reasoning.

2, 5, 8, 11, . . .

3, 6, 12, 24, . . .

4, 14, 24, 34, . . .

7, 12, 17, 22, . . .

16. **Building on the Essential Question** How can we extend patterns?

Name ________________________________

MY Homework

Lesson 6
Patterns

Homework Helper

Need help? connectED.mcgraw-hill.com

Tom is allowed to download 3 new songs each week. The table shows the total number of songs he can download for several weeks. Extend the pattern to find the next three terms.

Week	1	2	3	4
Number of Songs	3	6	9	12

3, 6, 9, 12, . . .

+ 3 + 3 + 3

Each term in the sequence can be found by adding 3 to the previous term.

12 + 3 = 15   15 + 3 = 18   18 + 3 = 21

The next three terms are 15, 18, and 21.

Practice

Algebra Identify the pattern. Then write the next three terms in each sequence.

1. 5, 10, 20, 40, ...

____

2. 63, 58, 53, 48, . . .

____

3. 192, 96, 48, 24, . . .

____

4. 4, 11, 18, 25, . . .

____

Problem Solving

5. **Mathematical PRACTICE 1 Make a Plan** Dino's Diner charges $4 for each sandwich. Carla's Café charges $8 for each sandwich. Write the total costs of 1, 2, 3, and 4 sandwiches for each restaurant. Then compare the total cost of 4 sandwiches at each restaurant.

$8? You're full of BOLOGNA!

$8

____

____

____

____

6. Luke always runs 1 lap to warm up for track practice. Alex always runs 3 laps to warm up for track practice. Write the number of total laps ran for 1, 2, 3, and 4 practices for each student. Then compare the total number of laps ran for 3 practices by both students.

____

____

Vocabulary Check 

7. Fill in each blank with the correct word to complete each sentence.

A sequence is a list of numbers that follow a specific __________.

Each number in the list is called a __________.

Test Practice

8. Which represents the next three terms in the sequence 8, 16, 24, 32, . . . ?

Ⓐ 36, 40, 44
Ⓑ 64, 128, 256
Ⓒ 40, 48, 56
Ⓓ 72, 216, 648

Name ....

Hands On
Map Locations

Lesson 7

ESSENTIAL QUESTION
How are patterns used to solve problems?

Draw It

You can use grid paper to represent locations on a map. From school, Marcia walks three blocks north to the library. Then she walks two blocks east to the park. Her home is located one block south of the park. Draw a map that shows these locations.

1. Use the blank grid to draw and label a dot in the lower left corner to represent the school.

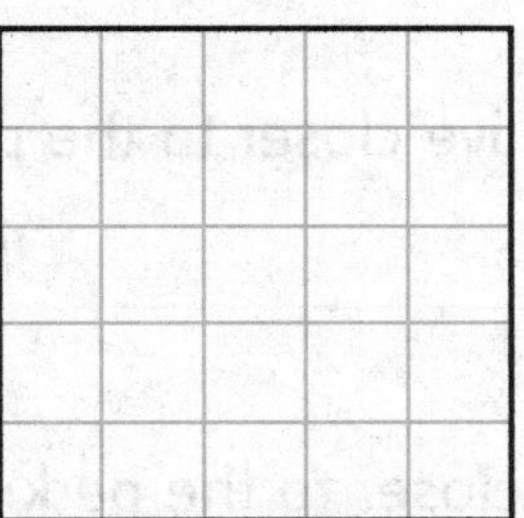

2. From the dot labeled "school", in what direction along the grid should you move to get to the library?

   ____

   How many units should you move to get to the library?

   ____

3. Draw and label a dot to represent the library's location. From the dot labeled "library", in what direction along the grid should you move to get to the park?

   ____

   How many units should you move to get to the park?

   ____

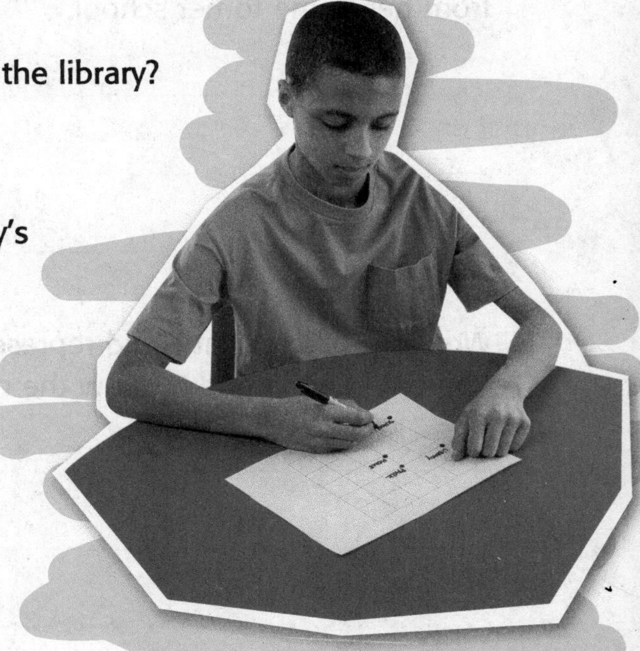

4 Draw and label a dot to represent the park's location. From the dot labeled "park", in what direction along the grid should you move to get to Marcia's home?

______

How many units should you move to get to Marcia's home?

______

Draw and label a dot to represent the location of Marcia's home.

Talk About It

1. On the grid provided, draw and label the locations from the Draw It Activity.

2. Does Marcia live closer to the park or to the library?

______

3. Is the library closer to the park or to Marcia's school?

______

4. Mathematical PRACTICE 1 **Make a Plan** Describe how Marcia could walk from her home to her school.

______

______

______

5. Write a problem that could represent locations of real-world objects. Use the grid to draw the map.

______

______

______

Name

Practice It

For Exercises 6–9, use the grid paper to draw a map of the given locations.

6. From the zoo entrance, Marco walks three units east to the gift shop. Then he walks four units north to the bear exhibit. The lion exhibit is located two units south and one unit west of the bear exhibit.

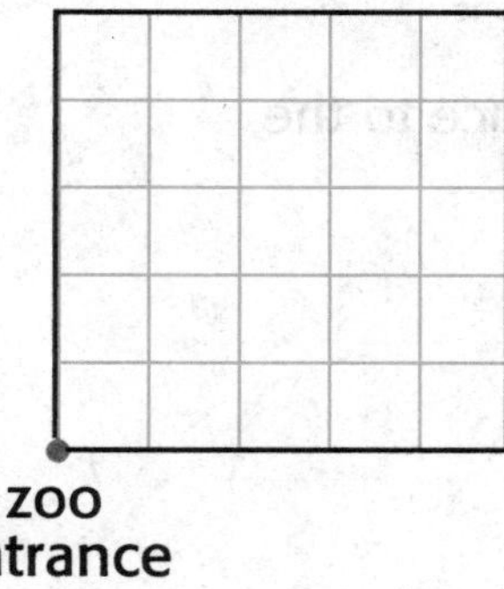

7. From the dining hall, a camper rides her bike four units north to the nature center. Then she rides her bike five units east and one unit south to her cabin. The campfire is one unit west and three units south of her cabin.

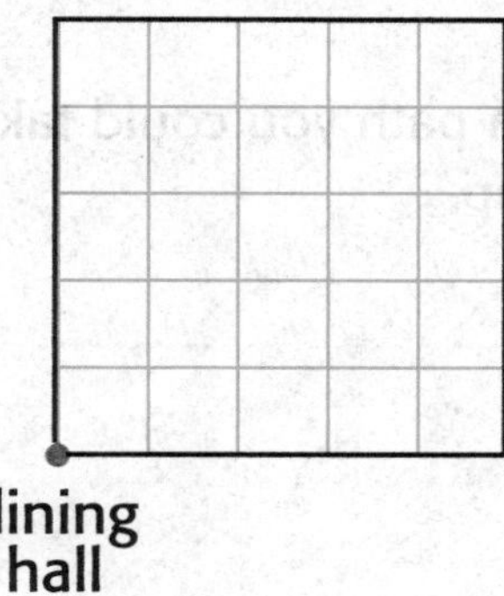

8. From the gymnasium entrance, Norah ran to the jump ropes that are located four units north and four units east. Then she ran two units west to the tumble mats. The skills challenge is one unit west and two units south of the tumble mats.

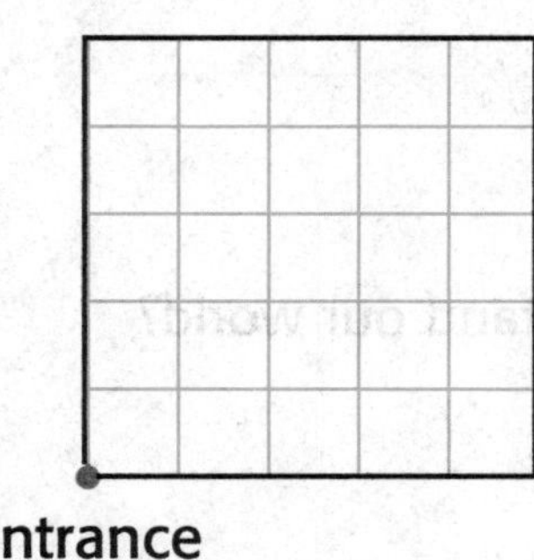

9. From the store entrance, Kyle walks to the toy section that is located four aisles north and two rows east. Then he walks two aisles south and two rows east to the boys' clothes. The cashier is three rows west and one aisle south of the boys' clothes.

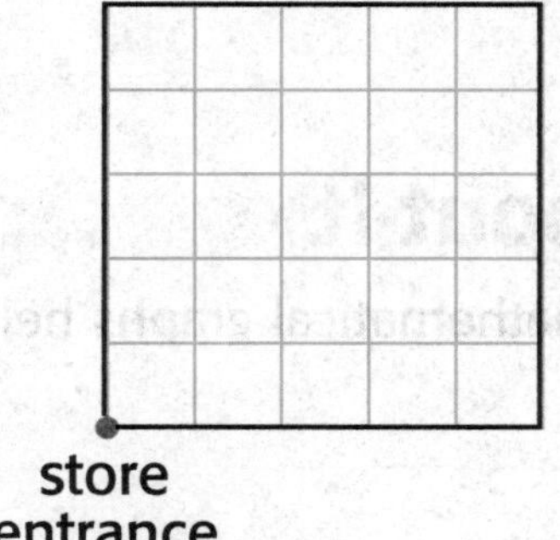

Apply It

Use the map of the amusement park below for Exercises 10–12. The walkways of the amusement park are represented by the vertical and horizontal lines on the map.

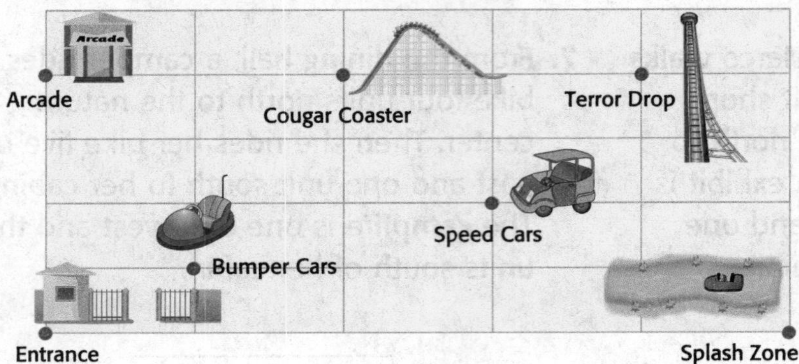

10. Describe a path you could take to get from the entrance to the Terror Drop.

11. After riding the Terror Drop, you decide to ride the Cougar Coaster. How many units do you need to move to get to the Cougar Coaster?

12. Mathematical PRACTICE 6 **Explain to a Friend** Lorenzo and Emily enter the park at the same time and head to two different attractions. Lorenzo walks to Speed Cars and Emily walks to the Cougar Coaster. Who walks farther? Explain to a friend.

Write About It

13. How do mathematical graphs help us better understand our world?

Name ______________________________

MY Homework

Lesson 7
Hands On: Map Locations

Homework Helper

Need help? connectED.mcgraw-hill.com

You can use grid paper to represent locations on a map. The map shows locations of animals at an aquarium. Describe how Dashiell can walk from the aquarium entrance to the giant octopus, penguins, and sharks, in that order.

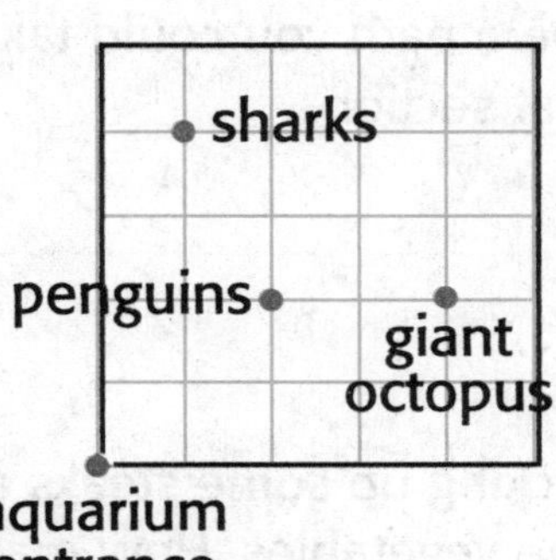

1 The aquarium entrance is located in the lower left corner of the map.

2 From the dot labeled "aquarium entrance", Dashiell walked 4 units to the right and then 2 units up to get to the giant octopus.

3 From the dot labeled "giant octopus", Dashiell walked 2 units to the left to get to the penguins.

4 From the dot labeled "penguins", Dashiell walked 2 units up and then 1 unit to the left to get to the sharks.

Practice

1. Refer to the Homework Helper. The squid exhibit is located three units east of the sharks. How many units north of the giant octopus is the squid exhibit?

____

Problem Solving

Use the map of Megamart below for Exercises 2–5. The aisles and rows of Megamart are represented by the vertical and horizontal lines on the map.

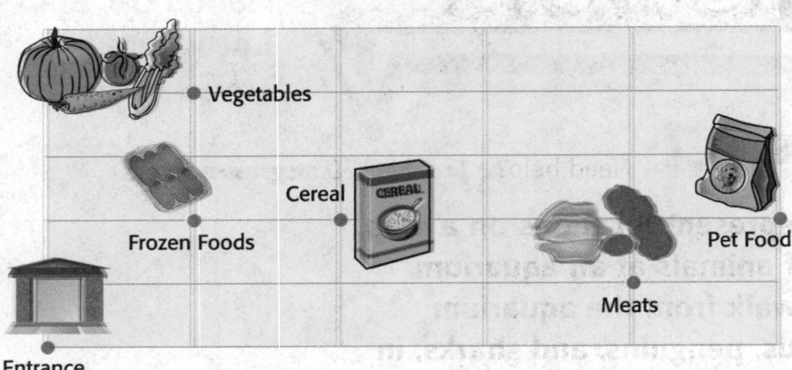

2. Describe a path you could take to get from the entrance to the meat section.

3. After picking up some steaks and hamburger, you decide to pick up some vegetables. How many total units do you need to move to get to the vegetables section? Explain.

4. Mathematical PRACTICE 5 **Use Math Tools** Rebecca and Lance enter Megamart and walk to two different sections. Rebecca walks to the cereal section and Lance walks to the frozen foods section. Who walks farther? Explain.

5. How many total units would you walk from the entrance to the pet food section? Explain.

Name ....

Ordered Pairs

Lesson 8
ESSENTIAL QUESTION
How are patterns used to solve problems?

A **coordinate plane** is formed when two perpendicular number lines intersect. One number line has numbers along the horizontal *x*-axis (across) and the other has numbers along the vertical *y*-axis (up). The point where the two axes intersect is the **origin.**

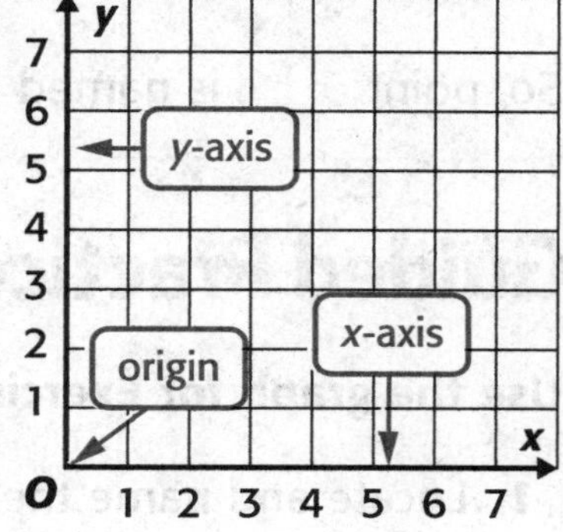

Math in My World

Example 1

Name the ordered pair for the location of Amy's house.

An **ordered pair** is a pair of numbers that is used to name a point.

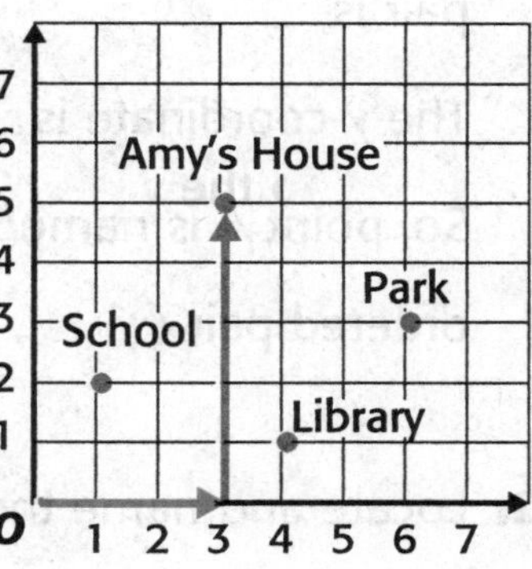

The first number is the ***x*-coordinate** and corresponds to a number on the *x*-axis.

(3, 5)

The second number is the ***y*-coordinate** and corresponds to a number on the *y*-axis.

1. Start at the origin (____, ____). Move right along the *x*-axis until you are under Amy's house. The *x*-coordinate of the ordered pair is ____.

2. Move up until you reach Amy's house. The *y*-coordinate is ____.

So, Amy's house is located at the ordered pair (____, ____).

Example 2 

Name the point for the ordered pair (2, 3).

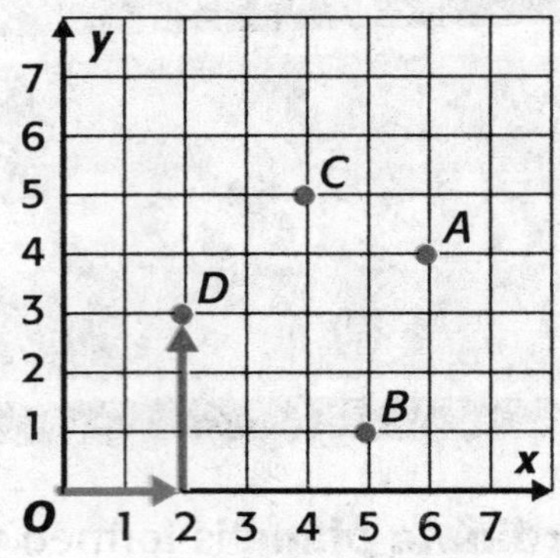

1 Start at the origin (____, ____). Move right along the *x*-axis until you reach ____, the *x*-coordinate.

2 Move up until you reach ____, the *y*-coordinate.

So, point ____ is named by the ordered pair (2, 3).

Guided Practice 

Use the graph for Exercises 1 and 2.

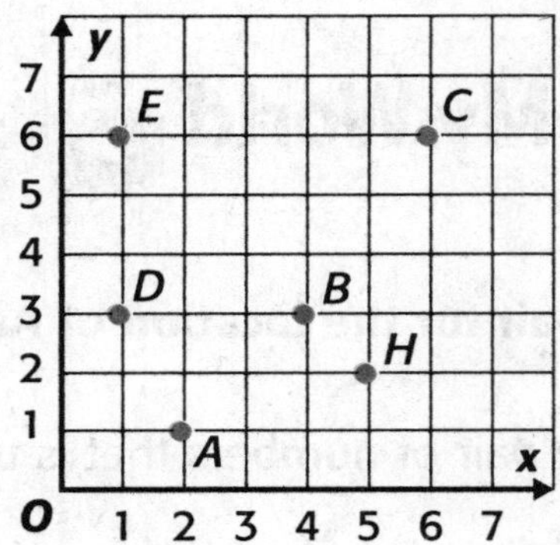

1. Locate and name the ordered pair for point *A*.

The *x*-coordinate of the ordered pair is ____.

The *y*-coordinate is ____.

So, point *A* is named by the ordered pair (____, ____).

2. Locate and name the point at (4, 3).

Move ____ units to the right.

Move up ____ units.

So, point ____ is named by the ordered pair (4, 3).

Talk MATH

Are the points at (3, 8) and (8, 3) in the same location? Explain your reasoning.

Name

Independent Practice

Use the graph for Exercises 3–14.

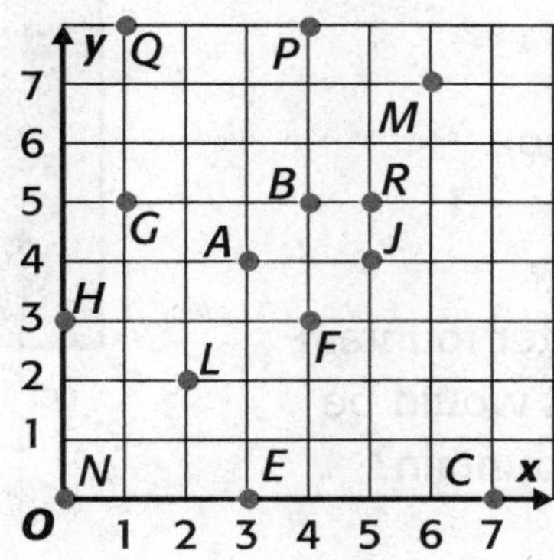

Locate and name each ordered pair.

3. *A* ________

4. *R* ________

5. *J* ________

6. *E* ________

7. *Q* ________

8. *N* ________

Locate and name each point.

9. (2, 2) ________

10. (0, 3) ________

11. (1, 5) ________

12. (6, 7) ________

13. (4, 8) ________

14. (7, 0) ________

Problem Solving

Use the map of the playground at the right for Exercises 15–20.

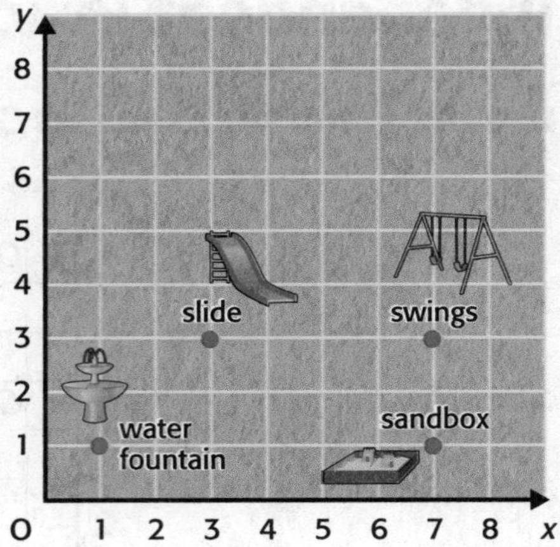

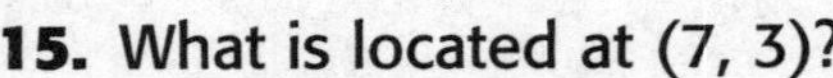

15. What is located at (7, 3)?

____

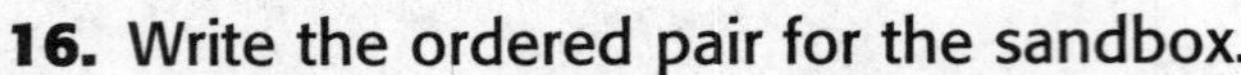

16. Write the ordered pair for the sandbox.

____

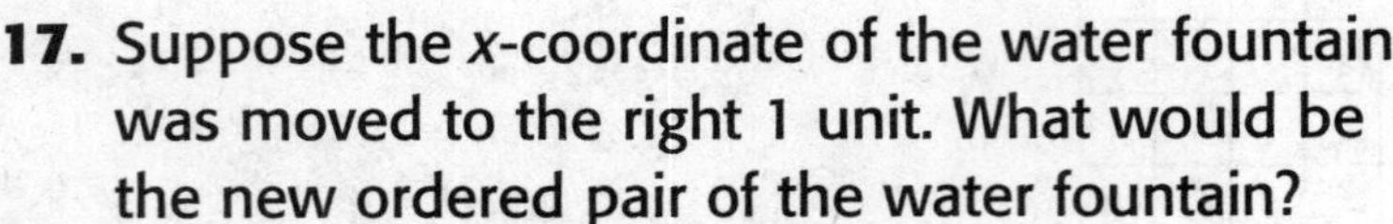

17. Suppose the *x*-coordinate of the water fountain was moved to the right 1 unit. What would be the new ordered pair of the water fountain?

____

18. If the *y*-coordinate of the slide was moved up 2 units, what would be the ordered pair of the slide?

____

19. Cam identified a point that was 4 units above the origin and 8 units to the right of the origin. What was the ordered pair?

____

20. Suppose point (6, 5) was moved 3 units to the left and moved 2 units down. Write the new ordered pair.

____

HOT Problems

21. Mathematical PRACTICE 1 **Make Sense of Problems** Name the ordered pair whose *x*-coordinate and *y*-coordinate are each located on an axis.

____

22. **Building on the Essential Question** How is the location of a point on a grid described?

____

____

____

Name ....

MY Homework

Lesson 8
Ordered Pairs

Homework Helper

Need help? connectED.mcgraw-hill.com

Name the ordered pair for point *A*.

Start at the origin (0, 0). Move right along the *x*-axis until you are under point *A*. The *x*-coordinate of the ordered pair is 3.

Move up until you reach point *A*. The *y*-coordinate is 5.

So, point *A* is named by the ordered pair (3, 5).

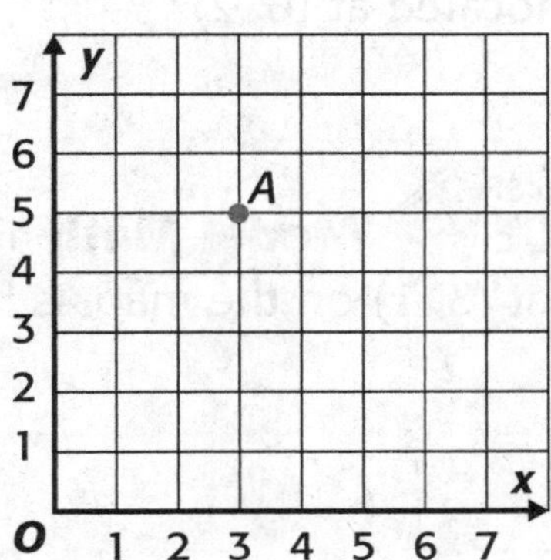

Practice

Use the graph for Exercises 1–6. Locate and name each ordered pair.

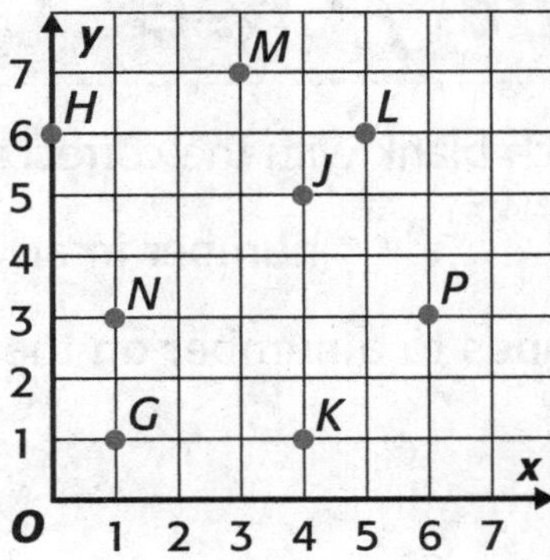

1. *M* ________

2. *P* ________

3. *J* ________

Locate and name each point.

4. (1, 3) ________

5. (5, 6) ________

6. (0, 6) ________

Problem Solving

Use the map for Exercises 7–10.

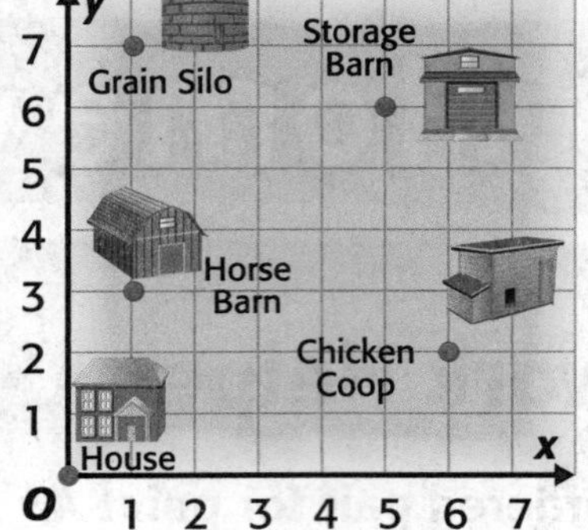

7. What ordered pair gives the location of the storage barn?

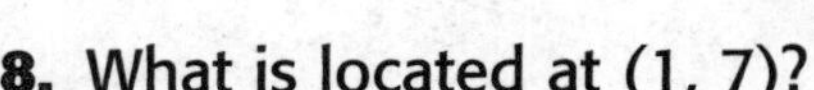

8. What is located at (1, 7)?

9. What is located at (6, 2)?

10. Mathematical PRACTICE 4 **Model Math** Jimmy says that the horse barn is located at (3, 1) on the map. Is his ordered pair correct? Explain.

Vocabulary Check

11. Fill in each blank with the correct word to complete the sentence.

The ____________ number in an ordered pair is the y-coordinate and corresponds to a number on the ____________.

Test Practice

12. What ordered pair represents point D on the coordinate grid?

Ⓐ (5, 7)

Ⓑ (5, 2)

Ⓒ (2, 5)

Ⓓ (3, 1)

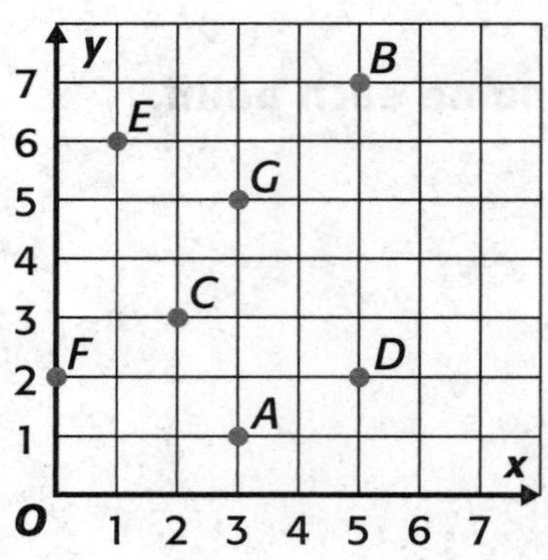

Name

Graph Patterns

Lesson 9

ESSENTIAL QUESTION
How are patterns used to solve problems?

We make a great pair!

Math in My World

Example 1

Tricia and her friends decided to rent bicycles to ride on their weekend trip. Bikes 'N More charges $5 for each hour and Adventure Bikes charges $10 for each hour. Find the cost of renting a bicycle from each store for 1, 2, 3, and 4 hours.

1 Complete the tables below.

Bikes 'N More				
Hours	1	2	3	4
Cost ($)				

Adventure Bikes				
Hours	1	2	3	4
Cost ($)				

2 Generate ordered pairs. Let each *x*-coordinate represent the number of ______. Let each *y*-coordinate represent the ______.

Bikes 'N More

(1, ____), (2, ____), (3, ____), (4, ____)

Adventure Bikes

(1, ____), (2, ____), (3, ____), (4, ____)

How much more would it cost to rent bicycles for 3 hours from Adventure Bikes than from Bikes 'N More? ______

Example 2

Refer to Example 1. Graph each set of ordered pairs on a coordinate plane. Label each set of ordered pairs. Does the difference in costs between the two stores increase or decrease as the number of hours increase?

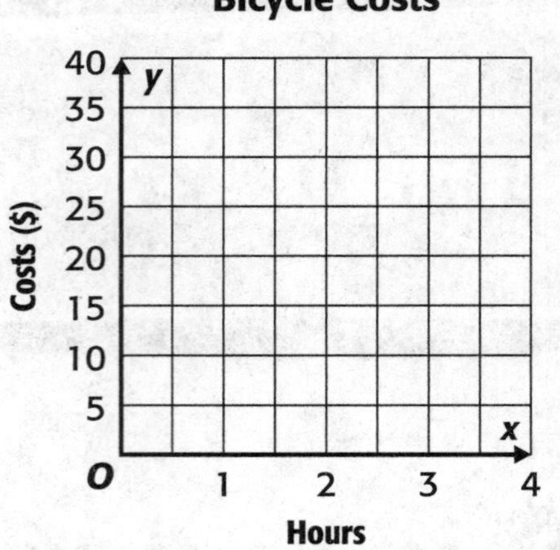

The graph shows that the difference in costs

between the two stores ________ as the number of hours increase.

Guided Practice

1. Birdseed is sold in 8-pound bags and 24-pound bags at the local store. Find the weight of buying 1, 2, 3, and 4 bags of both sizes of birdseed.

Complete the tables below.

8-Pound Bag				
Bags	1	2	3	4
Weight (lb)				

24-Pound Bag				
Bags	1	2	3	4
Weight (lb)				

Generate ordered pairs. Let each *x*-coordinate represent the number of ________. Let each *y*-coordinate represent the ________.

8-Pound Bag

(1, ____), (2, ____), (3, ____), (4, ____)

24-Pound Bag

(1, ____), (2, ____), (3, ____), (4, ____)

How many more pounds of birdseed would there be if you purchased 2 bags of 24-pound birdseed than 2 bags of 8-pound birdseed?

____

Talk MATH

Explain how you would graph two real-world patterns using ordered pairs.

Name

Independent Practice

2. Speedy Cab charges $2 per mile traveled while Purple Cab charges $4 per mile traveled. Find the costs of traveling 1, 2, 3, and 4 miles for both cab companies. Then graph the results as ordered pairs.

Speedy Cab				
Miles	1	2	3	4
Cost ($)				

Purple Cab				
Miles	1	2	3	4
Cost ($)				

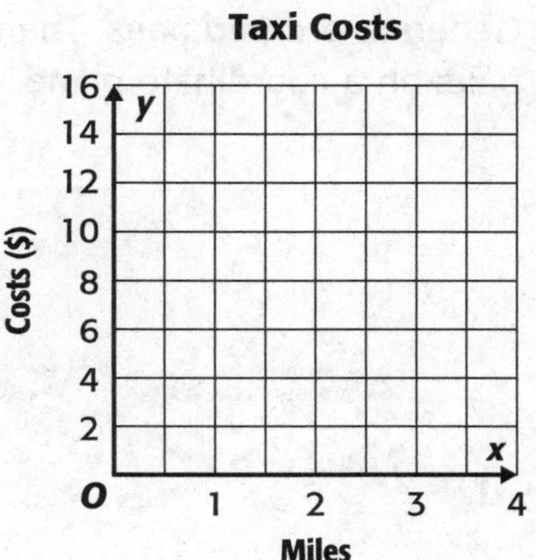

Does the difference in cost between the two taxi services increase or decrease as the number of miles increase?

____

3. Jennifer makes $9 per hour working for her neighbors after school each day. Carmen works for a local farmer and makes $3 per hour working after school each day. Find the total amount earned for each girl if they work 1, 2, 3, and 4 hours. Then graph the results as ordered pairs.

Jennifer's Hourly Wages				
Hours	1	2	3	4
Money Earned ($)				

Carmen's Hourly Wages				
Hours	1	2	3	4
Money Earned ($)				

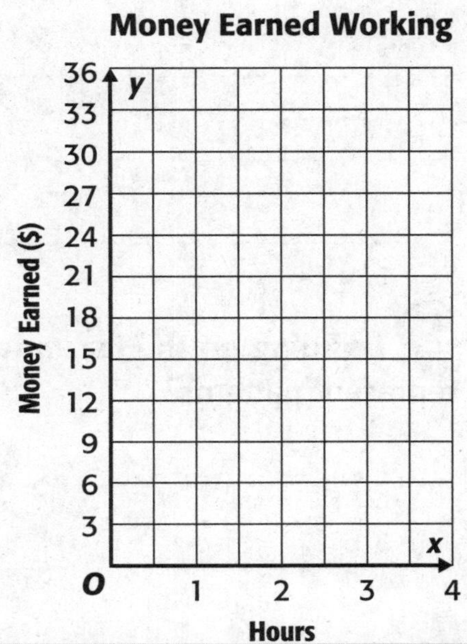

How much more money would Jennifer make if both girls worked 3 hours?

____

Problem Solving

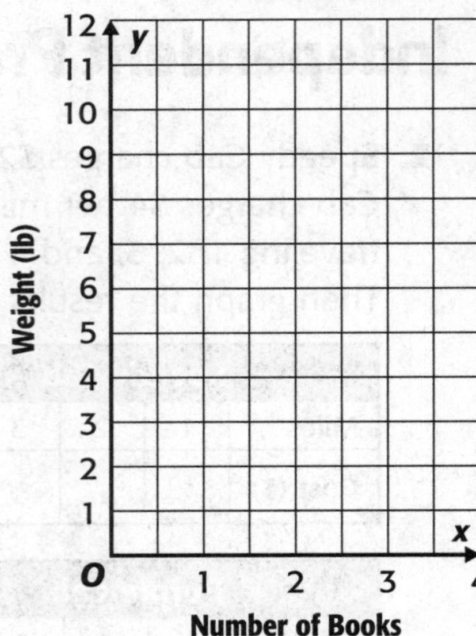

4. Jason places books in his book bag to take home. Each book weighs 2 pounds. Mason has books that weigh 3 pounds each. Find the weights of 1, 2, 3, and 4 books for both Jason and his brother. How many more pounds would Mason carry if they both carried 4 books? Generate ordered pairs. Then graph the ordered pairs on a coordinate plane.

____

HOT Problems

5. Mathematical PRACTICE 1 **Plan Your Solution** Write a real-world problem for which you could compare patterns by graphing ordered pairs.

____

____

____

6. **Building on the Essential Question** How are graphs used to represent patterns?

____

____

____

Name ....

MY Homework

Lesson 9
Graph Patterns

Homework Helper

Need help? connectED.mcgraw-hill.com

Lance is helping his dad remodel their house. He cuts boards with lengths of 2 feet and 8 feet. Find the amount of material needed for 1, 2, 3, and 4 boards of each length. Then graph the results as ordered pairs on a coordinate plane. How many more feet of 8-foot boards would there be if Lance cuts 3 boards of both lengths?

1 Complete the tables below.

2-Foot Board				
Number of Boards	1	2	3	4
Length (ft)	2	4	6	8

8-Foot Board				
Number of Boards	1	2	3	4
Length (ft)	8	16	24	32

2 Generate ordered pairs. Let each *x*-coordinate represent the number of boards. Let each *y*-coordinate represent the length in feet of each board.

2-Foot Board
(1, 2), (2, 4), (3, 6), (4, 8)

8-Foot Board
(1, 8), (2, 16), (3, 24), (4, 32)

3 Graph each set of ordered pairs on a coordinate plane.
If Lance cuts 3 boards of both lengths, he will have 6 feet of 2-foot boards and 24 feet of 8-foot boards.
$24 - 6 = 18$

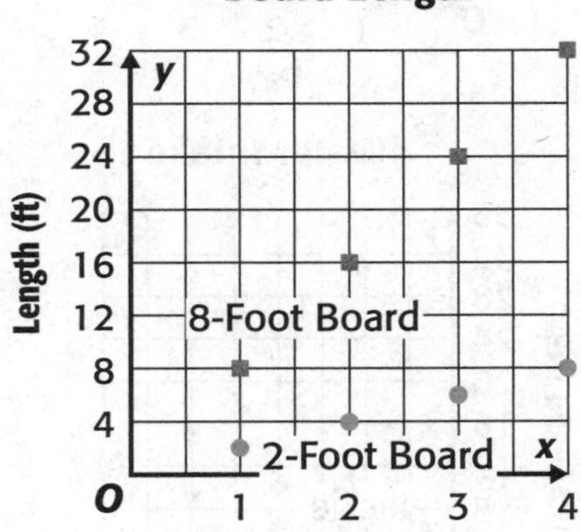

So, there will be 18 more feet of 8-foot boards if he cuts 3 boards of both lengths.

Problem Solving

1. Mathematical PRACTICE 4 **Model Math** Jarrett walks his puppy outside every day for 30 minutes. Angela walked her puppy every day for 90 minutes. Find the number of minutes that each puppy was walked for 1, 2, 3, and 4 days. Then graph the results as ordered pairs. How many more minutes does Angela spend walking her puppy over 2 days compared to Jarrett walking his puppy over 2 days?

____

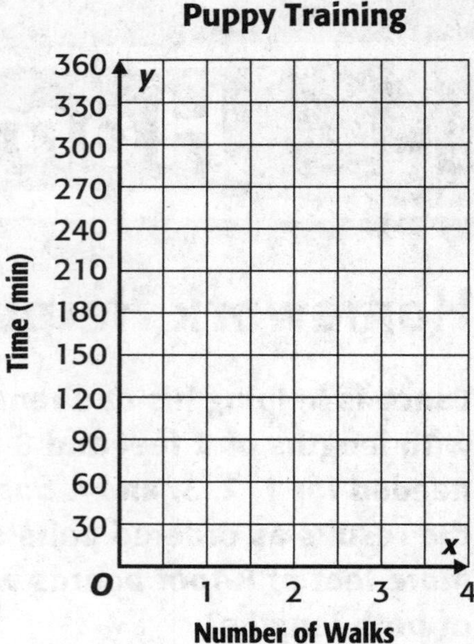

Test Practice

2. Emily drinks 6 cups of water every day, while Jackie drinks 8 cups of water every day. Which graph represents the total amount of water consumed by Emily and Jackie over a 4-day period?

Ⓐ
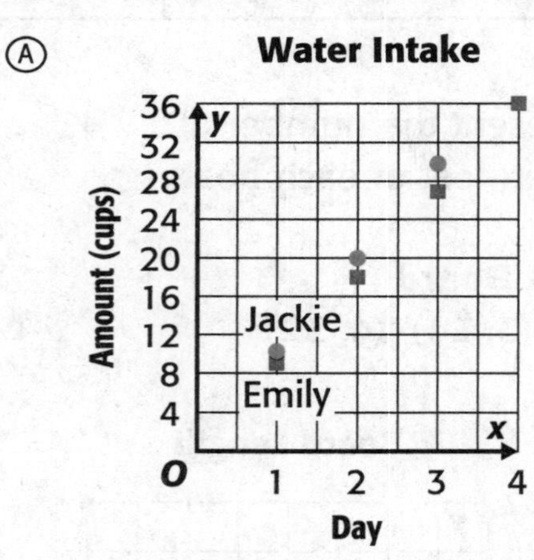

Ⓒ
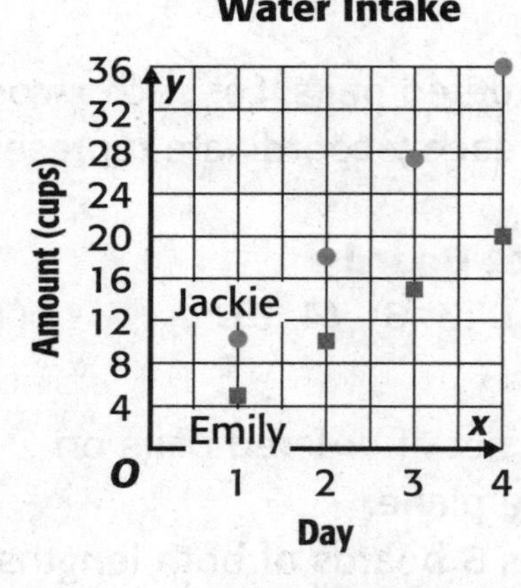

Ⓑ
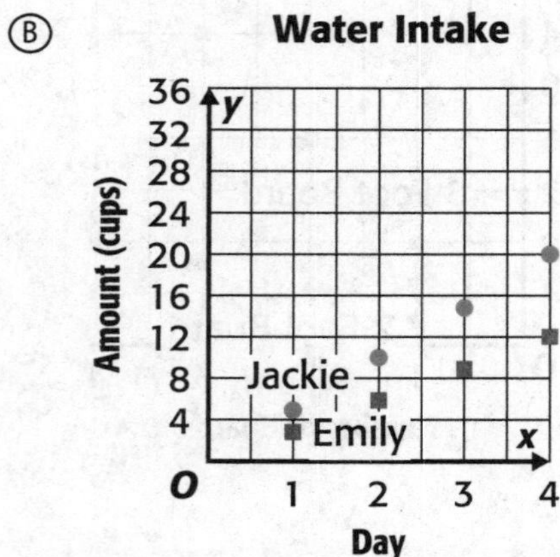

Ⓓ
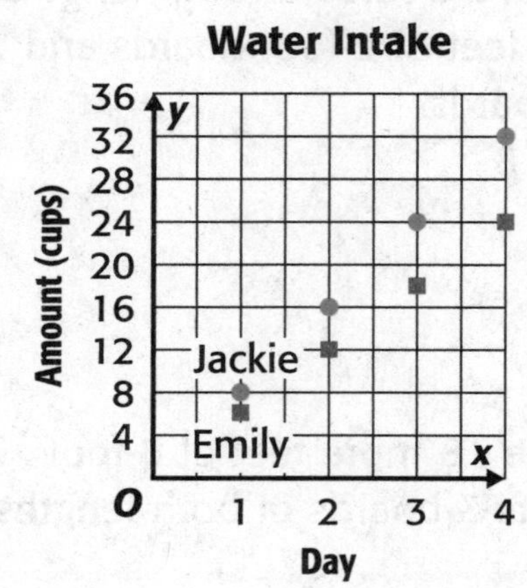

Review

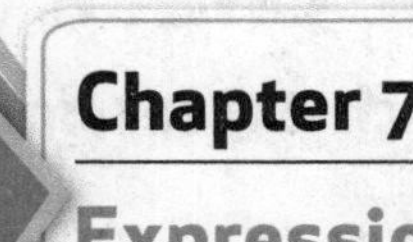

Chapter 7

Expressions and Patterns

Vocabulary Check

Use context clues to write a description for each boldfaced vocabulary word.

1. The **numerical expression** 5 + 12 represents the sum of 5 and 12.

2. Keira was asked why the **ordered pair** is important when plotting points on a graph.

3. Luis was given the expression 10 − (2 × 6) to evaluate and asked to identify the **order of operations**.

4. The **coordinate plane** can be used to name ordered pairs.

5. Majorie was asked to name the ***x*-coordinate** on the coordinate plane for point *A*.

Concept Check

Evaluate each expression.

6. $(2 \times 2^2) \times (4 + 7) =$ ______

7. $10 \times [(7^2 + 3) - 9] =$ ______

8. $(21 \div 3) + (17 - 7) =$ ______

9. $\{[(66 \div 11) + 3] \times 2\} =$ ______

10. $6 \times [5 \times (3^3 - 17)] =$ ______

11. $\{[(18 - 3) + 3^2] - 14\} \times 3 =$ ______

Write each phrase as a numerical expression.

12. divide 18 by 3, then add 9 ______

13. subtract 5 from 13 then add the product of 3 and 7 ______

14. Compare the pair of numerical expressions without evaluating them.

Expression 1	**Expression 2**
12×2	$(12 \times 2) \times 2$

Both expressions contain the same multiplication expression.

Write the expression. ______

In Expression 2, the product is multiplied by ______.

So, Expression 2 is ______ times as large as Expression 1.

Locate and name each ordered pair.

15. *A* ______ **16.** *B* ______ **17.** *C* ______

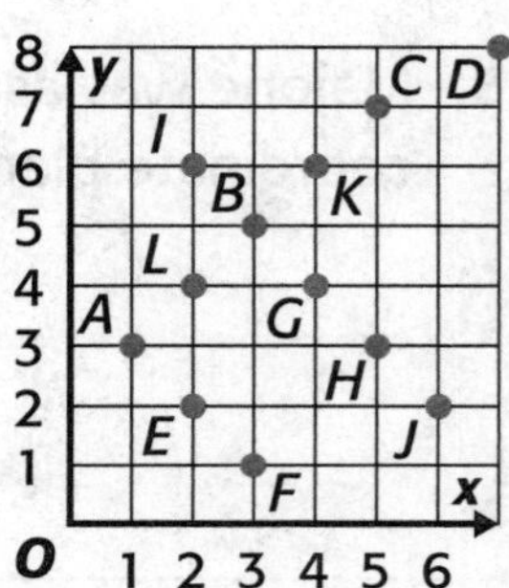

Locate and name each point.

18. (5, 3) ______ **19.** (4, 6) ______ **20.** (4, 4) ______

Name

Problem Solving

21. A store display will have 6 rows, with 12 containers in each row. If 39 containers has been set up so far, explain how to find the number of containers that still need to be set up. Then solve the problem.

22. Fashion for All has jeans on sale for \$13. Designer Pants has jeans on sale for \$26. Write the total costs of 1, 2, 3, and 4 pairs of jeans for each store. Then compare the total cost of 4 pairs of jeans.

23. Glenn swam 2 laps every morning for 7 days. In addition to the laps he swam each morning, he swam 3 laps with his friends on Tuesday and Thursday. Write the expression that shows the number of laps he swam during the week. Evaluate the expression to find the total number of laps he swam that week.

Use the graph for Exercises 24 and 25.

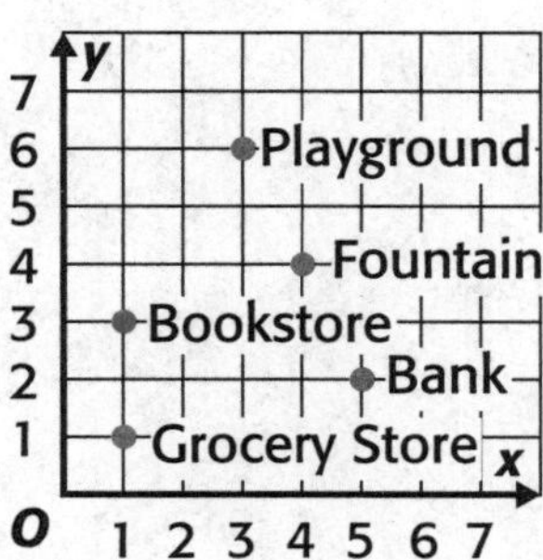

24. What is located at (3, 6)?

25. Write the ordered pair for the bookstore.

Test Practice

26. Refer to the graph for Exercises 24 and 25. If the *y*-coordinate of the grocery store was moved up 4 units, what would be the ordered pair of the grocery store?

Ⓐ (1, 5)

Ⓑ (5, 1)

Ⓒ (1, 7)

Ⓓ (5, 5)

Reflect

Chapter 7

Answering the ESSENTIAL QUESTION

Use what you learned about expressions and patterns to complete the graphic organizer.

Real-World Example

Graph the Patterns

Compare the Patterns

ESSENTIAL QUESTION

How are patterns used to solve problems?

Now reflect on the ESSENTIAL QUESTION  **Write your answer below.**

Glossary/Glosario

Go online for the eGlossary.

Go to the **eGlossary** to find out more about these words in the following 13 languages:

Arabic • Bengali • Brazilian Portuguese • Cantonese • English • Haitian Creole
Hmong • Korean • Russian • Spanish • Tagalog • Urdu • Vietnamese

English	Spanish/Español
acute angle An angle with a measure between 0° and 90°. 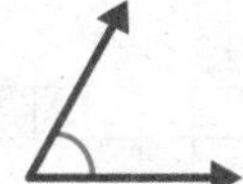	**ángulo agudo** Ángulo que mide entre 0° y 90°. 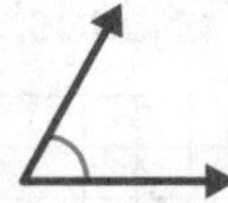
acute triangle A triangle with three acute angles.	**triángulo acutángulo** Triángulo con tres ángulos agudos.
algebra A branch of mathematics that uses symbols, usually letters, to explore relationships between quantities.	**álgebra** Rama de las matemáticas que usa símbolos, generalmente letras, para explorar relaciones entre cantidades.
angle Two rays with a common endpoint. 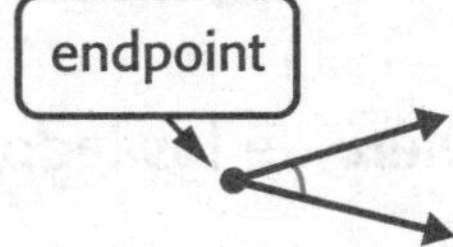	**ángulo** Dos semirrectas con un extremo común. 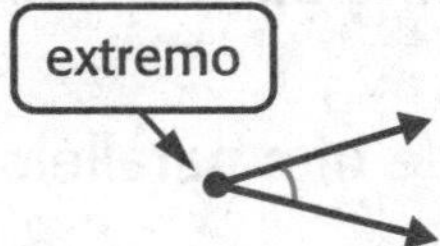
annex To place a zero to the right of a decimal without changing a number's value.	**agregar** Poner un cero a la derecha de un decimal sin cambiar el valor de un número.

Aa

area The number of square units needed to cover the surface of a closed figure.	**área** Cantidad de unidades cuadradas necesarias para cubrir la superficie de una figura cerrada.

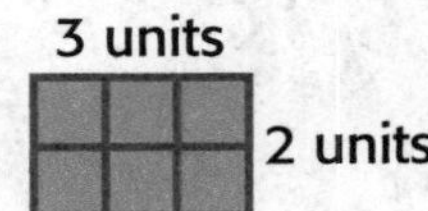

area = 6 square units

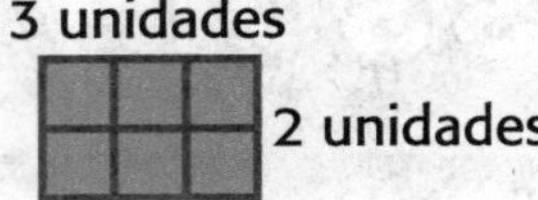

área = 6 unidades cuadradas

Associative Property Property that states that the way in which numbers are grouped does not change the sum or product.

propiedad asociativa Propiedad que establece que la manera en que se agrupan los números no altera la suma o el producto.

attribute A characteristic of a figure.

atributo Característica de una figura.

axis A horizontal or vertical number line on a graph. Plural is axes.

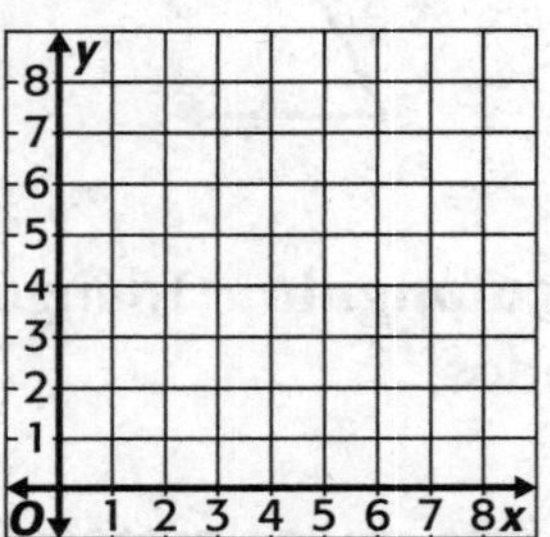

eje Recta numérica horizontal o vertical en una gráfica.

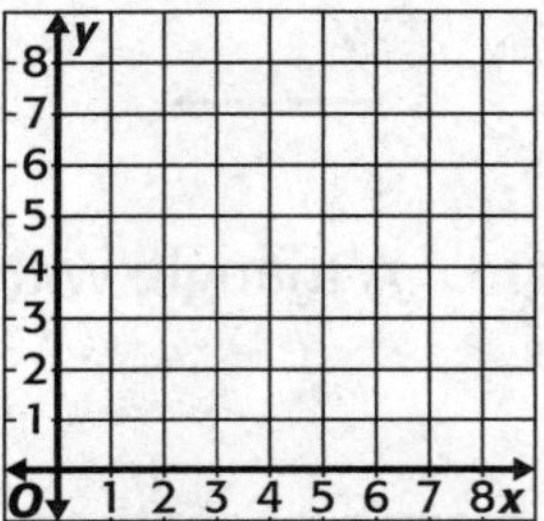

Bb

base In a power, the number used as a factor. In 10^3, the base is 10.

base En una potencia, el número que se usa como factor. En 10^3, la base es 10.

base Any side of a parallelogram.

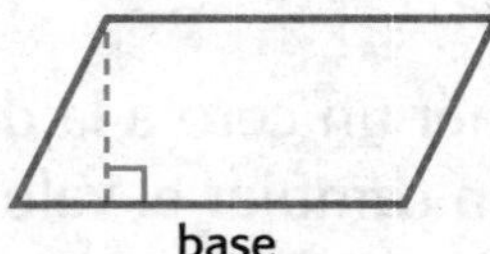

base Cualquiera de los lados paralelogramo.

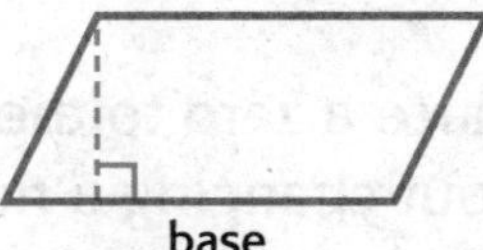

base One of the two parallel congruent faces in a prism.

base Una de las dos caras congruentes paralelas en un prisma.

Cc

capacity The amount a container can hold.

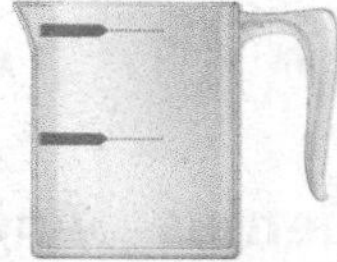

capacidad Cantidad que puede contener un recipiente.

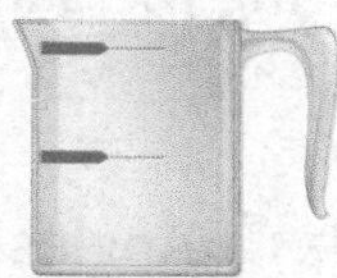

centimeter (cm) A metric unit for measuring length.

100 centimeters = 1 meter

centímetro (cm) Unidad métrica de longitud.

100 centímetros = 1 metro

common denominator A number that is a multiple of the denominators of two or more fractions.

denominador común Número que es múltiplo de los denominadores de dos o más fracciones.

common factor A number that is a factor of two or more numbers.

3 is a common factor of 6 and 12.

factor común Número que es un factor de dos o más números.

3 es factor común de 6 y 12.

common multiple A whole number that is a multiple of two or more numbers.

24 is a common multiple of 6 and 4.

múltiplo común Número natural múltiplo de dos o más números.

24 es un múltiplo común de 6 y 4.

Commutative Property Property that states that the order in which numbers are added does not change the sum and that the order in which factors are multiplied does not change the product.

propiedad conmutativa Propiedad que establece que el orden en que se suman los números no altera la suma y que el orden en que se multiplican los factores no altera el producto.

compatible numbers Numbers in a problem that are easy to work with mentally.

720 and 90 are compatible numbers for division because $72 \div 9 = 8$.

números compatibles Números en un problema con los cuales es fácil trabajar mentalmente.

720 ÷ 90 es una divisíon que usa son números compatibles porque $72 \div 9 = 8$.

composite figures A figure made up of two or more three-dimensional figures.

figura compuesta Figura conformada por dos o más figuras tridimensionales.

Cc

composite number A whole number that has more than two factors.

12 has the factors 1, 2, 3, 4, 6, and 12.

congruent Having the same measure.

congruent angles Angles of a figure that are equal in measure.

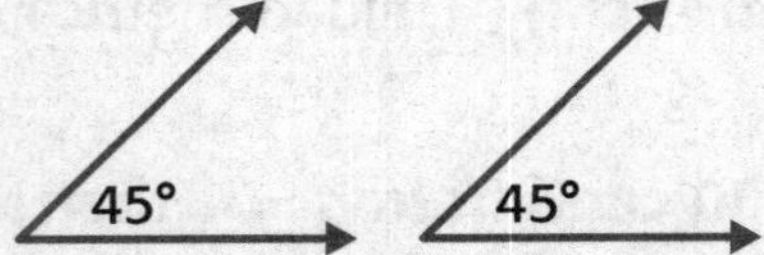

congruent figures Two figures having the same size and the same shape.

congruent sides Sides of a figure that are equal in length.

convert To change one unit to another.

coordinate One of two numbers in an ordered pair.

The 1 is the number on the *x*-axis, the 5 is on the *y*-axis.

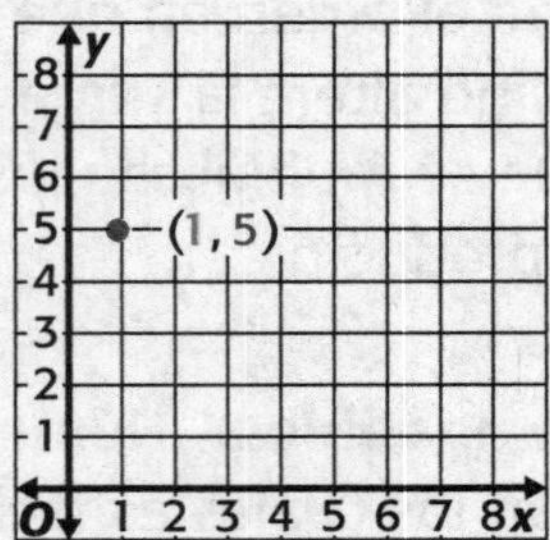

coordinate plane A plane that is formed when two number lines intersect.

número compuesto Número natural que tiene más de dos factores.

12 tiene a los factores 1, 2, 3, 4, 6 y 12.

congruentes Que tienen la misma medida.

ángulos congruentes Ángulos de una figura que tienen la misma medida.

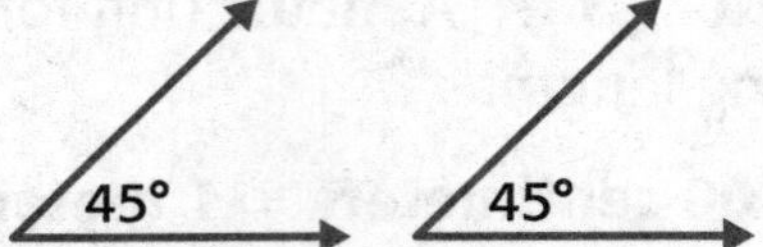

figuras congruentes Dos figuras que tienen el mismo tamaño y la misma forma.

lados congruentes Lados de una figura que tienen la misma longitud.

convertir Transformar una unidad en otra.

coordenada Cada uno de los números de un par ordenado.

El 1 es la coordenada *x* y el 5 es la coordenada *y*.

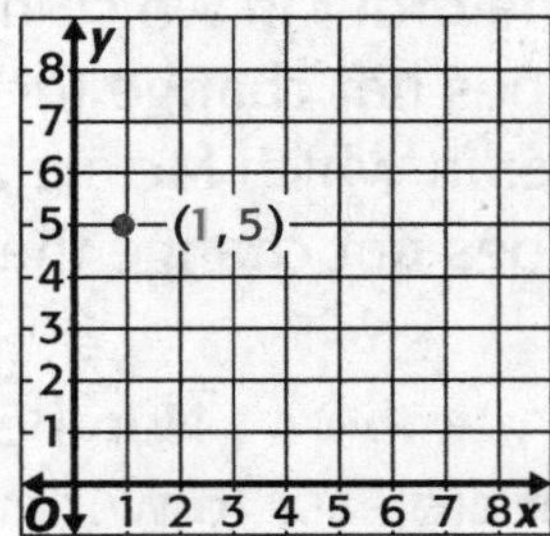

plano de coordenadas Plano que se forma cuando dos rectas numéricas se intersecan formando un ángulo recto.

cube A rectangular prism with six faces that are congruent squares.

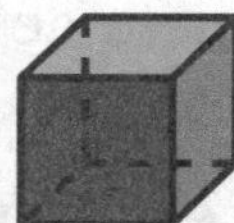

cubed A number raised to the third power; 10 × 10 × 10, or 10^3.

cubic unit A unit for measuring volume, such as a cubic inch or a cubic centimeter.

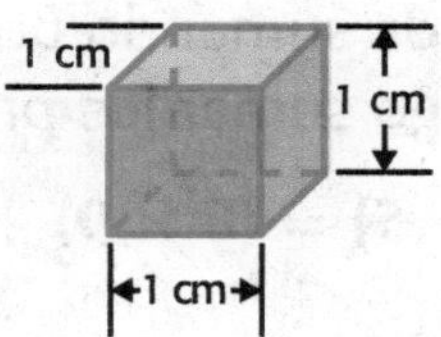

cup A customary unit of capacity equal to 8 fluid ounces.

customary system The units of measurement most often used in the United States. These include foot, pound, and quart.

cubo Prisma rectangular con seis caras que son cuadrados congruentes.

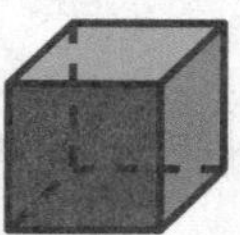

al cubo Número elevado a la tercera potencia; 10 × 10 × 10 o 10^3.

unidad cúbica Unidad de volumen, como una pulgada cúbica o un centímetro cúbico.

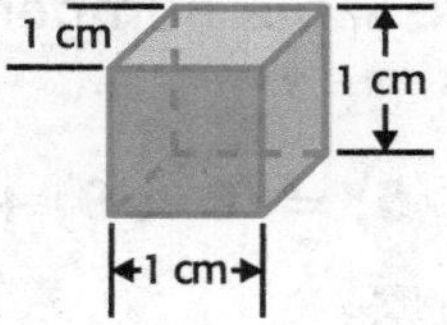

taza Unidad usual de capacidad que equivale a 8 onzas líquidas.

sistema usual Conjunto de unidades de medida de uso más frecuente en Estados Unidos. Incluyen el pie, la libra y el cuarto.

Dd

decimal A number that has a digit in the tenths place, hundredths place, and beyond.

decimal point A period separating the ones and the tenths in a decimal number.

0.8 or $3.77

degree (°) a. A unit of measure used to describe temperature. b. A unit for measuring angles.

decimal Número que tiene al menos un dígito en el lugar de las décimas, centésimas etcétera.

punto decimal Punto que separa las unidades y las décimas en un número decimal.

0.8 o $3.77

grado (°) a. Unidad de medida que se usa para describir la temperatura.
b. Unidad que se usa para medir ángulos.

Dd

denominator The bottom number in a fraction. It represents the number of parts in the whole.

In $\frac{5}{6}$, 6 is the denominator.

denominador Numero que se escribe debajo de la barra en una fracción. Representa el número de partes en que se divide un entero.

En $\frac{5}{6}$, 6 es el denominador.

digit A symbol used to write numbers. The ten digits are 0, 1, 2, 3, 4, 5, 6, 7, 8, and 9.

dígito Símbolo que se usa para escribir los números. Los diez dígitos son 0, 1, 2, 3, 4, 5, 6, 7, 8 y 9.

Distributive Property To multiply a sum by a number, you can multiply each addend by the same number and add the products.

$8 \times (9 + 5) = (8 \times 9) + (8 \times 5)$

propiedad distributiva Para multiplicar una suma por un número, puedes multiplicar cada sumando por ese número y luego sumar los productos.

$8 \times (9 + 5) = (8 \times 9) + (8 \times 5)$

divide (division) An operation on two numbers in which the first number is split into the same number of equal groups as the second number.

12 ÷ 3 means 12 is divided into 3 equal-size groups

dividir (división) Operación entre dos números en la cual el primer número se separa en tantos grupos iguales como indica el segundo número.

12 ÷ 3 significa que 12 se divide entre 3 grupos de igual tamaño.

dividend A number that is being divided.

$3\overline{)19}$ ← 19 is the dividend.

dividendo Número que se divide.

$3\overline{)19}$ ← 19 es el dividendo.

divisible Describes a number that can be divided into equal parts and has no remainder.

39 is divisible by 3 with no remainder.

divisible Describe un número que puede dividirse en partes iguales, sin residuo.

39 es divisible entre 3 sin residuo.

divisor The number that divides the dividend.

3 is the divisor. → $3\overline{)19}$

divisor Número entre el cual se divide el dividendo.

3 es el divisor. → $3\overline{)19}$

edge The line segment where two faces of a three-dimensional figure meet.

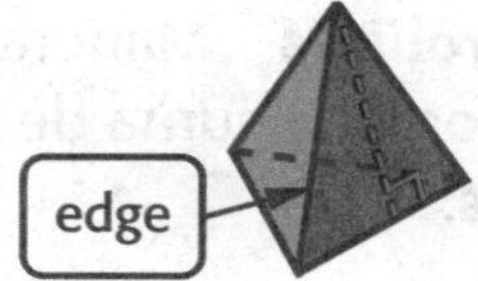

equation A number sentence that contains an equal sign, showing that two expressions are equal.

equilateral triangle A *triangle* with three *congruent* sides.

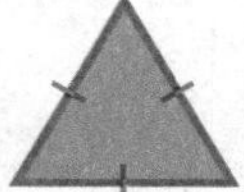

equivalent decimals Decimals that have the same value.

0.3 and 0.30

equivalent fractions Fractions that have the same value.

$$\frac{3}{4} = \frac{6}{8} = \frac{9}{12}$$

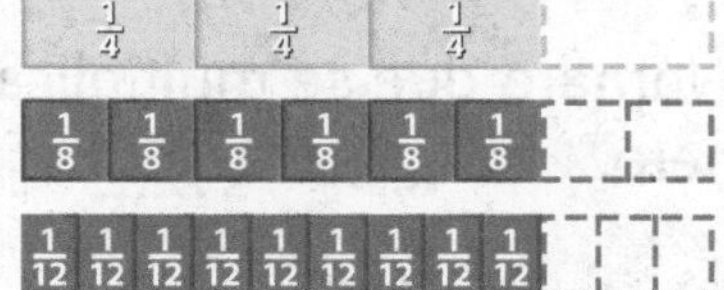

estimate A number close to an exact value. An estimate indicates about how much.

47 + 22 (round to 50 + 20)

The estimate is 70.

arista Segmento de recta donde se unen dos caras de una figura tridimensional.

ecuación Expresión numérica que contiene un signo igual y que muestra que dos expresiones son iguales.

triángulo equilátero *Triángulo* con tres lados *congruentes.*

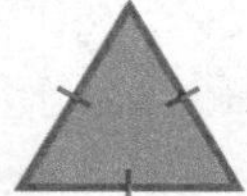

decimales equivalentes Decimales que tienen el mismo valor.

0.3 y 0.30

fracciones equivalentes Fracciones que tienen el mismo valor.

$$\frac{3}{4} = \frac{6}{8} = \frac{9}{12}$$

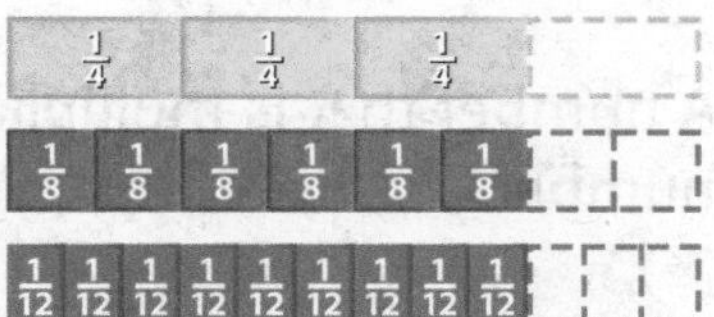

estimación Número cercano a un valor exacto. Una estimación indica una cantidad aproximada.

47 + 22 (se redondea a 50 + 20)

La estimación es 70.

Ee

evaluate To find the value of an expression by replacing variables with numbers.

evaluar Calcular el valor de una expresión reemplazando las variables por números.

even number A whole number that is divisible by 2.

número par Número natural divisible entre 2.

expanded form A way of writing a number as the sum of the values of its digits.

forma desarrollada Manera de escribir un número como la suma de los valores de sus dígitos.

exponent In a power, the number of times the base is used as a factor. In 5^3, the exponent is 3.

exponente En una potencia, el número de veces que se usa la base como factor. En 5^3, el exponente es 3.

expression A combination of numbers, variables, and at least one operation.

expresión Combinación de números, variables y por lo menos una operación.

face A flat surface.

A square is a face of a cube.

cara Superficia plana.

Cada cara de un cubo es un cuadrado.

fact family A group of related facts using the same numbers.

familia de operaciones Grupo de operaciones relacionadas que usan los mismos números.

factor A number that is multiplied by another number.

factor Número que se multiplica por otro número.

Fahrenheit (°F) A unit used to measure temperature.

Fahrenheit (°F) Unidad que se usa para medir la temperatura.

fair share An amount divided equally.

partes iguales Partes entre las que se divide equitativamente un entero.

fluid ounce A customary unit of capacity.

onza líquida Unidad usual de capacidad.

foot (ft) A customary unit for measuring length. Plural is feet.

1 foot = 12 inches

pie (pie) Unidad usual de longitud.

1 pie = 12 pulgadas

fraction A number that represents part of a whole or part of a set.

$\frac{1}{2}, \frac{1}{3}, \frac{1}{4}, \frac{3}{4}$

fracción Número que representa una parte de un todo o una parte de un conjunto.

$\frac{1}{2}, \frac{1}{3}, \frac{1}{4}, \frac{3}{4}$

gallon (gal) A customary unit for measuring capacity for liquids.

1 gallon = 4 quarts

galón (gal) Unidad de medida usual de capacidad de líquidos.

1 galón = 4 cuartos

gram (g) A metric unit for measuring mass.

gramo (g) Unidad métrica para medir la masa.

graph To place a point named by an ordered pair on a coordinate plane.

graficar Colocar un punto nombrado por un par ordenado en un plano de coordenadas.

Greatest Common Factor (GCF) The greatest of the common factors of two or more numbers.

The greatest common factor of 12, 18, and 30 is 6.

máximo común divisor (M.C.D.) El mayor de los factores comunes de dos o más números.

El máximo común divisor de 12, 18 y 30 es 6.

height The shortest distance from the base of a parallelogram to its opposite side.

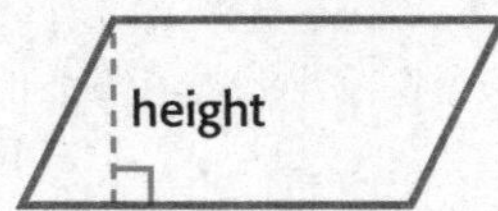

altura La distancia más corta desde la base de un paralelogramo hasta su lado opuesto.

Hh

hexagon A polygon with six sides and six angles.

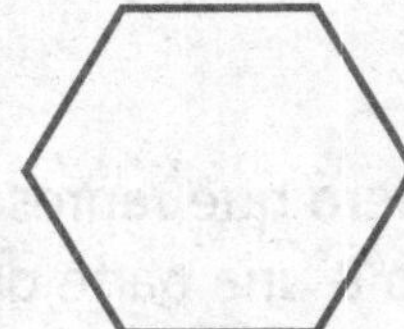

hexágono Polígono con seis lados y seis ángulos.

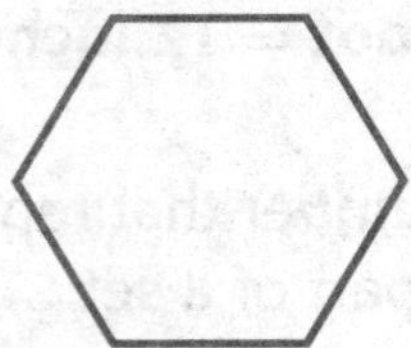

horizontal axis The axis in a coordinate plane that runs left and right (↔). Also known as the *x*-axis.

eje horizontal Eje en un plano de coordenadas que va de izquierda a derecha (↔). También conocido como eje *x*.

hundredth A place value position. One of one hundred equal parts. In the number 0.57, 7 is in the hundredths place.

centésima Valor posícional. Una de cien partes iguales. En el número 0.57, 7 está en el lugar de las centésimas.

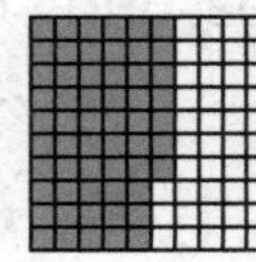

Ii

Identity Property Property that states that the sum of any number and 0 equals the number and that the product of any number and 1 equals the number.

propiedad de identidad Propiedad que establece que la suma de cualquier número y 0 es igual al número y que el producto de cualquier número y 1 es igual al número.

improper fraction A fraction with a numerator that is greater than or equal to the denominator.

$$\frac{17}{3} \text{ or } \frac{5}{5}$$

fracción impropia Fracción con un numerador mayor que él igual al denominador.

$$\frac{17}{3} \text{ o } \frac{5}{5}$$

inch (in.) A customary unit for measuring length. The plural is inches.

pulgada (pulg) Unidad usual de longitud.

inequality Two quantities that are not equal.

desigualdad Dos cantidades que no son iguales.

intersecting lines *Lines* that meet or cross at a common *point.*

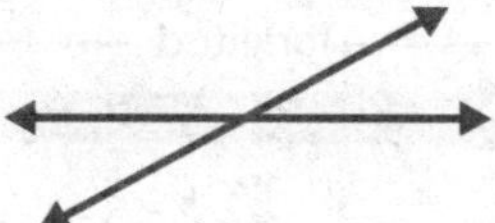

rectas secantes *Rectas* que se intersecan o se cruzan en un *punto* común.

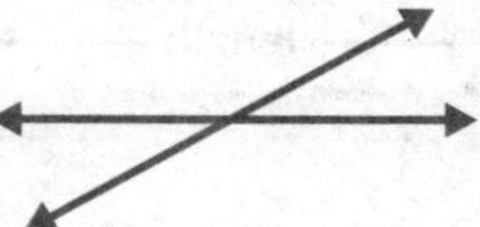

interval The distance between successive values on a scale.

intervalo Distancia entre valores sucesivos en una escala.

inverse operations Operations that undo each other.

operaciones inversas Operaciones que se cancelan entre sí.

isosceles triangle A *triangle* with at least 2 *sides* of the same *length.*

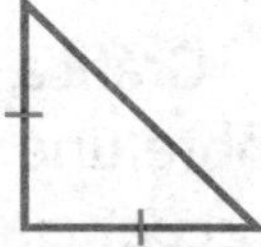

triángulo isósceles *Triángulo* que tiene por lo menos 2 *lados* del mismo largo.

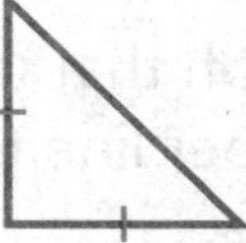

Kk

kilogram (kg) A metric unit for measuring mass.

kilogramo (kg) Unidad métrica de masa.

kilometer (km) A metric unit for measuring length.

kilómetro (km) Unidad métrica de longitud.

Ll

Least Common Denominator (LCD) The least common multiple of the denominators of two or more fractions.

$\frac{1}{12}, \frac{1}{6}, \frac{1}{8}$; **LCD is 24.**

mínimo común denominador (m.c.d.) El mínimo común múltiplo de los denominadores de dos o más fracciones.

$\frac{1}{12}, \frac{1}{6}, \frac{1}{8}$; **el m.c.d. es 24.**

Least Common Multiple (LCM) The smallest whole number greater than 0 that is a common multiple of each of two or more numbers.

The LCM of 2 and 3 is 6.

mínimo común múltiplo (m.c.m.) El menor número natural, mayor que 0, múltiplo común de dos o más números.

El m.c.m. de 2 y 3 es 6.

Ll

length Measurement of the distance between two points.

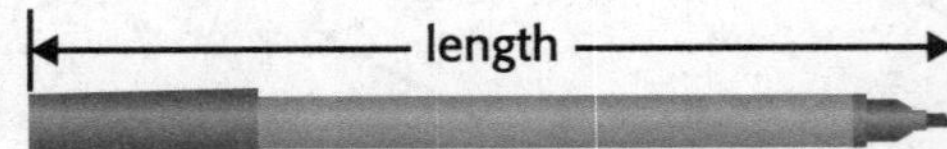

longitud Medida de la distancia entre dos puntos.

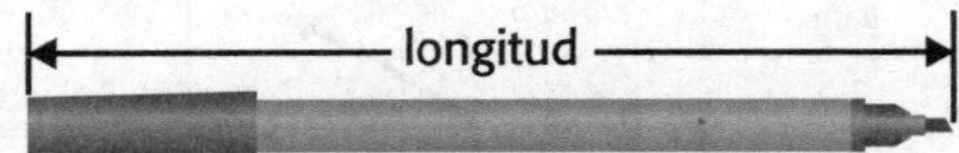

like fractions Fractions that have the same denominator.

$\frac{1}{5}$ **and** $\frac{2}{5}$

fracciones semejantes Fracciones que tienen el mismo denominador.

$\frac{1}{5}$ **y** $\frac{2}{5}$

line A set of *points* that form a straight path that goes on forever in opposite directions.

recta Conjunto de *puntos* que forman una trayectoria recta sin fin en direcciones opuestas.

line plot A graph that uses columns of Xs above a number line to show frequency of data.

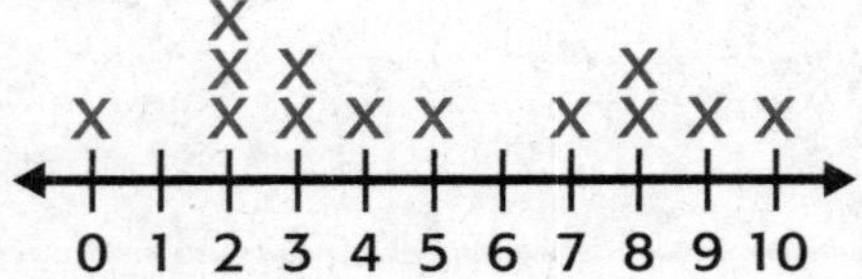

diagrama lineal Gráfica que usa columnas de X sobre una recta numérica para mostrar la frecuencia de los datos.

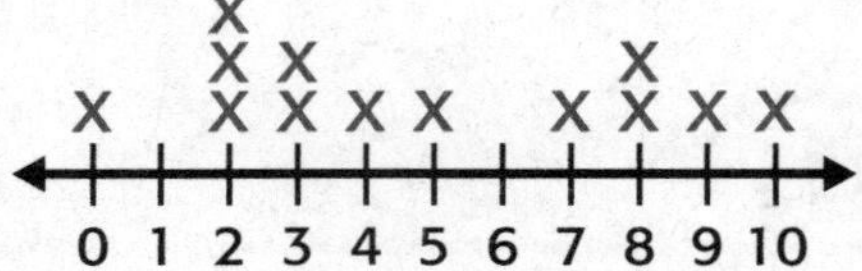

line segment A part of a *line* that connects two points.

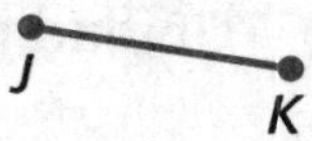

segmento de recta Parte de una *recta* que conecta dos puntos.

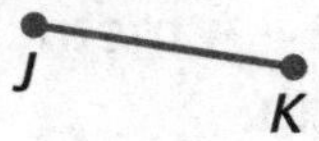

liter (L) A metric unit for measuring volume or capacity.

1 liter = 1,000 milliliters

litro (L) Unidad métrica de volumen o capacidad.

1 litro = 1,000 mililitros

mass Measure of the amount of matter in an object.

masa Medida de la cantidad de materia en un cuerpo.

meter (m) A metric unit used to measure length.

metro (m) Unidad métrica que se usa para medir la longitud.

metric system (SI) The decimal system of measurement. Includes units such as meter, gram, and liter.

sistema métrico (SI) Sistema decimal de medidas que incluye unidades como el metro, el gramo y el litro.

mile (mi) A customary unit of measure for length.

1 mile = 5,280 feet

milla (mi) Unidad usual de longitud.

1 milla = 5,280 pies

milligram (mg) A metric unit used to measure mass.

1,000 milligrams = 1 gram

miligramo (mg) Unidad métrica de masa.

1,000 miligramos = 1 gramo

milliliter (mL) A metric unit used for measuring capacity.

1,000 milliliters = 1 liter

mililitro (mL) Unidad métrica de capacidad.

1,000 mililitros = 1 litro

millimeter (mm) A metric unit used for measuring length.

1,000 millimeters = 1 meter

milímetro (mm) Unidad métrica de longitud.

1,000 milímetros = 1 metro

mixed number A number that has a whole number part and a fraction part. $3\frac{1}{2}$ is a mixed number.

número mixto Número formado por un número natural y una parte fraccionaria. $3\frac{1}{2}$ es un número mixto.

multiple (multiples) A multiple of a number is the product of that number and any whole number.

15 is a multiple of 5 because $3 \times 5 = 15$.

múltiplo Un múltiplo de un número es el producto de ese número por cualquier otro número natural.

15 es múltiplo de 5 porque $3 \times 5 = 15$.

multiplication An operation on two numbers to find their product. It can be thought of as repeated addition. 4×3 is another way to write the sum of four 3s, which is $3 + 3 + 3 + 3$ or 12.

multiplicación Operación entre dos números para hallar su producto. También se puede interpretar como una suma repetida. 4×3 es otra forma de escribir la suma de cuatro veces 3, la cual es $3 + 3 + 3 + 3$ o 12.

net A two-dimensional pattern of a three-dimensional figure.

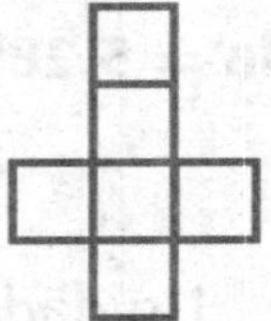

modelo plano Patrón bidimensional de una figura tridimensional.

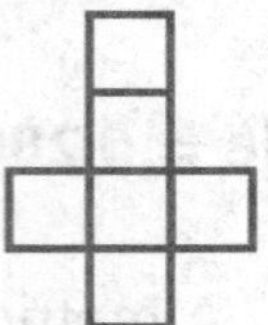

number line A line that represents numbers as points.

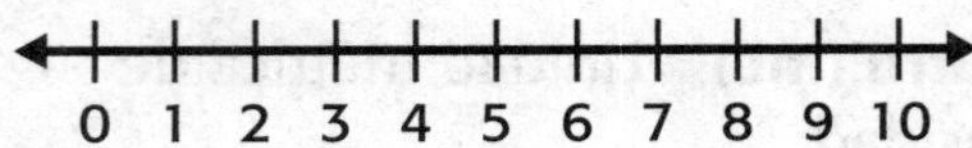

recta numérica Recta que representa números como puntos.

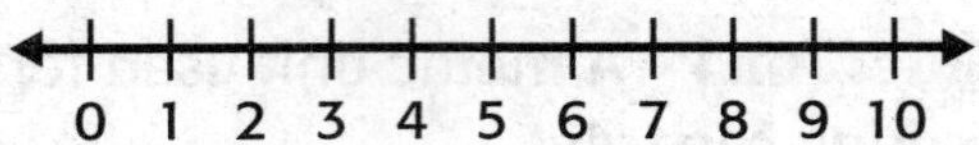

numerator The top number in a fraction; the part of the fraction that tells the number of parts you have.

numerador Número que se escribe sobre la barra de fracción; la parte de la fracción que indica el número de partes que hay.

numerical expression A combination of numbers and operations.

expresión numérica Combinación de números y operaciones.

obtuse angle An angle that measures between 90° and 180°.

ángulo obtuso Ángulo que mide entre 90° y 180°.

obtuse triangle A *triangle* with one *obtuse angle*.

triángulo obtusángulo *Triángulo* con un *ángulo obtuso*.

octagon A polygon with eight sides.

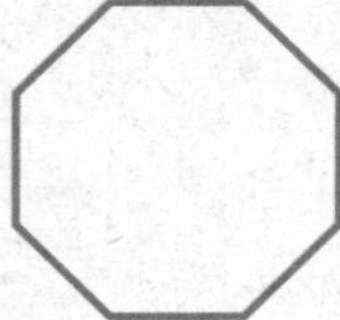

octágono Polígono de ocho lados.

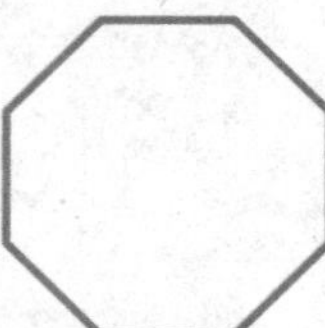

odd number A number that is not divisible by 2; such a number has 1, 3, 5, 7, or 9 in the ones place.

número impar Número que no es divisible entre 2. Los números impares tienen 1, 3, 5, 7, o 9 en el lugar de las unidades.

order of operations A set of rules to follow when more than one operation is used in an expression.

1. Perform operations in parentheses.
2. Find the value of exponents.
3. Multiply and divide in order from left to right.
4. Add and subtract in order from left to right.

orden de las operaciones Conjunto de reglas a seguir cuando se usa más de una operación en una expresión.

1. Realiza las operaciones dentro de los paréntesis.
2. Halla el valor de las potencias.
3. Multiplica y divide de izquierda a derecha.
4. Suma y resta de izquierda a derecha.

ordered pair A pair of numbers that is used to name a point on the coordinate plane.

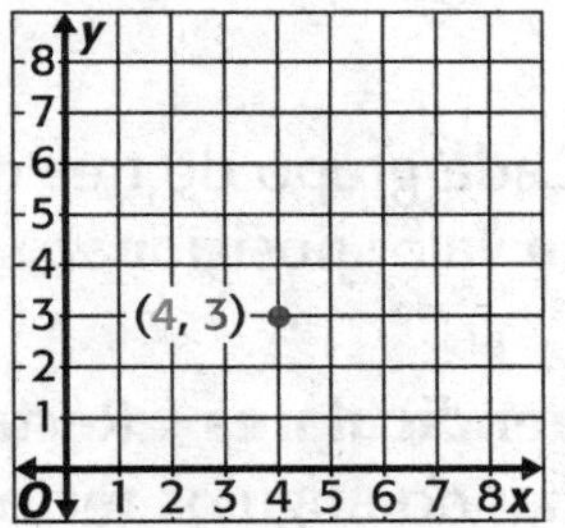

par ordenado Par de números que se usa para nombrar un punto en un plano de coordenadas.

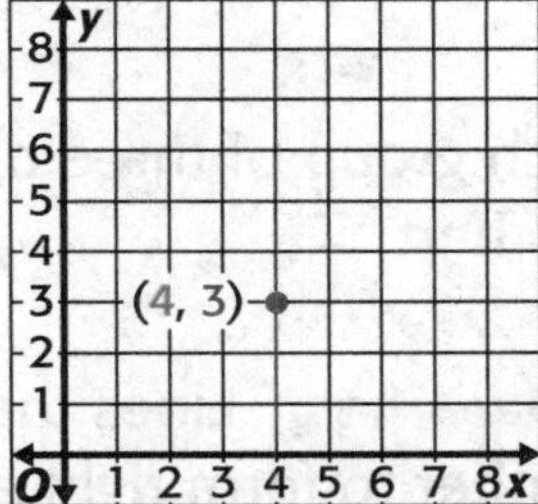

origin The point (0, 0) on a coordinate plane where the vertical axis meets the horizontal axis.

origen El punto (0, 0) en un plano de coordenadas donde el eje vertical interseca el eje horizontal.

ounce (oz) A customary unit for measuring weight or capacity.

onza (oz) Unidad usual de peso o capacidad.

parallel lines Lines that are the same distance apart. Parallel lines do not meet.

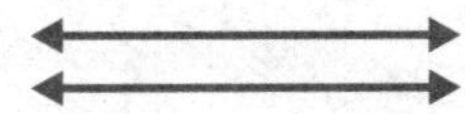

rectas paralelas Rectas separadas por la misma distancia en cualquier punto. Las rectas paralelas no se intersecan.

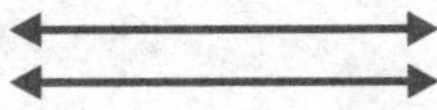

Pp

parallelogram A quadrilateral with four sides in which each pair of opposite sides are parallel and congruent.

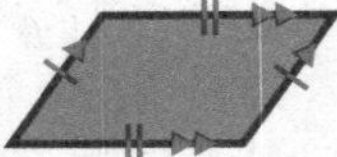

paralelogramo Cuadrilátero en el cual cada par de lados opuestos son paralelos y congruentes.

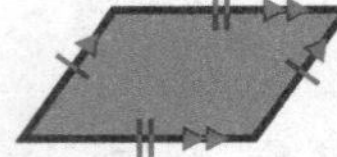

partial quotients A method of dividing where you break the dividend into sections that are easy to divide.

cocientes parciales Método de división por el cual se descompone el dividendo en secciones que son fáciles de dividir.

pentagon A polygon with five sides.

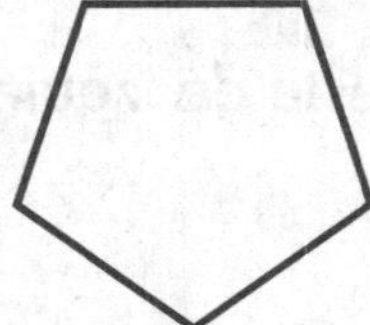

pentágono Polígono de cinco lados.

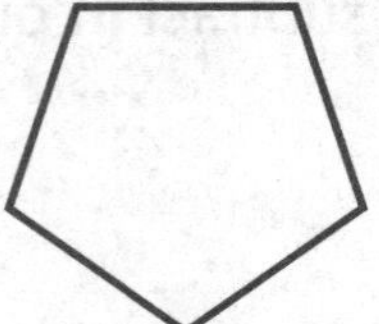

perimeter The *distance* around a polygon.

perímetro *Distancia* alrededor de un polígono.

period Each group of three digits on a place-value chart.

período Cada grupo de tres dígitos en una tabla de valor posicional.

perpendicular lines Lines that meet or cross each other to form right angles.

rectas perpendiculares Rectas que se cruzan formando ángulos rectos.

pint (pt) A customary unit for measuring capacity.

1 pint = 2 cups

pinta (pt) Unidad usual de capacidad.

1 pinta = 2 tazas

place The position of a digit in a number.

posición Lugar que ocupa un dígito en un número.

place value The value given to a digit by its position in a number.

valor posicional Valor dado a un dígito según su posición en el número.

place-value chart A chart that shows the value of the digits in a number.

tabla de valor posicional Tabla que muestra el valor de los dígitos en un número.

plane A flat surface that goes on forever in all directions.

plano Superficie plana que se extiende infinitamente en todas direcciones.

point An exact location in space that is represented by a dot.

punto Ubicación exacta en el espacio que se representa con una marca puntual.

polygon A closed figure made up of line segments that do not cross each other.

polígono Figura cerrada compuesta por segmentos de recta que no se intersecan.

positive number Number greater than zero.

número positivo Número mayor que cero.

pound (lb) A customary unit for measuring weight or mass.

1 pound = 16 ounces

libra (lb) Unidad usual de peso.

1 libra = 16 onzas

power A number obtained by raising a base to an exponent.

$5^2 = 25$ 25 is a power of 5.

potencia Número que se obtiene elevando una base a un exponente.

$5^2 = 25$ 25 es una potencia de 5.

power of 10 A number like 10, 100, 1,000 and so on. It is the result of using only 10 as a factor.

potencia de 10 Número como 10, 100, 1,000, etc. Es el resultado de solo usar 10 como factor.

prime factorization A way of expressing a composite number as a product of its prime factors.

factorización prima Manera de escribir un número compuesto como el producto de sus factores primos.

prime number A whole number with exactly two factors, 1 and itself.

7, 13, and 19

número primo Número natural que tiene exactamente dos factores: 1 y sí mismo.

7, 13 y 19

prism A three-dimensional figure with two parallel, congruent faces, called bases. At least three faces are rectangles.

prisma Figura tridimensional con dos caras congruentes y paralelas llamadas bases. Al menos tres caras son rectangulares.

product The answer to a multiplication problem.

producto Repuesta a un problema de multiplicación.

Pp

proper fraction A fraction in which the numerator is less than the denominator.

$\frac{1}{2}$

fracción propia Fracción en la que el numerador es menor que el denominador.

$\frac{1}{2}$

property A rule in mathematics that can be applied to all numbers.

propiedad Regla de las matemáticas que puede aplicarse a todos los números.

protractor A tool used to measure and draw angles.

transportador Instrumento que se usa para medir y trazar ángulos.

quadrilateral A polygon that has 4 sides and 4 angles.

square, rectangle, parallelogram, and trapezoid

cuadrilátero Polígono con 4 lados y 4 ángulos.

cuadrado, rectángulo, paralelogramo y trapezoide

quart (qt) A customary unit for measuring capacity.

1 quart = 4 cups

cuarto (ct) Unidad usual de capacidad.

1 cuarto = 4 tazas

quotient The result of a division problem.

cociente Resultado de un problema de división.

ray A line that has one endpoint and goes on forever in only one direction.

semirrecta Parta de una recta que tiene un extremo que se extiende infinitamente en una sola dirección.

rectangle A quadrilateral with four right angles; opposite sides are equal and parallel.

rectángulo Cuadrilátero con cuatro ángulo rectos; los lados opuestos son iguales y paralelos.

rectangular prism A prism that has six rectangular bases.

prisma rectangular Prisma que tiene bases rectangulares.

regular polygon A polygon in which all sides are congruent and all angles are congruent.

polígono regular Polígono que tiene todos los lados congruentes y todos los ángulos congruentes.

remainder The number that is left after one whole number is divided by another.

residuo Número que queda después de dividir un número natural entre otro.

rhombus A *parallelogram* with four *congruent sides.*

rombo *Paralelogramo* con cuatro *lados congruentes.*

right angle An angle with a measure of 90°.

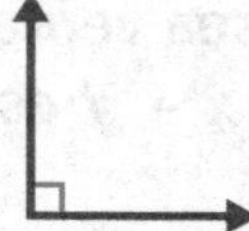

ángulo recto Ángulo que mide 90°.

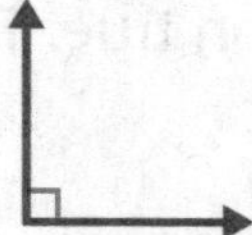

right triangle A *triangle* with one *right angle.*

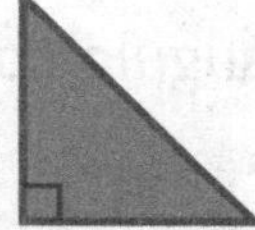

triángulo rectángulo *Triángulo* con un *ángulo recto.*

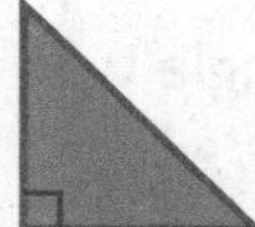

rounding To find the approximate value of a number.

6.38 rounded to the nearest tenth is 6.4.

redondear Hallar el valor aproximado de un número.

6.38 redondeado a la décima más cercana es 6.4.

Ss

scale A set of numbers that includes the least and greatest values separated by equal intervals.

escala Conjunto de números que incluye los valores menor y mayor separados por intervalos iguales.

scalene triangle A *triangle* with no *congruent sides.*

triángulo escaleno *Triángulo* sin *lados congruentes.*

Ss

scaling The process of resizing a number when it is multiplied by a fraction that is greater than or less than 1.

simplificar Proceso de redimensionar un número cuando se multiplica por una fracción que es mayor que o menor que 1.

sequence A list of numbers that follow a specific pattern.

secuencia Lista de números que sigue un patrón específico.

simplest form A fraction in which the GCF of the numerator and the denominator is 1.

forma simplificada Fracción en la cual el M.C.D. del numerador y del denominador es 1.

solution The value of a variable that makes an equation true. The solution of $12 = x + 7$ is 5.

solución Valor de una variable que hace que la ecualción sea verdadera. La solución de $12 = x + 7$ es 5.

solve To replace a variable with a value that results in a true sentence.

resolver Remplazar una variable por un valor que hace que la expresión sea verdadera.

square A rectangle with four *congruent sides.*

cuadrado Rectángulo con cuatro *lados congruentes.*

square number A number with two identical factors.

número al cuadrado Número con dos factores idénticos.

square unit A unit for measuring area, such as square inch or square centimeter.

unidad cuadrada Unidad de área, como una pulgada cuadrada o un centímetro cuadrado.

squared A number raised to the second power; 3×3, or 3^2.

al cuadrado Número elevado a la segunda potencia; 3×3 o 3^2.

standard form The usual or common way to write a number using digits.

forma estándar Manera usual o común de escribir un número usando dígitos.

straight angle An angle with a measure of 180°.

ángulo llano Ángulo que mide 180°.

sum The answer to an addition problem.

suma Respuesta que se obtiene al sumar.

tenth A place value in a decimal number or one of ten equal parts or $\frac{1}{10}$.

décima Valor posicional en un número decimal o una de diez partes iguales o $\frac{1}{10}$.

term A number in a pattern or sequence.

término Cada número en un patrón o una secuencia.

thousandth(s) One of a thousand equal parts or $\frac{1}{1,000}$. Also refers to a place value in a decimal number. In the decimal 0.789, the 9 is in the thousandths place.

milésima(s) Una de mil partes iguales o $\frac{1}{1,000}$. También se refiere a un valor posicional en un número decimal. En el decimal 0.789, el 9 está en el lugar de las milésimas.

three-dimensional figure A figure that has length, width, and height.

figura tridimensional Figura que tiene largo, ancho y alto.

ton (T) A customary unit for measuring weight. 1 ton = 2,000 pounds

tonelada (T) Unidad usual de peso. 1 tonelada = 2,000 libras

trapezoid A quadrilateral with exactly one pair of parallel sides.

trapecio Cuadrilátero con exactamente un par de lados paralelos.

triangle A polygon with three sides and three angles.

triángulo Polígono con tres lados y tres ángulos.

triangular prism A prism that has triangular bases.

prisma triangular Prisma con bases triangulares.

unit cube A cube with a side length of one unit.

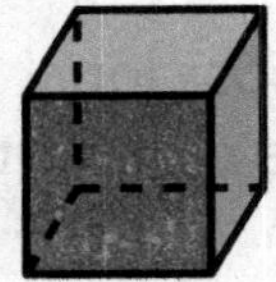

cubo unitario Cubo con lados de una unidad de longitud.

unit fraction A fraction with 1 as its numerator.

fracción unitaria Fracción que tiene 1 como su numerador.

unknown A missing value in a number sentence or equation.

incógnita Valor que falta en una oración numérica o una ecuación.

unlike fractions Fractions that have different denominators.

fracciones no semejantes Fracciones que tienen denominadores diferentes.

variable A letter or symbol used to represent an unknown quantity.

variable Letra o símbolo que se usa para representar una cantidad desconocida.

vertex The point where two rays meet in an angle or where three or more faces meet on a three-dimensional figure.

vértice a. Punto donde se unen los dos lados de un ángulo. b. Punto en una figura tridimensional donde se intersecan 3 o más aristas.

vertical axis A vertical number line on a graph (↕). Also known as the *y*-axis.

eje vertical Recta numérica vertical en una gráfica (↕). También conocido como eje *y*.

volume The amount of space inside a three-dimensional figure.

volumen Cantidad de espacio que contiene una figura tridimensional.

Ww

weight A measurement that tells how heavy an object is.

peso Medida que indica cuán pesado o liviano es un cuerpo.

Xx

***x*-axis** The horizontal axis (↔) in a coordinate plane.

eje *x* Eje horizontal (↔) en un plano de coordenadas.

***x*-coordinate** The first part of an ordered pair that indicates how far to the right of the *y*-axis the corresponding point is.

coordenada *x* Primera parte de un par ordenado; indica a qué distancia a la derecha del eje *y* está el punto correspondiente.

Yy

yard (yd) A customary unit of length equal to 3 feet or 36 inches.

yarda (yd) Unidad usual de longitud igual a 3 pies o 36 pulgadas.

***y*-axis** The vertical axis (↕) in a coordinate plane.

eje *y* Eje vertical (↕) en un plano de coordenadas.

***y*-coordinate** The second part of an ordered pair that indicates how far above the *x*-axis the corresponding point is.

coordenada *y* Segunda parte de un par ordenado; indica a qué distancia por encima del eje *x* está el punto correspondiente.

Name

Work Mat 1: Number Lines

0 1 2 3 4 5 6 7 8 9 10

0 1 2 3 4 5 6 7 8 9 10 11 12 13 14 15 16 17 18 19 20

Work Mat 2: Place-Value Chart (Billions to Ones)

Billions			Millions			Thousands			Ones		
hundreds	tens	ones	hundreds	tens	ones	hundreds	tens	ones	hundreds	tens	ones

Name

Work Mat 3: Centimeter Grid

Work Mat 4: Place-Value Chart (Hundreds to Thousandths)

Ones			Decimals		
hundreds	tens	ones	tenths	hundredths	thousandths

Name

Work Mat 5: Tenths and Hundredths Models

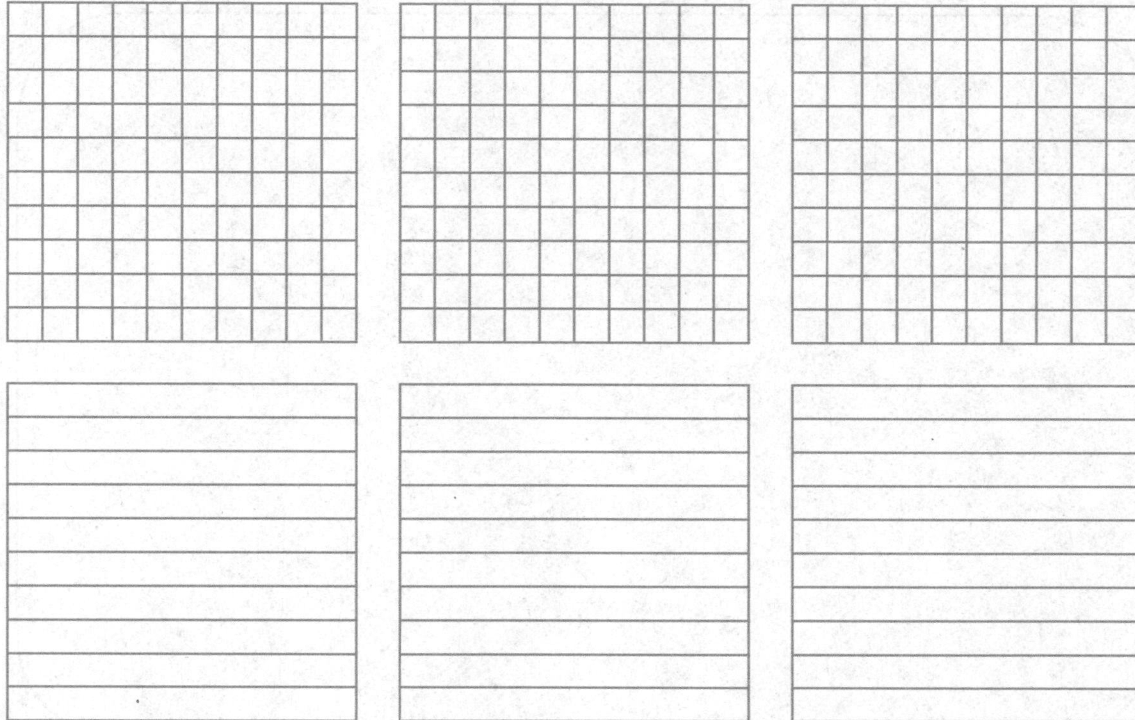

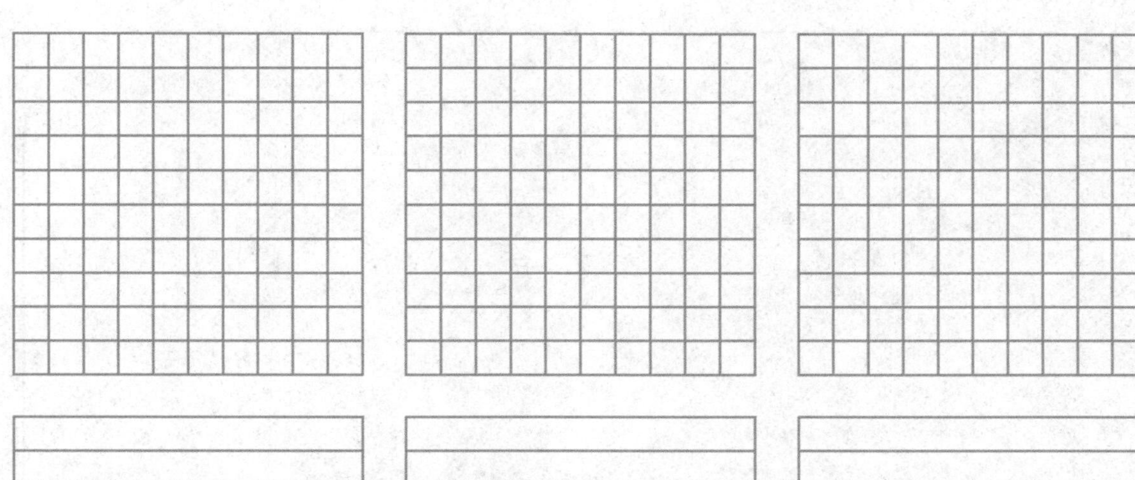

Work Mat 6: Algebra Mat

Name ....

Work Mat 7: First-Quadrant Grid

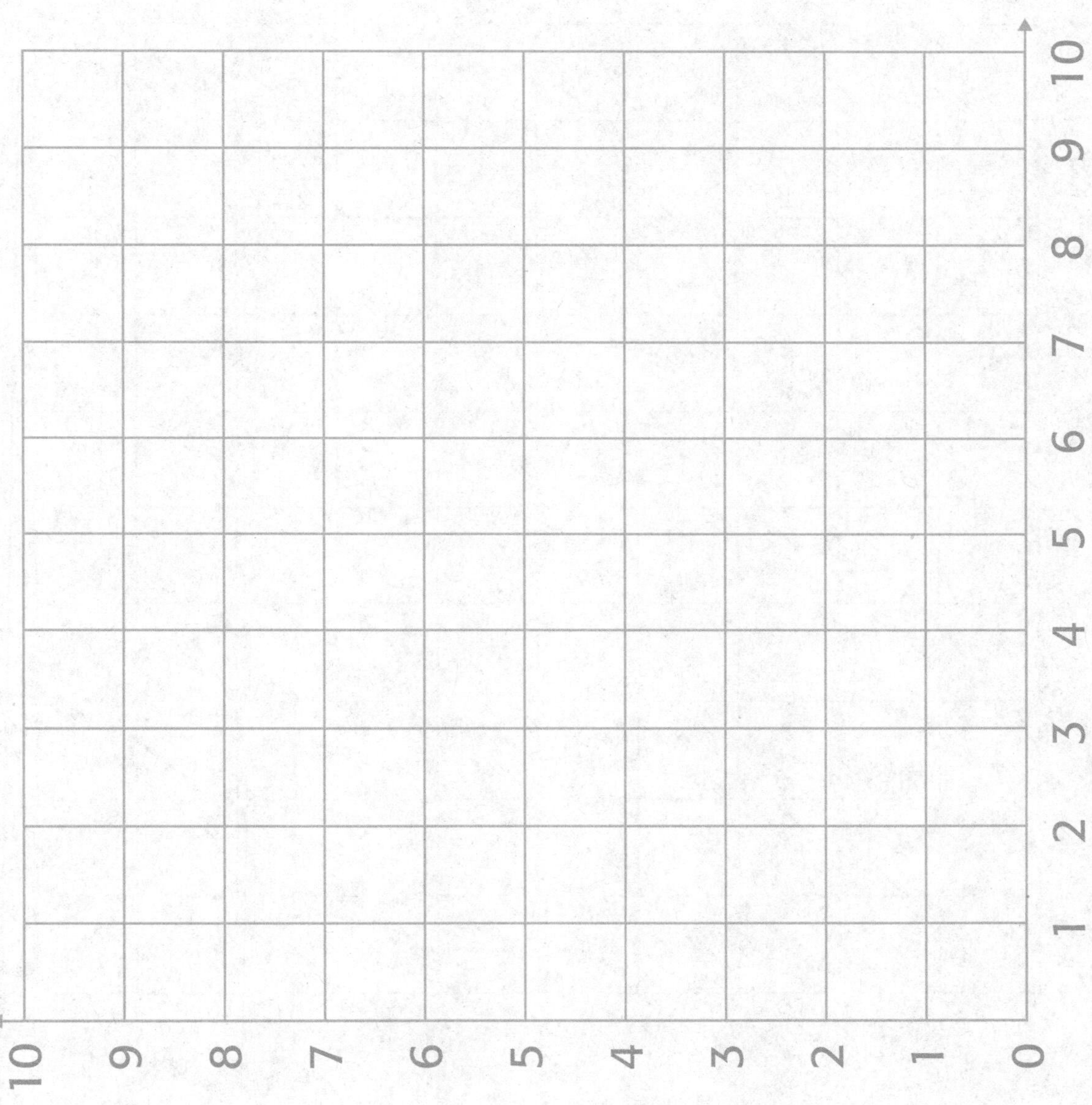

Work Mat 8: First-Quadrant Grid (blank)